Communications
in Computer and Information Scie ~~~~ ~~~

Editorial Board

Shahrul Azman Noah   Azizi Abdullah
Haslina Arshad   Azuraliza Abu Bakar
Zulaiha Ali Othman   Shahnorbanun Sahran
Nazlia Omar   Zalinda Othman (Eds.)

Soft Computing Applications and Intelligent Systems

Second International Multi-Conference
on Artificial Intelligence Technology, M-CAIT 2013
Shah Alam, August 28-29, 2013
Proceedings

 Springer

Volume Editors

Shahrul Azman Noah
Azizi Abdullah
Haslina Arshad
Azuraliza Abu Bakar
Zulaiha Ali Othman
Shahnorbanun Sahran
Nazlia Omar
Zalinda Othman

Universiti Kebangsaan Malaysia
Faculty of Information Science and Technology
Bangi, Selangor Darul Ehsan, Malaysia

E-mail: samn; azizi; has; aab; zao; shah; no; zalinda @ftsm.ukm.my

ISSN 1865-0929   e-ISSN 1865-0937
ISBN 978-3-642-40566-2   e-ISBN 978-3-642-40567-9
DOI 10.1007/978-3-642-40567-9
Springer Heidelberg New York Dordrecht London

Library of Congress Control Number: 2013946047

CR Subject Classification (1998): I.2, I.4, I.5, H.3, H.2.8

Typesetting: Camera-ready by author, data conversion by Scientific Publishing Services, Chennai, India

Printed on acid-free paper

Springer is part of Springer Science+Business Media (www.springer.com)

Preface

Welcome to the proceedings of the Second International Multi-Conference on Artificial Intelligence Technology (M-CAIT 2013). The theme of the conference was "The Next Wave of Computational Intelligence," which comprises a vast area of computational intelligence. M-CAIT 2013 was organized by the Center for Artificial Intelligence Technology (CAIT), a center of excellence at the Faculty of Information Science and Technology, Universiti Kebangsaan Malaysia.

M-CAIT 2013 was a conference hosting four artificial intelligence special tracks in a single event: The Special Track on Intelligence Computation on Pattern Analysis and Robotics (ICPAIR 2013), The Special Track on Data Mining and Optimization (DMO 2013), The Special Track on Semantic Technology and Information Retrieval (STAIR 2013), and The Special Track on Industrial Computing and Applied Informatics (ICOMP 2013). These special tracks were organized by the four research groups at the Center For Artificial Intelligence Technology (CAIT), namely, the Pattern Recognition, the Data Mining and Optimization Research Group (DMO), the Knowledge Technology Research Group (KT), and the Industrial Computing Research Group (IComp). M-CAIT 2013 was a platform to bring together researchers, developers, and practitioners from academia and industry working in all interdisciplinary areas of artificial intelligence to share their experience and exchange and cross-fertilize their ideas. The conference serves as a forum for the dissemination of state-of-the-art research, development, implementation, and applications within the four focus areas in CAIT: pattern recognition, data mining and optimization, knowledge technology, and industrial computing.

M-CAIT 2013 was delighted to host three fascinating keynote speakers: Dr. Benny Tjahjono, a senior lecturer in Manufacturing Systems Engineering, Cranfield University; Prof. Dr. Steffen Staab from the Institute for Computer Science, University of Koblenz, Germany; and Dr. Kenji Suzuki, Assistant Professor of Radiology, Medical Physics and Comprehensive Cancer, the University of Chicago.

We received 110 papers from 13 countries worldwide. The submitted papers went through a rigorous reviewing process and the 25 best papers were selected for inclusion in this CCIS vol. 378 under the common theme of "applications of soft computing and intelligent systems." We would like to express our greatest gratitude to the authors and reviewers of all paper submissions, as well as to all attendees, for their input and participation.

We are very grateful to a large of people for their effort in preparing this volume. We would like to acknowledge the financial support provided by Universiti Kebangsaan Malaysia. Our appreciation also goes to the members of the Program Committee, who spent a significant amount of time for not only ensuring the quality of the conference program, but also for the selection of papers that appear in this volume. In addition, we would like to offer a very special thank you to the supporting committee for their effort in editing and formatting the proceedings.

Finally, we would also like to thank the Faculty of Information Science and Technology, Bursary and Chancellery of Universiti Kebangsaan Malaysia for their valuable assistance.

August 2013   Abdul Razak Hamdan

M-CAIT 2013 Organization

Local Organizing Committee

Chair

Abdul Razak Hamdan

Co-chairs

Data Mining and Optimization

Salwani Abdullah

Pattern Recognition

Khairuddin Omar

Knowledge Technology

Nazlia Omar

Industrial Computing

Zainal Rasyid Mahayuddin

Secretary

Azmi Nasir

Treasurer

Syaimak Abd. Shukor

Publication Chairs

Azuraliza Abu Bakar
Shahrul Azman Mohd Noah
Haslina Arshad
Azizi Abdullah

Proceedings

Zalinda Othman
Siti Norul Huda Sh. Abdullah
Lailatul Qadri Zakaria
Hazura Mohamed

Publicity

Noor Faridatul Ainun Zainal
Elankovan Sundararajan
Saidah Saad
Hafiz Mohd Sarim

Logistics

Mohd Zamri Murah
Mohd Juzaiddin Abd Aziz
Mohd Zakree Ahmad Nazri
Azruhizam Shapii

Protocol

Shahnorbanun Sahran

Program

Shereena Mohd Arif
Bilal Bataineh
Nur Fazidah Elias
Suhaila Zainudin

ICPAIR 2013 Organization

Organizing Committee

General Chair
Othman A. Karim

Chair
Khairudin Omar

Publication
Azizi Abdullah

Local Organizing Committee and Secretariat
Secretary

Siti Norul Huda Sheikh Abdullah
Bilal Bataineh

Bursar

Mohammad Faidzul Nasrudin

Program

Shahnorbanun Sahran
Juhaida Abu Bakar

Publicity

Mir Shahriar Emami
Omar M. Wahdan

Proceedings

Anton Satria Prabuwono
Mohd Sanusi Azmi

Logistics

Zamri Murah

Protocol

Muhamad Shanudin Zakaria
Noor Faridatul Ainun Zainal

International Advisory Committee

Abdul Samad Hasan Basri	Universiti Teknikal Malaysia Melaka, Malaysia
Arief Syaichu Rohman	Institut Teknologi Bandung, Indonesia
Aswami Fadillah Bin Hj Mohd Ariffin	Cybersecurity, Malaysia
Azlinah Mohamed	Universiti Teknologi MARA, Malaysia
Christian Viard Gaudin	Ecole Polytechnique, France
Burhanudin Mohd Aboobaider	King Abdulaziz University, Saudi Arabia
Chan Chee Seng	Universiti Malaya, Malaysia
David Al-Dabass	Nottingham Trent University, UK
Fritz J. Neff	Karlsruhe University of Applied Sciences, Germany
Graham Kendall	University of Nottingham, UK
Ioannis Pitas	AIIA, University of Thessaloniki, Greece
Jan-Mark Geusebroek	The Netherlands
John-John Cabibihan	National University of Singapore
Kenji Suzuki	Universiti of Chicago, USA
Lukito Edi Nugroho	Gadjah Mada University, Indonesia
Marco A. Wiering	Groningen University, The Netherlands
Massudi Mahmuddin	Universiti Utara Malaysia
Mohamed Rawidean Mohd Kassim	MIMOS, Malaysia
Norhayati Ibrahim	Politeknik Sultan Salahuddin Abdul Aziz Shah, Malaysia
Nur Indrianti	University of Pembangunan Nasional "Veteran" Yogyakarta, Indonesia
Ramlan Mahmod	Universiti Putra Malaysia
Rubiyah Yusof	Universiti Teknologi Malaysia
Siti Maryam Shamsudin	Universiti Teknologi Malaysia
Siti Noridah Ali	Politeknik Sultan Salahuddin Abdul Aziz Shah, Malaysia
Taufik	Cal Poly State University, USA
Tengku Mohd Tengku Sembok	Universiti Islam Antarabangsa Malaysia
Wenzeng Zhang	Tsinghua University, China
Zaiki Awang	Universiti Teknologi MARA, Malaysia
Zuwairie Ibrahim	Universiti Teknologi Malaysia

DMO 2013 Organization

Local Organizing Committee

Chair

Salwani Abdullah — Fakulti Teknologi dan Sains Maklumat, UKM

Publication

Azuraliza Abu Bakar — Fakulti Teknologi dan Sains Maklumat, UKM

Secretary/Webmaster

Azmi Nasir — Fakulti Teknologi dan Sains Maklumat, UKM

Treasurer

Syaimak Abdul Shukor — Fakulti Teknologi dan Sains Maklumat, UKM

Proceedings Chair

Zalinda Othman — Fakulti Teknologi dan Sains Maklumat, UKM

Publicity

Hafiz Mohd Sarim — Fakulti Teknologi dan Sains Maklumat, UKM

Program

Suhaila Zainudin — Fakulti Teknologi dan Sains Maklumat, UKM

Logistics

Mohd Zakree Ahmad Nazri — Fakulti Teknologi dan Sains Maklumat, UKM

Protocol

Shahnorbanun Sahran — Fakulti Teknologi dan Sains Maklumat, UKM

STAIR 2013 Organization

Organizing Committee

Chair

Nazlia Omar — Universiti Kebangsaan Malaysia

Co-chair

Shahrul Azman Mohd Noah — Universiti Kebangsaan Malaysia

International Advisor

Fabio Crestani — University of Lugano, Switzerland
Alan Smeaton — Dublin City University, Ireland
Peter Willett — University of Sheffield, UK

Secretary

Shereena Mohd Arif — Universiti Kebangsaan Malaysia

Treasurer

Masnizah Mohd — Universiti Kebangsaan Malaysia

Proceedings

Lailatul Qadri Zakaria — Universiti Kebangsaan Malaysia

Protocol

Nazlena Mohamad Ali — Universiti Kebangsaan Malaysia
Akmal Aris — Universiti Kebangsaan Malaysia

Publicity

Sabrina Tiun — Universiti Kebangsaan Malaysia
Saidah Saad — Universiti Kebangsaan Malaysia

Logistics

Mohd Juzaiddin Abdul Aziz — Universiti Kebangsaan Malaysia

Program Committee

International

Adam Bermingham	Dublin City University, Ireland
Alan Smeaton	Dublin City University, Ireland
Apirak Hoonlor	Mahidol University, Thailand
Danica Damljanovic	Kuato Studios, UK
Eamonn Newman	Dublin City University, Ireland
Fabio Crestani	University of Lugano, Switzerland
Fernando Silva Parreiras	FUMEC University, Brazil
Gareth Jones	Dublin City University, Ireland
Ghassan Kanaan	Amman Arab University, Jordan
Hsin-Hsi Chen	National Taiwan University, Taiwan
James Lanagan	Technicolor
Jérôme Kunegis	University Koblenz-Landau, Germany
Jiao Tao	Rensselaer Polytechnic Institute, US
Marcin Grzegorzek	Universitat Siegen, Germany
Marcos Cintra	Sao Paolo University, Brazil
Marie Guégan	Technicolor
Mark Hughes	Dublin City University, Ireland
Michael Stewart Joy	University of Warwick, UK
Neil Ohare	Yahoo!, Barcelona
Peter Willett	University of Sheffield, UK
Quang Nhat Nguyen	Hanoi University of Science and Technology, Vietnam
Syed Noman Hasany	Qassim University, Saudi Arabia
Tassos Tombros	Queen Mary, University of London, UK
Thomas Gottron	University Koblenz Landau, Germany
Viorel Milea	Erasmus University Rotterdam, The Netherland

Local

Aida Mustapha	Universiti Putra Malaysia, Malaysia
Azreen Azman	Universiti Putra Malaysia, Malaysia
Juhana Salim	Universiti Kebangsaan Malaysia, Malaysia
Khalil Ben Mohamed	MIMOS, Malaysia
Lailatul Qadri Zakaria	Universiti Kebangsaan Malaysia, Malaysia
Lilly Suriani	Universiti Putra Malaysia, Malaysia
Maryati Mohd Yusof	Universiti Kebangsaan Malaysia, Malaysia
Mas Rina Mustapha	Universiti Putra Malaysia, Malaysia
Masnizah Mohd	Universiti Kebangsaan Malaysia, Malaysia
Mohd Juzaiddin Abdul Aziz	Universiti Kebangsaan Malaysia, Malaysia
Muthukkaruppan Annamalai	Universiti Teknologi MARA, Malaysia
Nazlena Mohamad Ali	Universiti Kebangsaan Malaysia, Malaysia
Nazlia Omar	Universiti Kebangsaan Malaysia, Malaysia
Nor Azan Mat Zain	Universiti Kebangsaan Malaysia, Malaysia

Noraini Seman	Universiti Teknologi MARA, Malaysia
Norwati Mustapha	Universiti Putra Malaysia, Malaysia
Nurfadhlina Mohd Sharef	Universiti Putra Malaysia, Malaysia
Nursuriati Jamil	Universiti Teknologi MARA, Malaysia
Rabiah Abdul Kadir	Universiti Putra Malaysia, Malaysia
Rohaya Latip	Universiti Putra Malaysia, Malaysia
Rusli Abdullah	Universiti Putra Malaysia, Malaysia
Shahrul Azman Mohd Noah	Universiti Kebangsaan Malaysia, Malaysia
Shereena Mohd Arif	Universiti Kebangsaan Malaysia, Malaysia
Syed Malek F.D	
Syed Mustapha	Asia e-University, Malaysia
Yusmadiah Ishak	Universiti Putra Malaysia, Malaysia
Zainab Abu Bakar	Universiti Teknologi MARA, Malaysia
Zuraidah Abdullah	Universiti Kebangsaan Malaysia, Malaysia

ICOMP 2013 Organization

Local Organizing Committee

Chair

Zainal Rasyid Mahayuddin

Secretary

Nur Fazidah Elias

Treasurer

Syaimak Abdul Shukor

Publicity and Promotion

Elankovan Sundararajan
Ang Mei Choo
Nazatul Aini Abd Majid
Mohamed Khatim Hasan
Amelia Natasya Abdul Wahab
Asmaul Husna Mohd Husaimi
Azyyati Zainal Abidin
Hendri Himawan
Md. Saifuddin Saif

Proceedings

Haslina Arshad
Anton Satria Prabuwono
Bahari Idrus
Hazura Mohamed

Logistics

Azrulhizam Shapii
Hairulliza Mohd Judi
Norazwin Buang
Riza Sulaiman

Protocol

Shahnorbanun Sahran
Muriati Mukhtar
Noraidah Sahari Ashaari
Ruzzakiah Jenal
Siti Aishah Hanawi

Table of Contents

Real World Coordinate from Image Coordinate Using Single Calibrated Camera Based on Analytic Geometry

Joko Siswantoro[1,2], Anton Satria Prabuwono[1], and Azizi Abdullah[1]

[1] Center For Artificial Intelligence Technology, Faculty of Information Science and
Technology, Universiti Kebangsaan Malaysia, 43600 UKM, Bangi, Selangor D.E., Malaysia
[2] Department of Mathematics and Sciences, Universitas Surabaya,
Jl. Kali Rungkut Tengilis, Surabaya, 60293, Indonesia
joko_siswantoro@ubaya.ac.id
{antonsatria,azizi}@ftsm.ukm.my

Abstract. The determination of real world coordinate from image coordinate has many applications in computer vision. This paper proposes the algorithm for determination of real world coordinate of a point on a plane from its image coordinate using single calibrated camera based on simple analytic geometry. Experiment has been done using the image of chessboard pattern taken from five different views. The experiment result shows that exact real world coordinate and its approximation lie on the same plane and there are no significant difference between exact real world coordinate and its approximation.

Keywords: real world coordinate, image coordinate, analytic geometry.

1 Introduction

The determination of image coordinate of a point in real word coordinate system can be easily calculated using a transformation after camera parameters that are used in image acquisition are known [1],[2] and [3]. Generally the reverse of this problem is cannot be performed, since the transformation from real world coordinate system to image coordinate system is not invertible. Information about the depth of position loses during the transformation. But under certain condition, such as the point in real world coordinate system lies on a ground plane, determination of world coordinate from a point in image coordinate system still can be performed [4]. The determination of real world coordinate has many applications in computer vision including robot positioning [5], object reconstruction [6], and measurement [4],[7] and [8]. Therefore, determination of the real world coordinate of a point from image coordinate is challenging problem in computer vision.

Common method to determine real world coordinate of a point from image coordinate is triangulation. Triangulation is problem of determining the real world coordinate of a point from a set of corresponding image locations and known camera parameters [3]. Mohamed et al [9] and Zhang [10] have used triangulation to determine the real world coordinate of a point using two corresponding image

S.A. Noah et al. (Eds.): M-CAIT 2013, CCIS 378, pp. 1–11, 2013.

coordinate acquired by two cameras in order to measure the accuracy of their proposed camera calibration method.

Many methods have been proposed to determine real word coordinate of a point in image coordinate system. Some of them do not apply camera parameters in determination of real world coordinate. Bucher [4] has proposed a decomposable image to world mapping where the transformation of vertical coordinate is independent from the horizontal position. Memony et al. [11] have proposed a multilayer artificial neural network model (ANN) to determine real world coordinate from matched pair of images. Vilaça et al. [8] have proposed method for determination of real world coordinate in a plane using two cameras and laser line. Polynomial was used to relate between image coordinate and real world coordinate. Xiaobo et al. [12] have used direct linear transformation and back propagation neural network in the determination of real world coordinate in a plane from image coordinate.

Currently the determination of camera parameters is not a hard problem. Using established camera calibration method, such as method proposed by Tsai [13] and Zhang [10], and supported by computer vision library that provides functions for camera calibration such as OpenCV [1] and camera calibration Toolbox for Matlab [14], camera parameters can be easily estimated. Therefore using estimated camera parameters and analytic geometry, the real world coordinate of a point on a plane can be easily obtained from image coordinate. Furthermore previews method can only be used to re-project a point in image coordinate system to its original position in real world coordinate system. They cannot be used to re-project a point to a plane that is different from its original position. This paper proposes the algorithm for determination of real world coordinate of a point on a plane using single calibrated camera base on analytic geometry. The algorithm can also be used to re-project a point in image coordinate system to any plane in real world coordinate system which is not perpendicular to image plane

2   Camera Model

Camera model is usually derived from simple pinhole camera based on collinearity principle. The origin of camera coordinate system is projection center. Each point in real world coordinate system is projected into image plane system by a line through projection center. The z-axis of the camera coordinate system is principle axis. This axis is perpendicular to image plane and intersects image plane at $z = f$, where f is the focal length of camera, as shown in Fig. 1. Projection from real world coordinate system onto image coordinate system in image plane consists of two transformations. The first one is transformation from real world coordinate system to camera coordinate system and the second one is transformation from camera coordinate system to image coordinate system.

Camera parameters are needed to construct the transformation. Camera parameters consist of extrinsic and intrinsic parameters. Extrinsic parameters are used to transform point in real world coordinate system to camera coordinate system. The parameters consist of rotation matrix **R** and translation vector **T** as the following form.

$$\mathbf{R} = \begin{pmatrix} r_{11} & r_{12} & r_{13} \\ r_{21} & r_{22} & r_{33} \\ r_{31} & r_{32} & r_{33} \end{pmatrix}, \mathbf{T} = \begin{pmatrix} t_{11} \\ t_{21} \\ t_{31} \end{pmatrix}$$

Rotation matrix is the product of three rotation matrices $\mathbf{R}_x(\psi), \mathbf{R}_y(\varphi)$ and $\mathbf{R}_z(\theta)$. Where $\mathbf{R}_x(\psi), \mathbf{R}_y(\varphi)$ and $\mathbf{R}_z(\theta)$ are rotation matrices around x-, y-, and z- axis with respective rotation angles ψ, φ and θ. Translation vector is a shift from real world coordinate system to camera coordinate system

Intrinsic parameters are used to transform point in camera coordinate system to image coordinate system. The parameters consist of the focal length of camera f and the center of image plane coordinates $\mathbf{C}(c_x, c_y)$. All camera parameters are obtained from camera calibration process.

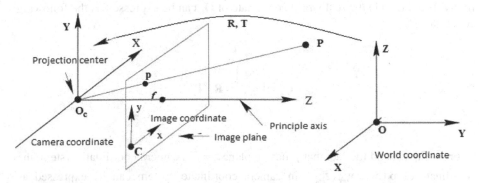

Fig. 1. The geometry of camera model based on pinhole camera model

Suppose $\mathbf{P}(x_w, y_w, z_w)$, $\mathbf{P_c}(x_c, y_c, z_c)$, and $\mathbf{p}(x_{im}, y_{imw})$ are coordinate of a point in real world coordinate system, camera coordinate system and image coordinate system respectively. By assuming that camera used in image acquisition has very small distortion such that its distortion coefficients can be neglected, then transformation from real world coordinate system to image coordinate system can be describe in the following equations [1],[2] and [3].

$$\begin{pmatrix} x_c \\ y_c \\ z_c \end{pmatrix} = \mathbf{R} \begin{pmatrix} x_w \\ y_w \\ z_w \end{pmatrix} + \mathbf{T} \tag{1}$$

$$\begin{pmatrix} x_{im} \\ y_{im} \\ 1 \end{pmatrix} = \begin{pmatrix} f_x & 0 & c_x \\ 0 & f_y & c_y \\ 0 & 0 & 1 \end{pmatrix} \begin{pmatrix} x_c/z_c \\ y_c/z_c \\ 1 \end{pmatrix} \qquad (2)$$

Where $f_x = s_x f$, $f_y = s_y f$ are focal length in x, y direction respectively and s_x, s_y are the size of individual imager elements in x, y direction respectively.

3   Real World Coordinate from Image Coordinate

In order to determine the real world coordinate of point $P(x_w, y_w, z_w)$ from its image coordinate $p(x_{im}, y_{imw})$, camera calibration is firstly performed to estimate camera parameters. Suppose P is located on plane A or we want to re-project point p to point P on plane A, where A is not parallel to line l through projection center O_c and image point p. Under this assumption, the coordinate of P is intersection of plane A and line l, as shown in Fig. 2. Therefore real world coordinate of O_c and p must be determined firstly. From Eq. (1) the real world coordinate of O_c can be expressed as the following equation.

$$O_c = \begin{pmatrix} o_{c1} \\ o_{c2} \\ o_{c3} \end{pmatrix} = R^{-1}T \qquad (3)$$

From Eq. (2) and the fact that p lies on plane $z = f$ in camera coordinate system, the coordinate of $p(x_{imc}, y_{imc}, z_{imc})$ in camera coordinate system can be expressed as follow

$$x_{imc} = \frac{x_{im} - c_x}{s_x} \qquad (4)$$

$$y_{imc} = \frac{y_{im} - c_y}{s_y} \qquad (5)$$

$$z_{imc} = f \qquad (6)$$

Using Eq. (1), (4), (5), and (6), the coordinate of $p(x_{imw}, y_{imw}, z_{imw})$ in real world coordinate can be expressed as follow

$$\begin{pmatrix} x_{imw} \\ y_{imw} \\ z_{imw} \end{pmatrix} = \mathbf{R}^{-1} \left(\begin{pmatrix} x_{imc} \\ y_{imc} \\ z_{imc} \end{pmatrix} - \mathbf{T} \right) \tag{7}$$

According to geometry analytic [15], a line in 3D space is determined by a point and a direction vector of the line. Direction vector of a line is a vector that parallel to the line, as shown in Fig. 2. The vector

$$\mathbf{v} = \overrightarrow{\mathbf{O_c p}} = \left(x_{imw} - o_{c1} \right) \mathbf{i} + \left(y_{imw} - o_{c2} \right) \mathbf{j} + \left(z_{imw} - o_{c3} \right) \mathbf{k} \tag{8}$$

is parallel to line l. Therefore the equation of line l is given by

$$x = o_{c1} + t\left(x_{imw} - o_{c1} \right), \ y = o_{c2} + t\left(y_{imw} - o_{c2} \right), \ z = o_{c3} + t\left(z_{imw} - o_{c3} \right), \ t \in \mathbb{R} \tag{9}$$

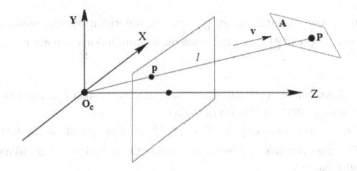

Fig. 2. Line l is line through $\mathbf{O_c}$ and $\mathbf{p}$. Vector $\mathbf{v}$ is direction vector of line l. The real coordinate of $\mathbf{P}$ is intersection between plane $\mathbf{A}$ and line l.

Suppose the equation of plane $\mathbf{A}$ is given by following equation

$$ax + by + cz = d \tag{10}$$

where $a\left(x_{imw} - o_{c1} \right) + b\left(y_{imw} - o_{c2} \right) + c\left(z_{imw} - o_{c3} \right) \neq 0$. This condition will guarantee that plan $\mathbf{A}$ and line l are not parallel and have an intersection point. The point intersection of plane $\mathbf{A}$ and line l is obtained by substituting Eq. (9) to Eq. (10) and found

$$t = \frac{d - \left(ao_{c1} + bo_{c2} + co_{c3} \right)}{a\left(x_{imw} - o_{c1} \right) + b\left(y_{imw} - o_{c2} \right) + c\left(z_{imw} - o_{c3} \right)}. \tag{11}$$

Substitute back t in Eq. (11) to Eq. (9) and the coordinate of **P** in real world coordinate system is approximated by

$$x_{wapp} = o_{c1} + \frac{\left[d - \left(ao_{c1} + bo_{c2} + co_{c3}\right)\right]\left(x_{imw} - o_{c1}\right)}{a\left(x_{imw} - o_{c1}\right) + b\left(y_{imw} - o_{c2}\right) + c\left(z_{imw} - o_{c3}\right)} \tag{12}$$

$$y_{wapp} = o_{c2} + \frac{\left[d - \left(ao_{c1} + bo_{c2} + co_{c3}\right)\right]\left(y_{imw} - o_{c2}\right)}{a\left(x_{imw} - o_{c1}\right) + b\left(y_{imw} - o_{c2}\right) + c\left(z_{imw} - o_{c3}\right)} \tag{13}$$

$$z_{wapp} = o_{c3} + \frac{\left[d - \left(ao_{c1} + bo_{c2} + co_{c3}\right)\right]\left(z_{imw} - o_{c3}\right)}{a\left(x_{imw} - o_{c1}\right) + b\left(y_{imw} - o_{c2}\right) + c\left(z_{imw} - o_{c3}\right)} \tag{14}$$

From the above explanation, we propose the algorithm for real world coordinate determination of point $\mathbf{P}\left(x_w, y_w, z_w\right)$ on a plane from its image point $\mathbf{p}\left(x_{im}, y_{imw}\right)$ as follow:

Step 1. Perform camera calibration to obtain extrinsic camera parameters **R**, **T** and intrinsic camera parameters f_x, f_y, c_x, c_y .

Step 2. Find the coordinate of point $\mathbf{p}\left(x_{im}, y_{im}\right)$ in image coordinate system.

Step 3. Find the coordinate of projection center $\mathbf{O_c}$ in real world coordinate system using Eq. (3).

Step 4. Find the coordinate of $\mathbf{p}\left(x_{imw}, y_{imw}, z_{imw}\right)$ in real world coordinate system using Eq. (7).

Step 5. Approximate the coordinate of **P** in real world coordinate system using Eq. (12), (13), and (14).

4   Experiment and Result

Experiment was performed in the laboratory to validate proposed algorithm. Proposed algorithm was implemented in C++ using OpenCV library. The methodology used in this experiment consists of camera calibration, real world coordinate system construction, image acquisition, corner detection, real world coordinate approximation, and error analysis. Fig. 3 shows the methodology used in the experiment.

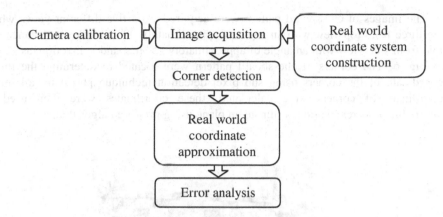

Fig. 3. Methodology used in the experiment

Web camera Logitech C270 was used for image acquisition in the experiment. The camera acquired image with dimension 640 × 480 pixels and resolution 96 dpi in both vertical and horizontal directions. The camera has very small distortion both in radial and tangential distortions. Therefore the distortion coefficients can be neglected. Camera calibration was performed base on Zhang's method [10] using OpenCV library. A 9 × 6 corners flat chessboard pattern with 24.65 mm × 24.65 mm in each square was used in camera calibration, as shown in Fig. 4. Ten views of the pattern were acquired to estimate intrinsic camera parameters and one view was acquired to estimate extrinsic camera parameters.

Fig. 4. Chessboard pattern used in camera calibration

Fifty four inner corners of chessboard pattern were used as points in real world coordinate system by assuming the points lie on plane $z = 0$ for simplicity. Therefore the real world coordinate of the corners have the form $(x_i, y_i, 0)$, $i = 1, 2, ..., 54$. For the construction of real world coordinate system, the following assumptions were used. The center of real world coordinate system is located at top left corner of chessboard pattern. Positive x-axis lies along top left corner to bottom left corner and positive y-axis lies along top left corner to top right corner. Positive z axis is perpendicular to x axis and y axis according to right hand rule. Fig. 5 shows constructed real world coordinate system used in the experiment.

The images of chessboard pattern were acquired from five different views which are three from top view with an angle of approximately 90° in different distance and two from side view with an angle of approximately 45° in x and y direction, as shown in Fig. 6. The corners of chessboard pattern were located to determine the image coordinate of the corners using sub pixel detection technique [1]. After all image coordinate of corners were calculated, these coordinates were then used to approximate its real world coordinate on plane using proposed algorithm.

Fig. 5. Constructed real world coordinate system used in the experiment

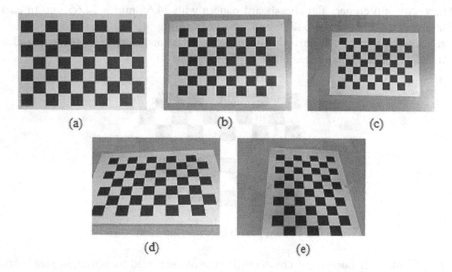

Fig. 6. The images of chessboard pattern from five different views: (a), (b), and (c) top view with an angle of approximately 90° and different distance (d) side view with an angle of approximately 45° in x direction (e) side view with an angle of approximately 45° in y direction.

The accuracy of proposed algorithm was measured using absolute error between exact real world coordinate and its approximation in x, y, and z direction using the following equation.

$$E_x = \left| x_w - x_{wapp} \right| \tag{15}$$

$$E_y = \left| y_w - y_{wapp} \right| \tag{16}$$

$$E_z = \left| z_w - z_{wapp} \right| \tag{17}$$

Where $X_w = (x_w, y_w, z_w)$ and $X_{west} = (x_{wapp}, y_{wapp}, z_{wapp})$ are exact real world coordinate and its approximation respectively.

The result of experiment shows that the approximations of real world coordinate of 54 corners are also on plane $z = 0$. The exact coordinate of 54 corners and its approximation from image coordinate using proposed algorithm on plane $z = 0$ are shown in Fig. 7. From Fig. 7, it can be seen that there are no significant different between the exact coordinate of 54 corners and its approximations. The absolute errors between real world exact coordinate and its approximation are summarized in Table 1.

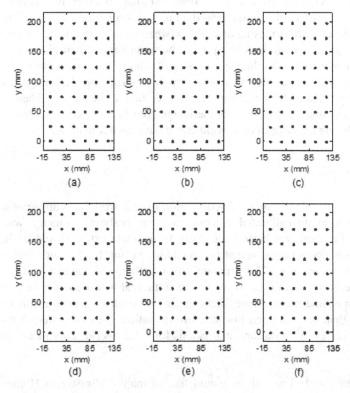

Fig. 7. (a) The exact coordinate of 54 corners on plane $z = 0$ and its approximations from image coordinate using proposed algorithm: (b) – (f) from image in Fig. 5 (a) – (e) respectively

Table 1. Absolute error between exact real world coordinate and its approximation in x, y, and z direction (in mm)

View	Mean E_x	Std. dev. E_x	Mean E_y	Std. dev. E_y	Mean E_z	Std. dev. E_z
View a	0.2188	0.1456	0.2117	0.1371	0.0000	0.0000
View b	0.2839	0.1990	0.2874	0.1732	0.0000	0.0000
View c	0.3271	0.2265	0.3616	0.2168	0.0000	0.0000
View d	0.3206	0.2052	0.2786	0.1879	0.0000	0.0000
View e	0.2399	0.1673	0.3381	0.2730	0.0000	0.0000

Since the approximation of real world coordinates of 54 corners lie on $z = 0$, therefore the absolute errors in z direction are zeros. In x and y directions the largest absolute error means are 0.3271 mm and 0.3616 mm respectively occurred in approximation using image from view c. The standard deviation of absolute error is less than 0.27 mm for all views and all direction. It indicates that the entire absolute error tend to be very close to the mean. Therefore proposed algorithm gives a good approximation for real world coordinate from image coordinate.

In this experiment error can be occurred due to error in camera parameters estimation and inaccuracy in locating the image coordinate of corners. Absolute error mean increases with increasing distance between camera and plane, as shown in view a, b, and c. In addition, decreasing angle between camera and plane $z = 0$ also has impact on increasing absolute error mean. In both case, increasing absolute error mean may be occurred due to inaccuracy in determining the image coordinate of corner when the distance from camera to image increases. Therefore camera parameters estimation and locating the image coordinate of corners play an important role in reducing the error of real world coordinate approximation.

5   Conclusion

In this paper the algorithm for determination of real world coordinate of a point on a plane using single calibrated camera base on analytic geometry was proposed. Camera calibration was performed firstly to estimate extrinsic and intrinsic camera parameters including rotation matrix, translation vector, focal length, and the center of image plane coordinate. These parameters together with camera model and simple analytic geometry were used to approximate the real world coordinate of a point on a plane from its image coordinate. The experiment result shows the absolute errors have mean less than 0.37 mm and low standard deviation. It can be inferred that proposed algorithm gives a good approximation for real world coordinate from image coordinate.

Acknowledgment. The authors would like to thanks Ministry of Higher Education Malaysia and Universiti Kebangsaan Malaysia for providing facilities and financial support under Grants No. FRGS/1/2012/SG05/UKM/02/12 and PTS-2011-054.

References

1. Bradski, G., Kaehler, A.: Learning OpenCV: Computer Vision with the OpenCV Library. O'Reilly Media, Inc. (2008)
2. Heikkila, J., Silven, O.: A four-step camera calibration procedure with implicit image correction. In: Proc. IEEE Computer Society Conf. Computer Vision and Pattern Recognition, pp. 1106–1112 (1997)
3. Szeliski, R.: Computer Vision Algorithms and Applications. Springer (2011)
4. Bucher, T.: Measurement of Distance and Height in Images based on easy attainable Calibration Parameters. In: Proc. IEEE Intelligent Vehicles Symposium, pp. 314–319 (2000)
5. Neves, A.J.R., Martins, D.A., Pinho, A.J.: Obtaining the distance map for perspective vision systems. In: Proc. of the ECCOMAS Thematic Conference on Computational Vision and Medical Image Processing (2009)
6. Mulayim, A.Y., Yilmaz, U., Atalay, V.: Silhouette-based 3-D model reconstruction from multiple images. IEEE Transactions on Systems, Man, and Cybernetics, Part B: Cybernetics 33, 582–591 (2003)
7. Lee, D.J., Xu, X., Eifert, J., Zhan, P.: Area and volume measurements of objects with irregular shapes using multiple silhouettes. Optical Engineering 45, 027202 (2006)
8. Vilaça, J.L., Fonseca, J.C., Pinho, A.M.: Calibration procedure for 3D measurement systems using two cameras and a laser line. Optics & Laser Technology 41, 112–119 (2009)
9. Mohamed, R., Ahmed, A., Eid, A., Farag, A.: Support Vector Machines for Camera Calibration Problem. In: 2006 IEEE International Conference on Image Processing, pp. 1029–1032 (2006)
10. Zhang, Z.: A flexible new technique for camera calibration. IEEE Transactions on Pattern Analysis and Machine Intelligence 22, 1330–1334 (2000)
11. Memony, Q., Khanz, S.: Camera calibration and three-dimensional world reconstruction of stereo-vision using neural networks. International Journal of Systems Science 32, 1155–1159 (2001)
12. Xiaobo, C., Hanghang, F., Yinghua, Y., Shukai, Q.: The research of camera distortion correction basing on neural network. In: 2011 Chinese Control and Decision Conference (CCDC), pp. 596–601 (2011)
13. Tsai, R.Y.: A versatile camera calibration technique for high-accuracy 3D machine vision metrology using off the shelf TV cameras and lenses. IEEE Journal of Robotics and Automation RA-3, 323–344 (1987)
14. Bouguet, J.-Y.: http://www.vision.caltech.edu/bouguetj/calib_doc/
15. Weir, M.D., Thomas, G.B., Hass, J.: Thomas' Calculus. Addison-Wesley (2010)

Articulated Human Motion Tracking
with Online Appearance Learning

Wei Ren Tan[1], Chee Seng Chan[1], Hernan E. Aguirre[2], and Kiyoshi Tanaka[2]

[1] University of Malaya, Centre of Image and Signal Processing,
Faculty of Comp. Sci. and Info. Tech., Kuala Lumpur, 50603, Malaysia
`willtwr@siswa.um.edu.my`, `cs.chan@um.edu.my`
[2] Shinshu University, Faculty of Engineering, 4-17-1 Wakasato,
Nagano, 380-8553, Japan
`{ktanaka,ahernan}@shinshu-u.ac.jp`

Abstract. Automatic tracking of the articulations of human from a video sequence is a difficult task due to complex motions of the limbs, dynamic background, and varieties of poses. These challenges make it difficult to train a generative motion and appearance model to be used in different scenarios. In our work, we employ particle swarm optimization framework to avoid the need of motion model. Particularly, we propose a novel appearance learning strategy to learn the appearance of each body part in real time. Furthermore, we also propose an appearance model to represent the shape of each body part. Samples from UIUC dataset had been used in experiments. The results had shown that our method performed well on complex activities without motion model and online appearance training. It also showed the robustness of our method to recover from tracking failure in an occluded video.

Keywords: Visual Tracking, Online Learning, PSO, Human Motion Analysis.

1 Introduction

Articulated human motion tracking is a task that automatic tracks the articulations of human from a video sequence. It is useful in computer vision applications such as sport analysis, gait analysis, human-computer interaction, surveillance, etc. Although many works [1], [2], [3], [4], [5], [6], [7] and [8] had been done, articulated human motion tracking remains a challenging problem due to the complex motions of the limbs, dynamic background, and varieties of poses and cloths.

Many previous works had been focusing on training motion model [9] and [10], and appearance model [11] and [12]. However, offline training is time consuming and high computational cost. **Motion Model:** Although motion model helps in predicting the pose in next frame, movements of a person in a video are usually hard to predict because human tends to have complex actions (e.g. walk, run, jump, etc.) in real life. Moreover, a trained motion model does not guarantee to work all the time [13] and [27]. Therefore, as oppose to conventional methods, we want to avoid using motion model. We adopted the framework in [5], which uses Particle Swarm Optimization

S.A. Noah et al. (Eds.): M-CAIT 2013, CCIS 378, pp. 12–26, 2013.
© Springer-Verlag Berlin Heidelberg 2013

(PSO) to estimate the pose of a human without the need of motion model. However, as contrasted with their work, we learn the appearance model in real time.

Appearance Model: Appearance model interprets an image that contains complex structure such as head or limb of a human using model-based approach. Different features had been used for appearance modeling, such as HOG [15], color-based [16] and [17], edges [12], and silhouette [18]. Unlike the conventional method, we intend to avoid training the appearance model offline. Balan and Black [19] proposed an adaptive appearance model based on WSL framework [20] to cope with changes in appearance including occlusion. Kalal et al. [21] proposed a Training-Learning-Detection (TLD) method for online learning method by learning face appearance in different view. Both methods have a collective set of appearance model with different changes. One of the appearance models from the collection will be chosen based on a strategy to track the target in an observed image. Inspired by their work, we propose a novel appearance learning strategy to learn a generalized appearance model for each body part. New generalized appearance model is learnt in each frame. Then, the learnt new generalized appearance model will be used to track new pose in the next frame.

The experiments had been carried out using sample videos from UIUC dataset [10]. This dataset contains complex activities where the subject changes the actions from time to time. The results are shown and compared qualitatively and quantitatively. Both results had shown that our method performed well without the need of motion model and offline appearance training. Moreover, we also occluded the sample video by adding a black region to test the robustness of our method. Empirically, it also shown that our method is capable to recover from tracking failure in the occluded video.

The rest of the paper is structured as follows. Section 2 describes the framework of articulated human motion tracking. We will explain the learning strategy in Section 3. Experimental results and discussions are detailed in Section 4 and 5, respectively. Finally, conclusion is drawn in Section 6.

2   Proposed Framework

This section will discuss the overview framework of our articulated human motion tracking. The framework is initialized at first frame with manual annotation of a set points, $P = \{p_j\}$, where $p_j = (x_j, y_j, \theta_j)$ and $j = 1, 2, ..., J$. J is the number of points. θ_j is the angle of rotation at position (x_j, y_j). In this work, we had defined 15 points as illustrated in Fig. 1.

Next, features are extracted from the frame. The features are: (i) edges, I^e extracted using Canny edge detector; (ii) silhouette, I^s extracted from static background subtraction [22]. Then, initial appearance model is computed from the I^e and details of the appearance model are discussed in Subsection 2.1. Following that, we estimate each point in P using the PSO where the extracted features are used to compute the cost function. The cost function measures how well the generalized appearance model fits into the features. Subsection 2.2 shows how PSO estimates the position of the points in detail. Cost function will be discussed in detail in Subsection 2.3. Once all the points are estimated, new generalized appearance model will be learnt based on the appearance learning strategy and will be used to estimate new P in next frame. Algorithm 1 shows the pseudocode of the proposed framework.

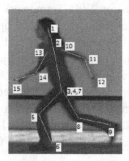

Fig. 1. Sample of Points and Skeleton

Algorithm 1. Pseudocode for Articulated Human Motion Tracking

---

Begin Algorithm
 - Manual initialization for the first frame
 - Store current position, $P = \{p_1, p_2, \ldots, p_j\}$
 - Extract features, I^e and I^s
 - Compute initial appearance model, G (Eq. 8 or 9)
         for each consecutive frame, f **do**
                 Features extraction, I_f^e and I_f^s
                 for each body part, j **do**
                         Begin PSO
                         - Swarm initialization, $X = \{i_{1d}, i_{2d}, \ldots, i_{Nd}\}$
                         - Velocity initialization, $V = \{v_1, v_2, \ldots, v_N\} = 0$
                         - Cost function evaluation, Σ (Eq. 6)
                         - Store *lbest* and *gbest*
                         while stopping condition not fulfilled **do**
                         - Update X and Y as to Eq. 2 and 3, respectively
                         - Cost function evaluation, Σ (Eq. 6)
                         - Update *lbest* and *gbest*
                         - Check stopping condition
                         end while
                         - Update p_j
                         End PSO
                 end for
                 - Update appearance model, G_f (Eq. 8 or 9)
         end for
End Algorithm

2.1   Appearance Model

Appearance model is essential for successful tracking by matching the appearance model onto the features. First, local edges are extracted for each body part. Then, we need to remove the noises and extract the information from the edge to represent the shape of each body part. Fig. 2 illustrates the appearance model for the human body parts. It consists of two parts: outline-based and line-detection-based. Outline-based is used to represent the outline of the head and torso because they have curve shape. While, line-detection-based is used to represent the limbs using lines.

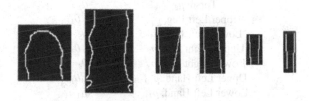

Fig. 2. Appearance Model (from left to right): Head, torso, upper-leg, lower-leg, upper hand, lower hand

Outline-Based - The space between edge pixels are filled to compute a local silhouette. Then, Canny edge detector is used to get the outline shape. The process is illustrated in Fig. 3.

Fig. 3. Outline-based: local edge to local silhouette to appearance

Line-Detection-Based - Edge image is divided into two parts: left half and right half. One line will be extracted from both sides by using Hough transform [23] on the edge, respectively. Fig. 4 illustrates the process for this part.

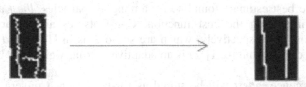

Fig. 4. Line-detection-based: local edge to appearance

2.2   Particle Swarm Optimization

Similar to [8], PSO is employed to optimize the parameters listed in Table 1. The remaining parameters in each body part, p_j that are not optimized will be calculated using the homogeneous transform as to Eq. 1.

Table 1. Parameters to be optimized for each body part

Body parts	Parameters
Head, p_1	x_1, y_1, θ_1
Torso, p_2	θ_2
Upper Left Leg, p_4	x_4, θ_4
Lower Left Leg, p_5	θ_5
Upper Right Leg, p_7	x_7, θ_7
Lower Right Leg, p_8	θ_8
Upper Left Hand, p_{10}	x_{10}, θ_{10}
Lower Left Hand, p_{11}	θ_{11}
Upper Right Hand, p_{13}	x_{13}, θ_{13}
Lower Right Hand, p_{14}	θ_{14}

$$\begin{pmatrix} x_{j+1} \\ y_{j+1} \end{pmatrix} = \begin{pmatrix} cos\,(\theta_j) & -sin\,(\theta_j) \\ sin\,(\theta_j) & cos\,(\theta_j) \end{pmatrix} \begin{pmatrix} x_j \\ y_j \end{pmatrix}, j = 1, 2, \dots, 15 \tag{1}$$

In PSO, X is a set of solutions that consists of N number of particles, i_{nd}, $n = 1, 2, \dots,$ N. d is the number of parameters to be optimized in each body part. The swarm of particles in the next frame is initialized by sampling a Gaussian distribution center at the current best estimation with covariance set to 1. X and velocity, V are updated in each iteration, t using Eq. 2-3, respectively.

$$X_{t+1} = X_t + V_{t+1} \tag{2}$$

$$V_{t+1} = \omega V_t + c_1 rand_1 (lbest_t - X_t) + c_2 rand_2 (gbest_t - X_t) \tag{3}$$

Local best, $lbest$ is the best estimate encounter so far for each particle. While, global best, $gbest$ is the best estimate found so far among all particles. $lbest$ and $gbest$ are computed by minimizing the cost function. Constants, c_1 and c_2 are social and cognition components, respectively, which are set to 2 as in [14]. $rand$ is a uniform random number drawn from [0, 1]. ω is an adaptive inertial weight [7] which change over time.

Upon convergence, $gbest$ will be stored as final solution. Convergence happens when: (i) average cost function value does not improve for 5 iterations, or (ii) it had reached 100 iterations.

2.3   Cost Function

Cost function is used to measure how well a pose fits an observed frame. Appearance model will be projected onto the features extracted from the observed frame based on the parameters in the particle. The projected appearance model of nth particle is referred as projected model points, M_n. Cost function consists of two parts [6], [8]:

Edge-Based Part - I_f^e is first convolved with a Gaussian kernel to create an edge distance map, $I_f^{e'}$. Appearance model is projected onto $I_f^{e'}$ and edge-based cost function, Σ^e is computed using Sum of Squared Difference (SSD):

$$\Sigma^e(M_n, I_f^{e'}) = \frac{1}{N_e} \sum_{i=1}^{N_e} (1 - p_i^e(M_n, I_f^{e'}))^2 \tag{4}$$

where $p_i^e(M_n, I_f^{e'})$ represents the value of the edge distance map at projected model points.

Silhouette-Based Part - I_f^s is a pixel map constructed with foreground set to 1 and background set to 0. Appearance model is projected onto I_f^s and SSD is computed for silhouette-based cost function, Σ^s.

$$\Sigma^s(M_n, I_f^s) = \frac{1}{N_s} \sum_{i=1}^{N_s} (1 - p_i^s(M_n, I_f^s))^2 \tag{5}$$

where $p_i^s(M_n, I_f^s)$ represents the value of the pixel map at projected model points.

Then, the overall cost function will be computed by combining the edge and silhouette parts:

$$\Sigma(M_n) = \Sigma^e(M_n, I_f^{e'}) + \Sigma^s(M_n, I_f^s) \tag{6}$$

3   Online Learning Strategy

The framework learns and updates new generalized appearance model, G from a collection of local appearance model, L. G will be used as the appearance model for next frame to compute cost function. Fig. 5 shows the learnt appearance model at each frame. Based on the appearance model as discussed in Section 2.1, the learning strategy has 2 parts: outline-based and line-detection-based:

Outline-Based - L for this part is the local silhouette computed from edge as shown in Fig. 3. First, L is collected from first frame until current frame, f' by using Eq. 7. Then, H is normalized and then threshold to vote if a pixel is belonging to the body part. Lastly, canny edge detector is used to extract the outline to produce G. Eq. 8 represents the process of G.

$$H_{f'} = \sum_{f=1}^{f'} L_f \tag{7}$$

$$G_{f'} = edge(\frac{H_{f'}}{\max H_{f'}} \geq th) \tag{8}$$

where th is a threshold value.

Line-detection-based - L for this part is the appearance in Fig. 4. Similarly, L will be collected from first frame to current frame, f' from Eq. 7. Then, H is converted into binary image and G will be computed by applying Hough transform line detection using Eq. 9.

$$G_{f'} = hough(H_{f'} > 0) \tag{9}$$

Fig. 5. Learnt appearance model: (From left to right) Frame 1, 5, 10, 15, 20, and 25

3.1 Collection Filtering

One must note that there is no guarantee that the estimation or tracking done are correct all the time. Hence, using all local appearance to learn the generalized appearance should be avoided. In order to pick the suitable local appearance, we do a filtering process when collecting the local appearance. In this paper, we check the cost function value in order to filter the local appearance. Therefore, Eq. 7 can be rewritten into:

$$H_{f'} = H_{f'-1} + L_{f'} \tag{10}$$

where current H is calculated by the sum of previous H and current L. During the learning step in each frame, cost function of $L_{f'}$ will be checked thoroughly. If it is below the learning threshold, th_l, then it will be accumulated into the collection:

$$H_{f'} = \begin{cases} H_{f'-1} + L_{f'} & if \ \Sigma_{L_{f'}} < th_l \\ H_{f'} - 1 & otherwise \end{cases} \tag{11}$$

4   Experimental Result

We had used sample videos from UIUC dataset [10] for experiments. UIUC dataset consists of complex human activities where the subject changes the actions from time to time which make the dataset is challenging. As an example, the subject was initially walking and then started to run in the middle of the video.

In this paper, the quantitative measurement is done by calculating the distance error [5] between the estimated points and ground truth of the points for each frame:

$$\sigma(P_f, \tau_f) = \frac{\Sigma_{j=1}^{J} \left\| p_f^j - \tau_f^j \right\|}{J} \tag{12}$$

where τ is the ground truth points. Lower $\sigma(P_f, \tau_f)$ means the estimated points are closer to the actual points. Average distance error can be calculated by averaging over all distance error of the frames:

$$\sigma_{avg}(P, \tau) = \frac{\Sigma_{f=1}^{F} \sigma(P_f, \tau_f)}{F} \tag{13}$$

where F is the number of frames in the video.

4.1   Result Evaluations

In this subsection, we will evaluate the performance of our method with [10], [19]. Though both work are able to cope with complex human activities, however, in their work, a set of motion models are trained beforehand. Then, they built a large range of complex queries to a collection of complex motion and activity. In our work, these steps are exempted and we still able to obtain comparable results as shown in Fig. 6.

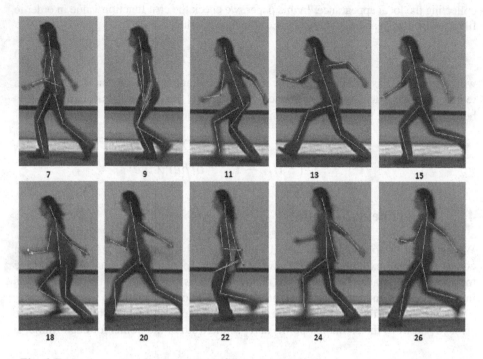

Fig. 6. Example results of walk-run: the numbers below the images represent frame number

In Fig. 6, the subject was walking and started to run at frame 13. By visual inspection, our framework is able to trace the articulations of the human successfully. During running, fast movement of the limbs might cause some motion blur. Even in this situation, our method still able to perform well in most of the frame without the need of motion model and offline appearance training.

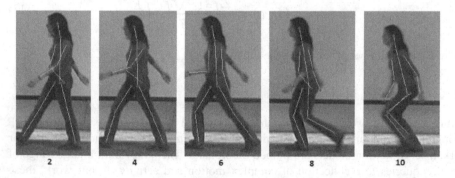

Fig. 7. Example result of tracking error (walk-run)

Fig. 7 shows an example of tracking failure. In most of the cases, the failures are only occur to the limbs. One of the main problems is due to self-occlusion, where other body parts occlude the hand or leg. It makes the tracking of the occluded body part almost impossible because there is limited way of extracting the features from the occluded body parts. However, our proposed method is able to recover from tracking failure when the occluded body parts became visible, e.g. frame 10. Unfortunately, it fails to recover the hand in frame 4 to 6. This might be caused by the bounded search space in PSO, which is used to avoid the particles from searching the solutions in an overly large search space. More discussion about bounded search space will be highlighted in Section 5.

Table 2. Different combination of actions

Actions	Average, σ_{avg} (mm)	Max, σ_{max} (mm)
walk-run	09.2656	16.4727
walk-reach-jump-walk	14.3167	36.1980
walk-crouch-walk	18.4054	38.1283

Table 2 shows the average and maximum distance error between different combinations of actions. Based on the results, the average and maximum distance error is higher when the subject jumps or crouches. The reason is that the shape of certain body parts changes significantly during the actions. For instance, when jumping, the hair oats and causes change of the shape of head as shown in Fig. 8. In Fig. 9, the subject crouches and the neck are blocked by the shoulder. This caused occlusion scenario. Such occlusion will change the shape of the head. Another problem is due to error propagation. As shown in both figures, the error from the head propagates to torso and then to limbs. Such errors are usually accumulated from parts to parts causing high average distance error.

Fig. 8. Jumping

Fig. 9. Crouching

4.2   Population Size in PSO

Determining the number of particles used in PSO is always a challenge. In this subsection, we compare the performance with different population size. Fig. 10 shows the comparisons between different numbers of particles. Firstly, small population size (10 particles) has the highest distance error. Small population size in PSO always suffers higher chance of not converging into optimum solution. From the graph, the performance of 30 particles is very stable by maintaining at flow distance error in most of the frame. However, due to the randomness of PSO, high distance error might occur randomly, e.g. frame 18 to 19. Admittedly, with higher population size, the performance is able to gain more stability but higher computational cost will be required. Therefore, empirically, we will be using 50 particles for the rest of the experiments.

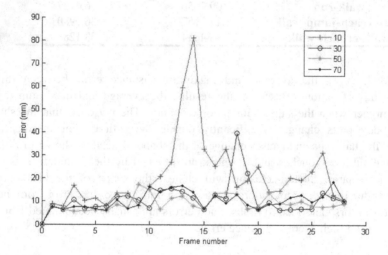

Fig. 10. Distance error graph of walk-run for comparison between 10, 30, 50, and 70 particles

4.3   Learning Threshold

The learning threshold determines the need of updating the appearance model at a particular frame. Although learning strategy is applied, the outcome may become worse without proper learning threshold value. Experiments had been done and comparison between different learning threshold and without learning will be discussed in this subsection.

Fig. 11 shows the comparison between different threshold values, th_l and without learning. While, Table 3 shows the average and maximum distance error. The performance is relatively stable with lowest average distance error without learning. The reason is that with manual annotation, the appearance model computed at first

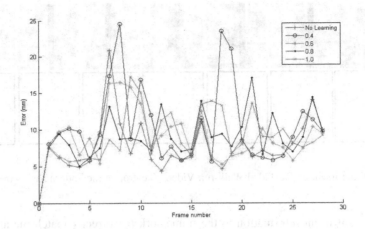

Fig. 11. Distance error graph of walk-run for comparison between different threshold values, th_l

Table 3. Different threshold value (walk-run)

Threshold	Average, σ_{avg} (mm)	Max, σ_{max} (mm)
No Learning	7.9627	20.8440
0.4	10.5668	24.5115
0.6	9.2656	16.4727
0.8	9.3695	17.0610
1.0	9.3136	17.2106

frame is significant to be used throughout whole tracking process. However, the maximum distance error is quite high compare to others. This is because without learning, it lacks of adaptability in clothing-invariant, where the shape of the cloth changes from time to time.

The framework will learn the appearance model more frequently with higher learning threshold, but higher computational cost will be required. In order to ensure the stability, minimizing the maximum distance error and average distance error will be the requirement for choosing a suitable learning threshold. Meanwhile, we would want to reduce the computational cost by using as low learning threshold as possible. Therefore, we had chosen 0.6 as our learning threshold.

4.4   Robustness

Robustness is important in evaluating the ability of the framework to recover from tracking failure cause by occlusions or noises. We occluded the sample video by adding in black region to test the robustness of our method. Fig. 12 shows the result of the experiment on the occluded video. When the subject is passing through the occluded space, the tracking fails as expected. This is because the occluded features

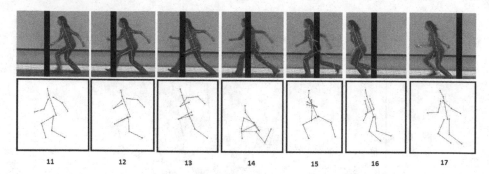

Fig. 12. Experiment on Occluded Walk-run Video. Skeleton at the bottom shows the estimated pose

provide insufficient information to the framework to correctly match the appearance model onto the features. However, as the subject leaves the black region, the tracking is able to slowly recover. In frame 14, the estimated locations are completely messed up. Due to bounded search space, fully recovery from tracking failure at next few frames is not possible. Overall, the experiment shows that our method is robust to recover when failure occurs without the need of motion model and offline appearance training.

5   Discussions and Future Works

One of the key problems in articulated human motion tracking is the reliability of cost function. During the experiments, we found that failures in tracking are always caused by weak features. For instance, some crucial edges are not able to be detected because the color of the upper-hand and torso are very similar due to cloth's color. In future, other features such as optical flow [24] and HOG [15] will be investigated for better features and cost function.

In PSO, the bounded search space [25], [26] is a constraint that limits the search space of the particles. In articulated human motion tracking, allowing bounded search space is justifiable because motion between two consecutive frames is small due to high frame rate of the video (15Hz). Such constraint helps to increase the convergence rate and reduce the computational cost. However, it is hard to determine the size of the bounded search space. Thus, it would be worth investigating a dynamic bounded search space for a flexible search space size.

Likewise, population size affects the convergence rate, computational cost, and the ability to converge to optimum solution. Fixing the size does not guarantee PSO to perform well all the time. Recently, dynamic population size [27] had become a topic of interest for researchers. In future, dynamic population size should be considered for better performance.

Initialization of the position of body globally is important in articulated tracking because it helps in constraint for the search space of the articulations of 14 the human to reduce the time consumption of tracking. In [8], initialization was started by localizing the torso of the subject. While, in our paper, we initialize by localizing head's position. We found that the hair of the subject always causes problem in initializing the head's position correctly. In future, it is important to determine a better way of global localization of the subject.

Currently, the algorithm is initialized by manual annotation. For future work, we will apply an automatic initialization for a more intelligent system. In order to do so, a fixed initial appearance model that cooperates with online learning will be considered. In this case, we will also investigate the human body topology for computing a good initial appearance. Last but not least, more sample videos will be included for experiments in future.

6   Conclusions

As a conclusion, we had presented a framework of articulated human motion tracking using an nature inspired approach, PSO without the need of motion model and offline appearance training. An online learning strategy is introduced to learn a generalized appearance model for tracking human body parts in a video. We also presented a new method to automatically extract the appearance of body parts from the features.

Empirically, the framework is able to track the articulations of the human in a complex human activities video. The results also show the robustness of our framework to be able to recover from tracking failure in an occluded video. Although many other works are able to achieve same result, but our framework do not require any motion model and offline training.

We had also discussed how the parameters in PSO and online learning strategy affect the system performance. Currently, the parameters are mostly set to a fixed value based on the experiments. For more intelligent system, dynamicity of the parameters is important to ensure that the framework can adapt to any situation automatically. Investigating methods for dynamic parameters will be one of our tasks in future.

References

1. Bregler, C., Malik, J.: Tracking People with Twists and Exponential Maps. In: CVPR, pp. 8–15 (1998)
2. Sidenbladh, H., Black, M.J., Fleet, D.J.: Stochastic Tracking of 3D Human Figures using 2D Image Motion. In: Vernon, D. (ed.) ECCV 2000, Part II. LNCS, vol. 1843, pp. 702–718. Springer, Heidelberg (2000)
3. Balan, A.O., Sigal, L., Black, M.J.: A Quantitative Evaluation of Video-based 3D Person Tracking. In: ICCCN, pp. 349–356 (2005)
4. Deutscher, J., Reid, I.: Articulated Body Motion Capture by Stochastic Search. Int. Journal of Computer Vision 61, 185–205 (2005)
5. Chan, C.S., Liu, H., Brown, D.J., Kubota, N.: A Fuzzy Qualitative Approach to Human Motion Recognition. In: FUZZ-IEEE, pp. 1242–1249 (2008)
6. Ivekovic, S., John, V., Trucco, E.: Markerless Multi-view Articulated Pose Estimation using Adaptive Hierarchical Particle Swarm Optimisation. In: Di Chio, C., et al. (eds.) EvoApplicatons 2010, Part I. LNCS, vol. 6024, pp. 241–250. Springer, Heidelberg (2010)

7. John, V., Trucco, E., Ivekovic, S.: Markerless Human Articulated Tracking using Hierarchical Particle Swarm Optimisation. Image and Vision Computing 28, 1530–1547 (2010)
8. Chan, C.S., Liu, H.: Fuzzy Qualitative Human Motion Analysis. IEEE Transactions on Fuzzy Systems 17(4), 851–862 (2009)
9. Sidenbladh, H., Black, M.J., Sigal, L.: Implicit Probabilistic Models of Human Motion for Synthesis and Tracking. In: Heyden, A., Sparr, G., Nielsen, M., Johansen, P. (eds.) ECCV 2002, Part I. LNCS, vol. 2350, pp. 784–800. Springer, Heidelberg (2002)
10. Ikizler, N., Forsyth, D.: Searching for Complex Human Activities with No Visual Examples. Int. Journal of Comp. Vision 80, 337–357 (2008)
11. Yang, Y., Ramanan, D.: Articulated Human Detection with Flexible Mixtures of Parts. IEEE Transactions on Pattern Anal. and Machine Intelligence (in press)
12. Ramanan, D., Forsyth, D., Zisserman, A.: Tracking People by Learning Their Appearance. IEEE Transactions on Pattern Anal. and Machine Intelligence 29, 65–81 (2007)
13. Cifuentes, C.G., Sturzel, M., Jurie, F., Brostow, G.: Motion Models that Only Work Sometimes. In: BMVC, Surrey, pp. 55.1–55.12 (2012)
14. Dalal, N., Triggs, B.: Histograms of Oriented Gradients for Human Detection. In: CVPR, San Diego, pp. 886–893 (2005)
15. Czyz, J., Ristic, B., Macq, B.: A Particle Filter for Joint Detection and Tracking of Color Object. Image and Vision Computing 25, 1271–1281 (2007)
16. Tan, W.R., Chan, C.S., Yogarajah, P., Condell, J.: A Fusion Approach for Efficient Human Skin Detection. IEEE Transactions on Industrial Informatics 8(1), 138–147 (2012)
17. Plänkers, R., Fua, P.: Model-based Silhouette Extraction for Accurate People Tracking. In: Heyden, A., Sparr, G., Nielsen, M., Johansen, P. (eds.) ECCV 2002, Part II. LNCS, vol. 2351, pp. 325–339. Springer, Heidelberg (2002)
18. Balan, A.O., Black, M.J.: An Adaptive Appearance Model Approach for Model-based Articulated Object Tracking. In: CVPR, New York, pp. 758–765 (2006)
19. Chan, C.S., Liu, H., Lai, W.K.: Fuzzy Qualitative Complex Actions Recognition. In: FUZZ-IEEE, Barcelona, pp. 1–8 (2008)
20. Jepson, A., Fleet, D., El-Maraghi, T.: Robust Online Appearance Models for Visual Tracking. IEEE Transactions on Pattern Anal. and Machine Intelligence 25, 1296–1311 (2003)
21. Kalal, Z., Mikolajczyk, K., Matas, J.: Face-tld: Tracking-learning-detection Applied to Faces. In: ICIP, Hong Kong, pp. 3789–3792 (2010)
22. Piccardi, M.: Background Subtraction Techniques: A Review. In: IEEE SMC, The Hague, vol. 4, pp. 3099–3104 (2004)
23. Chutatape, O., Guo, L.: A Modified Hough Transform for Line Detection and Its Performance. Pattern Recognition 32, 181–192 (1999)
24. Beauchemin, S.S., Barron, J.L.: The Computation of Optical Flow. ACM Computing Surveys 27, 433–466 (1995)
25. Helwig, S., Wanka, R.: Particle Swarm Optimization in High-dimentional Bounded Search Spaces. In: SIS, Honolulu, pp. 198–205 (2007)
26. Lim, M.K., Chan, C.S., Monekosso, D., Remagnino, P.: SwATrack: A Swarm Intelligence-based Abrupt Motion Tracker. In: IAPR MVA, Kyoto (2013)
27. Lu, H., Yen, G.: Dynamic Population Size in Multiobjective Evolutionary Algorithms. In: CEC, Hawaii, vol. 2, pp. 1648–1653 (2002)

Arabic-Jawi Scripts Font Recognition
Using First-Order Edge Direction Matrix

Bilal Bataineh[1], Siti Norul Huda Sheikh Abdullah[2],
Khairuddin Omar[3], and Anas Batayneh[3]

Pattern Recognition Research Group
Center For Artificial Intelligence Technology,
Faculty Of Information Science And Technology,
Universiti Kebangsaan Malaysia, 43600, Bangi, Selangor, Malaysia
Department of Computer Engineering
Faculty of Hijjawi for Engineering Technology,
Yarmouk University, 21163, Irbid, Jordan
b.btnh@yahoo.com, {mimi,ko}@ftsm.ukm.my,
anasb86@hotmail.com

Abstract. Document image analysis and recognition (DIAR) techniques are a primary application of pattern recognition. OFR is one of the most important DIAR techniques. The information about font type indicates important information to support human knowledge and other document analysis and recognition techniques. In this paper, a new optical font recognition method for Arabic scripts is proposed based on the First order edge direction matrix, which is an effected simple feature extraction method for binary images. The proposed methods based on several recent methods in pre-processing and feature extraction stages. The performance of the proposed method is compared with the previous OFR methods that based on texture analysis methods in the feature extraction stage. The results show that the proposed method presents the best performance than of other methods in terms of computation time and accuracy.

Keywords: Arabic scripts, documents, font recognition, feature extraction, EDMS, GLCM, Gabor, LBP.

1   Introduction

Document Image Analysis and Recognition (DIAR) techniques present a major branch in artificial intelligence. DIAR techniques aim to extract information from document images to support human knowledge and computer processes (Jain et al. 2000). In this side, the font types and styles present important information about document images such as the document structure, tasks, parts, contents, and history. Therefore, recognition the font types and styles by machine is presented as important domain in DIAR, which call optical font recognitions (OFR) [1]. OFR is one of the main challenges of DIAR. The available OFR techniques deal with the recent documents and fonts types. Most of OFR implemented with printed Latin and Chinese

S.A. Noah et al. (Eds.): M-CAIT 2013, CCIS 378, pp. 27–38, 2013.

characters. They are neglected other important scripts such as Arabic script, which used in many usual languages such as Arabic, Jawi, Persian, and much more. Moreover, there are ignoring to the historical font styles such as calligraphy types.

Based on the general structure of pattern recognition systems, the OFR framework includes three main stages; document image pre-processing, feature extraction and recognition stages [2] and [3]. Usually, document images have several properties that affected on the performance in feature extraction stage. Those properties presented noises, multiple grey-scale or colour levels, different size of texts, skew, spaces between words, lines and boundaries. Therefore, image preparing processes such as binarization, skew correction, and filtering are required in the pre-processing stage. Feature extraction is one of the most critical issues in pattern recognition applications. Xudong [3] claimed that the effectiveness of each feature extraction technique is highly associated with distinguishing the similarity of patterns that belong to each identical class from other patterns or noise. Basically, the texture analysis feature extraction methods deal with overall or sub-region of the natural image to extract global properties of an input image for the recognition stage [4]. Last, the recognition stage is applied to classify the targets based on the input features.

This work presents a proposed optical font recognition system for Arabic calligraphy (OFR-AC). The proposed OFR-AC is based on recent methods in document image binarization, generating a texture image of document images and feature extraction. Based on an enhanced binarization method, better quality images are produced at the pre-processing stage. Then, texture blocks of the text are generated before passing to the post-processing stages. Next, the features are extracted based on edge direction matrixes (EDM_1). In the classification stage, the back-propagation neural network is applied to identify the font type of the calligraphy. The work structure as following: section 2 is state of art, section 3 the proposed method, section 4 the experiment and section 5 the conclusion.

2   State of the Arts

Usually, documents consist of different font types based on several conditions such as text aim and document structure. OFR is used in various applications, such as document characterisation and classification [5], document layout analysis [6], improvement of multi-font optical character recognition (OCR) [7] and reprinting documents [8].

Moreover, most of OCR systems have been implemented on texts based on a single font type. The performance of OCR is affected negatively on multi-fonts, where each character has varied forms based on the font type. OFR helps to solve the above problem by developing multi-font optical character recognition systems. This goal is accomplished by using OFR as a pre-processing stage in an OCR system to match the database with the font type used. That reduces the likelihood of conflict between two characters, reducing the processing time by reducing the amount of data used in training the recognition stage [7] and [9]. Moreover, the information about fonts helps to reprint documents in the original input image format [8].Only a few research has focused on the importance of the font for the process of OCR. Manna et al.[7] and

Arjun et al.[9] proposed a multi-font numeral recognition system. Arjun et al. [9] claimed that this method needs less time to be implemented and gives higher performance than other methods.

2.1 The Previous OFR Works

Abuhaiba [6] has presented an optical font recognition algorithm for three fonts of printed text based on different styles, sizes and weights. In total, 36 classes are considered. He used the most common 100 words in Arabic langue to determine the font type of the text. This determination was achieved by assuming that the font type of the common words within the text represented the font type. A set of 48 geometrical features were extracted from words used in the decision tree classification method to recognise the font type. However, the main disadvantage of this work is that it failed when the common Arabic words did not exist in the text.

Sun [10], presented a method for Chinese and English fonts. He used a stroke template technique as input features into a decision tree identifier. This method uses forty fonts (twenty English and twenty Chinese). Here, a mathematical representation instead of an image representation of stroke templates is used to accelerate the process. Conversely, the stroke cannot be generalised to different languages. It requires modification to insert any new font. Also, the stroke method is based on local analysis approach that requires a private segmentation process, pre-processing and template matching before the font recognition process can be executed. An algorithm for font recognition implemented on a single unknown Chinese character has been presented by Ding et al. [5]. They employed a wavelet transform for feature extraction to recognise a set of 7 fonts.

Zhang et al. [11] presented a method for English italic fonts. They used strokes patterns as input in a 2-D wavelet decomposition method. This method uses 22,384 common word images from four fonts (Arial, Comic Sans MS, Courier and Times New Roman) in both normal and italic styles. However, they did not address the use of recognition techniques. This method is used for most common words; therefore, the process fails as missing words in the dataset.

The global texture analysis approach has been applied in many document image analysis and recognition researches. It is based on applying Gabor texture analysis methods to text as a texture. Yong et al. [4] have described a new texture analysis approach to handle font recognition for Chinese and English fonts. They employ the text block as a texture block, where each font has a special texture. In the pre-processing stage, they prepared the document images to generate a texture block. The 2-D Gabor filter was used to extract the features from a font's texture blocks. A set of 32 features were extracted based on the mean and standard deviation values of the 16 filtered images.

Ma and Doermann [12] have proposed a new method based on grating cell operator. It extracts the orientation texture features of the text images. This method was compared to the isotropic Gabor filter, and it classified five fonts of three scripts. The pre-processing stage of [12] is similar to that of [4]. A weighted Euclidean distance classifier and a back-propagation neural network classifier were applied in this work. They conducted the experiment using three alphabets, Latin, Greek and

Cyrillic. Ramanathan et al. [13] proposed English font recognition based on global texture analysis. The Gabor filters were applied to extract the features to classify six font types. They used the features in their proposed SVM classifier. In [14], they applied the same method to the Tamil language. In general, [4], [12], [13] and [14] focused on the Gabor filter feature extraction method. Each work imitated the previous work using different datasets. They generated sixteen different filtered images (by Gabor filter) and computed their mean and standard deviation as their input features. However, [15] and [16] claimed that the Gabor filter consumes more processing time than other feature extraction methods. This method is also sensitive to image noise and ignores the image's structural properties.

2.2   The Arabic Calligraphy

Arabic is an international language. It is the primary language spoken in twenty-two countries. It has also been used as a second language in many else regions. The Arabic alphabet is one of the most widely used alphabets in the world. It is used in many widespread languages such as Arabic, Persian, Jawi, Pashto, and Urdu. In general, more than one billion people use Arabic alphabets in written documents around the world [17]. There are three styles of written forms of Arabic: printed, handwritten and calligraphy. The oldest form is Arabic calligraphy (Fig. 1). This most beautiful form is considered an Arabic hallmark and an Islamic art and has been created by professional writers or calligraphers. Initially, the calligraphy type implied a particular area, history, official functions and special daily transactions. However, nowadays the calligraphy type appears in the art of engineering, ornamental publications and decoration. It is often found in mosques, historic buildings and museums. There are seven major types of Arabic calligraphy: Kufi, Thuluth, Naskh, Roqaa, Diwani, Persian and Maghrebi (Andalusi). Fig. 1 shows examples of different types of calligraphy.

(a)   (b)   (c)

(d)   (e)   (f)

(g)

Fig. 1. Examples of the Arabic calligraphy datasets for (a) Kufi, (b) Naskh, (c) Roqaa, (d) Diwani, (e) Thuluth, (f) Parisian, and (g) Andalusi classes

3   The Proposed Method

The framework of the OFR-AC is depicted in Fig. 2. There are three main stages in this framework: pre-processing, feature extraction and recognition. In pre-processing, the document images of calligraphy are prepared to produce the edge texture blocks. That process involves several steps: (1) document image binarization, (2) preparing the texture of the document image and (3) extracting the edges between the black and white levels. The results of the pre-processing stage are used in the feature extraction stage. The feature extraction stage aims to extract the features of the document image of Arabic calligraphy. This work employed global feature extraction based on a texture analysis approach, using the extended EDMS namely EDM_1 method. Eventually, the extracted features are used in the recognition stage. The back-propagation neural network is used to recognize the type of calligraphy. The details of OFR-AR are explained fowling.

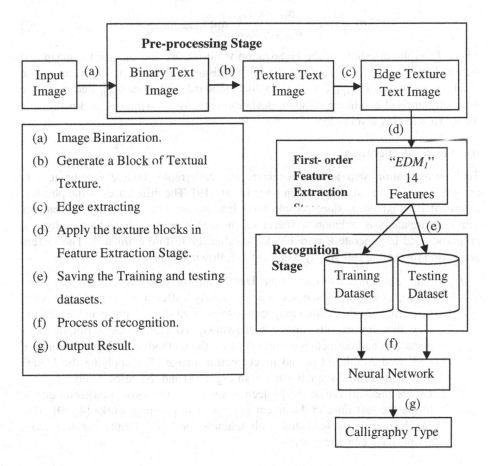

Fig. 2. The proposed framework of the OFR-AC

3.1   Pre-processing Stage

Usually, the input document images have different properties. Therefore, the properties are unformatted in the pre-processing stage to give a constant affection in the feature extraction stage. In this stage, images are prepared using binarization, text normalisation, and edge extraction as explained following:

A) Document Image Binarization

The binarization process aims to remove unwanted information to keep only the text information by classify the pixels into two values, 0 for the text and 1 for the background. The binarization thresholding method of [18]* has used in this step.

$$T_W = m_W - \frac{m_W^2 * \sigma_W}{(m_g + \sigma_W)(\sigma_{Adaptive} + \sigma_W)} \tag{1}$$

$$\sigma_{Adaptive} = \frac{\sigma_W - \sigma_{min}}{\sigma_{max} - \sigma_{min}} max_{level} \tag{2}$$

where T_W is the threshold of the binarization window, m_W and σ_W are the mean and standard deviation of the pixels of the binarization window, m_g is the mean of the pixels in all image, $\sigma_{Adaptive}$ is the adaptive standard deviation, σ_{min} and σ_{max} are the minimum and maximum standard deviation value of all windows, and max_{Level} is the maximum gray level (=255).

B) Texture Generation

Text normalisation step aims to generate the calligraphy texture for the texture analysis method in feature extraction stage [4, 16, 19]. The differences in the physical nature of the text (size, slop, complement) lead to weaknesses in finding features using texture analysis techniques. Therefore, the text is normalised to fix the physical properties and to generate texture blocks suitable for feature extraction. The texture generation for Arabic calligraphy requires the following steps:

i.   *Skew deduction and correction*: Due to an improper scanning process, i.e., wrong alignment, some script images may suffer a skew problem. Apart from that, Arabic calligraphy can also be written and presented in artistic way that intrinsically involves skewness. To realign the script, a skew deduction and correction is applied using the method of [20]. An example of a skewed Arabic script and its correction image after applying the Hough transform technique are illustrated in Fig. 3.(a) and (b), respectively.

ii.   *Locate the Text Lines*: the projection profile is the most used technique to locate the text lines in document images as in previous works [4, 19]. This can be easily implemented with Chinese or Latin scripts because those scripts have consistent height.

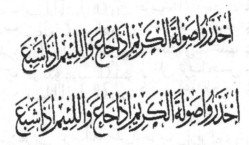

Fig. 3. (Left) The source Thuluth calligraphy image, and (right) the resulting image after applying skew correction using the Hough transform technique

iii.   *Line Normalisation:* Next, each individual sub-word has been determined vertically. The vertical projection profile is used to calculate the gaps between words. Usually, the gaps between some words are inconsistent. Therefore the gaps' sizes are uniformed based on two pixels distance, and each line is normalised to fit in a 512×100 pixel image.

iv.   *Text Block Normalisation:* Each normalised line is taken to build a text block (texture of the text). Each block contained five prepared lines. The texture block is built by padding each line to other to get 512×512 block. Two pixels are inserted as space within each two lines to avoid connecting between text patterns. The previous steps produce a text block representing a texture of Arabic calligraphy as shown in Fig. 4, while Fig. 5 shows a selected input image and the final texture image result after applying the above steps in the pre-processing stage.

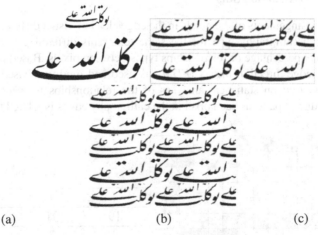

(a)   (b)   (c)

Fig. 4. (a) The original binary image of Diwani calligraphy, (b) its result after line localization and normalization, and (c) its texture block results after text normalisation

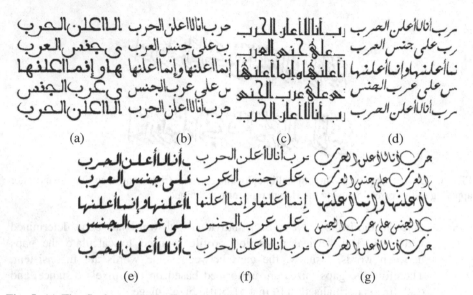

(a)   (b)   (c)   (d)

(e)   (f)   (g)

Fig. 5. (a) The final texture blocks of a same text by (a) Andalusi, (b) Naskh, (c) Kufi, (d) Persian, (e) Roqaa, (f) Thuluth, and (g) Diwani calligraphies

C) Edge Detection

The adopted feature extraction method (edge direction matrix (EDMS)) implements with the edge of the pattern. The edges of the texts in the block image are detected by keep the black pixels are adjacent with white pixels; else the pixel will be removed.

3.2   Feature Extraction Stage

The different arrangements of the pixels in the edges lead to visual difference among the textures of each calligraphy type for the same text. This difference is caused by the difference in the percentage of relationships types between pixels. Based on above note, the first order edge direction matrix (EDM_1) is used and modified based on this work [21]. EDM_1 is based on statistical analyses of the relationships between the adjacent pixels in the edge of pattern. Each pixel in the calligraphy edges is related to other pixels

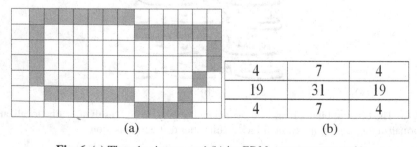

(a)   (b)

Fig. 6. (a) The edge image, and (b) its EDM_1 occurrences results

by special relationships. These relationships are represented as occurrences. The number of occurrences is stored in the related cell in EDM_1 (Fig. 6). Then, the symmetrical in EDM_1 is ignored and only the values of 0° to 135° take to extract the features.

Those values are used in statistical methods to extract several features. In this work, the features are correlation, homogeneity, pixel regularity, weights and edge. The method generates 14 statistical features of the texture of the fonts.

$$\text{Correlation } (\boldsymbol{\theta}) = EDM_1 (x,y)/\sum \text{all } EDM_1(x,y) \qquad (3)$$

$$\text{Homogeneity } (\boldsymbol{\theta}) = EDM_1 (x,y)/ EDM_1(2,2) \qquad (4)$$

$$\text{Regularity } (\boldsymbol{\theta}) = EDM_1 (x,y)/ (\sum EDM_1(x,y) - EDM_1(2,2)) \qquad (5)$$

$$\text{Weight} = EDM_1 (2,2)/ \sum (I(x,y) = \text{black}) \qquad (6)$$

$$\text{Edges Direction} = \text{Max} (EDM_1 (x,y)) \qquad (7)$$

where θ represents 0°, 45°, 90° and 135°, θ_* represents 0°, 45°, 90° and 135°, $EDM_1(x,y)$ and presents the occurrences value in the relative position, and $I(x,y)$ presents the original image.

3.3   Recognition Stage

The back-propagation neural network (BPNN) classification technique has been used in this work. The back-propagation neural network is one of the most used neural networks. It consists of an input layer, a hidden layer and an output layer. The hidden layer contains complex relationships between the input and output layers (Daniel et al. 2008). In this research, 14 features are used to identify one of the 7class types of Arabic calligraphy. Based on the training experiments, the BPNN used 14 nodes in its input layer. It has 13 nodes in the hidden layer, and 7 nodes represent the Andalusi, Naskh, Kufi, Parisian, Roqaa, Thuluth and Diwani calligraphies in the output layer. Please draw the BPNN network scheme here???

4   Experiments and Results

In this section, the experiments and performances of the proposed OFR-AC are presented. About 100 documents of each calligraphy type have been collected from different resources. In total, 700 images are presented the Kufi, Diwani, Persian, Roqaa, Thuluth, Naskh and Andalusi are used in this work. This dataset is split into 60% training and 40% testing datasets. The performance of the proposed OFR method is compared with the previous OFR methods such as [12-14] which based on Gabor filter method, [19] which based on GLCM, [21] which based on EDMS, and Local Binary Pattern (LBP) feature extraction method.

Table 1. The accuracy rate of the OFR method of each of the proposed, EDMS, Gabor, LBP and GLCM on the Arabic calligraphy Dataset and Neural Network

	EDM_1	GLCM	LBP	Gabor	EDMS
Neural Network	96.79%	89.49%	84.8%	97.2%	93.67%

Table 2. The required runtimes of the OFR method of each of proposed, EDMS, Gabor, LBP and GLCM on the Arabic calligraphy Dataset and Neural Network

	EDM_1	GLCM	LBP	Gabor	EDMS
Process time/ N	$18N^2$	$3N^2$	$9N^2$	$480N^2$	$27N^2$
Process time/ minutes	52	9	26	1400	79

Based on the experimental results in Table 1, the Gabor filter and the EDM_1 obtained the higher accuracy rates were about 97.2% and 96.79% respectively. While the GLCM, LBP and EDMS obtained about 89.49%, 84.8% and 93.67% respectively. However, as shown in Table 2; Gabor filter required the longest process times to extract the features of all the images in the dataset compared with EDM_1. That shows a significant weakness at Gabor filter performance, its average time was around 23 hours. While the required time to extract the futures from 700 images of the dataset by EDM_1 was around 52 minutes in the same conditions (Intel(R) core (TM)2 Duo CPU), 200GHz, and VB.NET). Moreover, the GLCM, LBP and EDMS required better process time compared with Gabor filter about 9, 26 and 79 minutes. In conclusion, the proposed OFR method given the best performances based on both of accuracy rate and process time compared with other methods. The interface of the proposed method prototype is shown in Fig. 7.

Fig. 7. The interface of the OFR-AC prototype

5   Conclusion

In this work, a method of font recognition for Arabic calligraphy is proposed. The recent methods for binarization, generating the texture of the Arabic calligraphy and EDM_1 feature extraction using BPNN as the classifier are used. In image pre-processing, the proposed binarization process is applied on Arabic calligraphy document images. Then, the text textures are generated. Finally, the edges of the generated texture blocks are extracted. Then, the EDM_1 is applied as the feature extraction method. It is used to extract 14 features of the texture of the seven calligraphy types. Finally, the BPNN is used in the recognition stage to classify the calligraphy font type. Several experiments were conducted to find the performance accuracy of the proposed method. The results show that the proposed method presented the better performances than [12], [14], [19] and [21], and Local Binary Pattern (LBP) feature extraction method. And it is able to add new font types without major modification.

Acknowledgments. In [18] detected an error in its binarization equation (equation (1)) whereby the numerator $m_W^2 - \sigma_W$, should be $m_W^2 * \sigma_W$. The authors would like to thank the Faculty of Information Science and Technology and Center for Research and Instrumentation Management of the, Universitiy Kebangsaan Malaysia, for providing facilities and financial support under Exploration Research Grant Scheme Project No. ERGS/1/2011/STG/UKM/01/18 entitled "Calligraphy Recognition in Jawi Manuscripts using Palaeography Concepts Based on Perception Based Model" and Fundamental Research Grant Scheme No. FRGS/1/2012/SG05/UKM/02/8 entitled "Generic Object Localization Algorithm for Image Segmentation".

References

1. Marinai, S.: Introduction to Document Analysis and Recognition. SCI, vol. 90, pp. 1–20 (2008)
2. Joshi, G.D., Garg, S., Sivaswamy, J.: A generalised framework for script identification. International Journal on Document Analysis and Recognition (IJDAR) 10, 55–68 (2007)
3. Xudong, J.: Feature extraction for image recognition and computer vision. In: 2nd IEEE International Conference on Computer Science and Information Technology "ICCSIT 2009", Beijing, pp. 1–15 (2009)
4. Yong, Z., Tieniu, T., Yunhong, W.: Font recognition based on global texture analysis. IEEE Transactions on Pattern Analysis and Machine Intelligence 23, 1192–1200 (2001)
5. Ding, X., Chen, L., Wu, T.: Character Independent Font Recognition on a Single Chinese Character. IEEE Transactions on Pattern Analysis and Machine Intelligence 41, 153–159 (2007)
6. Abuhaiba, I.: Arabic Font Recognition using Decision Trees Built from Common Words. Journal of Computing and Information Technology 13, 211–223 (2005)
7. Manna, S.L., Sperduti, A., Colla, A.M.: Optical Font Recognition for Multi-Font OCR and Document Processing. In: Proceedings of the 10th International Workshop on Database & Expert Systems Applications, Florence, Italy, pp. 549–553. IEEE Computer Society (1999)

8. Zramdini, A., Ingold, R.: Optical Font Recognition Using Typographical Features. IEEE Transaction on Pattern Analysis and Machine Intelligence 20, 877–882 (1998)
9. Arjun, N.S., Navaneetha, G., Preethi, G.V., Babu, T.K.: Approach to Multi-Font Numeral Recognition. In: IEEE Region 10 Conference in Approach to Multi-Font Numeral Recognition, pp. 1–4 (2007)
10. Sun, H.: Multi-Linguistic Optical Font Recognition Using Stroke Templates. In: 18th International Conference on Pattern Recognition (ICPR 2006), Hong Kong, pp. 889–892 (2006)
11. Zhang, L., Lu, Y., Tan, C.L.: Italic Font Recognition Using Stroke Pattern Analysis on Wavelet Decomposed Word Images. In: Proceedings of the 17th International Conference on Pattern Recognition (ICPR 2004), Cambridge, UK, pp. 835–838. IEEE Computer Society (2004)
12. Ma, H., Doermann, D.S.: Font identification using the grating cell texture operator. In: Document Recognition and Retrieval XII, pp. 148–156. SPIE, San Jose (2005)
13. Ramanathan, R., Thaneshwaran, L., Viknesh, V., Soman, K.P.: A Novel Technique for English Font Recognition Using Support Vector Machines. In: International Conference on Advances in Recent Technologies in Communication and Computing, pp. 766–769 (2009)
14. Ramanathan, R., Ponmathavan, S., Thaneshwaran, L., Nair, A.S., Valliappan, N., Soman, K.P.: Tamil Font Recognition Using Gabor Filters and Support Vector Machines. In: International Conference on Advances in Computing, Control, and Telecommunication Technologies (ACT 2009), pp. 613–615 (2009)
15. Tuceryan, M., Jain, A.K.: Texture Analysis. In: The Handbook of Pattern Recognition and Computer Vision, pp. 207–248. World Scientific Publishing Co., Singapore (1999) ISBN: 9810230710
16. Petrou, M., Sevilla, G.P.: Image Processing, Dealing with Texture. John Wiley & Sons, Ltd., Chichester (2006) ISBN: 0-470-02628-6
17. Bataineh, B., Abdullah, S.N.H.S., Omar, K.: Generating an Arabic Calligraphy Text Blocks for Global Texture Analysis. International Journal on Advanced Science, Engineering and Information Technology 1, 150–155 (2011)
18. Bataineh, B., Abdullah, S.N.H.S., Omar, K.: An Adaptive Local Binarization Method for Document Images Based on a Novel Thresholding Method and Dynamic Windows. Pattern Recognition Letters 32, 1805–1813 (2011)
19. Busch, A., Boles, W., Sridharan, S.: Texture for Script Identification. IEEE Transactions on Pattern Analysis and Machine Intelligence 27, 1720–1732 (2005)
20. Singh, C., Bhatia, N., Kaur, A.: Hough transform based fast skew detection and accurate skew correction methods. Pattern Recognition 41, 3528–3546 (2008)
21. Bataineh, B., Abdullah, S.N.H.S., Omer, K.: A Novel Statistical Feature Extraction Method for Textual Images: Optical Font Recognition. Expert Systems With Applications 39, 5470–5477 (2012)

A New Initialization Algorithm for Bees Algorithm

Wasim A. Hussein[1,*], Shahnorbanun Sahran[2], and Siti Norul Huda Sheikh Abdullah[3]

Pattern Recognition Research Group,
Center of Artificial Intelligence Technology,
Faculty of Information Systems and Technology,
Universiti Kebangsaan Malaysia,
43650 Bandar Baru Bangi, Malaysia
wassimahmed@yahoo.com, {shah,mimi}@ftsm.ukm.my

Abstract. The Bees Algorithm (BA) is a swarm- based metaheuristic algorithm inspired by the foraging behavior of honeybees. This algorithm is very efficient, simple and natural algorithm. In this paper, two natural aspects, namely the patch environment and Levy motion are employed to propose a novel initialization algorithm to initialize the population of bees in the Bees Algorithm. Thus, an improved version of Bees Algorithm is adopted based on the proposed initialization procedure. This initialization algorithm is more natural modeling the patch environment in nature and Levy motion that is believed to characterize the foraging patterns of bees in nature. Experimental results prove the effectiveness of the proposed initialization algorithm. The obtained results confirm that the improved Bees Algorithm employing the proposed initialization algorithm outperforms the standard Bees Algorithm in terms of convergence speed and success rate.

Keywords: bees algorithm, population initialization, levy flight, patch environment.

1 Introduction

Metaheuristic algorithms are often nature-inspired algorithms imitating the most successful behaviors in nature. These metaheuristics could be divided into two classes: population-based algorithms and trajectory-based algorithms. The trajectory-based algorithms use a single agent through the search space such as Simulated Annealing (SA), whereas population-based algorithms use a population and set of agents through the search space such as Genetic Algorithm (GA) [1]. This category of population-based metaheuristics includes techniques inspired by modeling the collective intelligent behaviours of swarm of animals and insects such as fishes, birds, bacteria, ants and fireflies. As a result, the swarm intelligence emerged in the field of artificial intelligence and algorithms such as Ant Colony Optimization (ACO), Particle Swarm Optimization (PSO), Bacterial Foraging Optimization (BFO) and Firefly Algorithm (FA) have been developed.

---

[*] Corresponding author.

S.A. Noah et al. (Eds.): M-CAIT 2013, CCIS 378, pp. 39–52, 2013.

In the last decade, the collective intelligent behaviors of swarm of bees have attracted much attention of researchers to develop intelligent search algorithms such as Honey Bee Optimization (HBO), Artificial Bee Colony (ABC), Beehive (BH) and Bee Colony Optimization (BCO). One of the recent algorithms inspired by the behaviors of bees is the Bees Algorithm (BA). The Bees Algorithm is a population-based search algorithm proposed by Pham et al. [2] and inspired by the foraging behavior of the honeybees to search for the good food sources. In its basic version, the algorithm performs a kind of exploitatory local (neighborhood) search combined with an exploratory global search. The Bess Algorithm has been successfully applied to both combinatorial (discrete) optimization [3], [4] and [5] and functional (continuous) optimization [2] problems.

The Bees Algorithm has enjoyed a significant interest of researchers as it has been proven to be robust and efficient optimization tool. In addition, it is very simple and more intuitive and natural. The Bees Algorithm can be divided into four parts: the parameter tuning and setting part, the initialization part, the local search (exploitation) part and the global search (exploration) part. Several studies have been done to improve the Bees Algorithm and to enhance the performance of the Bees Algorithm. Some of these studies focused on the parameter tuning and setting part. The improvements in these studies were by reducing the number of tunable parameters [6] or by investigating other strategies to define some control parameters [7] and [8]. In addition, some of these studies aimed at developing methods for tuning all parameters of the Bees Algorithm [9]. Other studies focused on developing other concepts and strategies for the local (neighborhood) search part either for general purpose [7], [8] and [10] or for specific problems [9].

However, limited attention has been paid to the study of initialization and global search parts. In initialization step, the foragers or searchers fly at random to initial food sources. The initial location of foragers or searchers relative to optimal source (target) may affect the degree of optimality of other algorithm steps. As a result, the initialization part is a critical part since it can significantly affect the quality of the resource reached and the speed of convergence to the optimal target. In the Bees Algorithm, this part is performed by distributing the bees uniformly at random to food sources. In addition, there is no clear exploitation for the fact that food sources usually found in patches in the initialization part, even though, the concept of patch is used clearly in the neighborhood search part. However, flowers in nature are usually distributed in patches which regenerate and are rarely completely depleted [11]. Furthermore, a scout honeybee searches for new flowers by flying away from the hive and moving randomly throughout the space according to a Levy flight pattern [12], [13] and [14].

Hence, in this paper, an improved version of Bees Algorithm is proposed in which the initialization part is enhanced and improved to be very close to nature. In this version of Bees Algorithm, the search space which represents the environment is dived into segments that represent patches so that the food sources are clearly patchily distributed. Then, the bees are distributed into this search space at random according to Levy flight distribution that is believed to approximate bee flight patterns in nature.

The remaining of paper is organized as follows. Section 2 reviews the basic Bees Algorithm. Section 3 describes the Levy flight. Section 4 presents the proposed algorithm. Section 5 displays the performance evaluation and experimental results obtained by the improved Bees Algorithm and compares these results to the corresponding ones produced using the original algorithm. Finally, section 6 concludes the paper.

2   Standard Bees Algorithm

The Bees Algorithm (BA) is a swarm intelligence-based optimization algorithm inspired by the food foraging behavior of honeybees. Honeybees can exploit a large number of flower patches as food sources by extending their foraging over more than 10 km and in multiple directions [2]. During the harvesting, the foraging process begins in a colony by deploying scout bees to search for useful flower patches with rich pollen or nectar and that are close to the hive. The scout bees fly randomly from one patch to another. When they return to the hive, they deposit their collected nectar or pollen. Then these returned bees go to the dance floor to report their foraging results to other bees in the hive by performing the so-called waggle dance. This dance is very important for colony to communicate three pieces of information regarding the food source, including its direction, distance and quality rating which represent the fitness of the source. As a result, this waggle dance is essential in evaluating the food sources simultaneously and enables the colony to recruit follower bees following the dancers to the promising food sources precisely. More follower bees are sent to more promising patches. This helps the colony to search food quickly and efficiently. Moreover, during the whole harvesting season, a colony continues its exploration, keeping a percentage of the population as scout bees.

The Bees Algorithm is a population-based search algorithm proposed by Pham et al. [2] and inspired by the natural foraging behavior of the honeybees to search for good food sources. In its basic version, during the selection of the best solutions, the algorithm performs fine balance between an exploitatory local or neighborhood search and an exploratory global search. Both kinds of search implement random search. In the global search, the scout bees are distributed uniformly randomly to different areas of the search space to scout for potential solutions. In the neighborhood search, follower bees are recruited uniformly at random to patches found by scout bees to be more promising to exploit these patches. To conduct the local search, two processes are required which are the selection and recruitment processes. In the selection process, the patches found to be more promising are chosen whereas in the recruitment operation follower bees are recruited to those promising patches where more bees are recruited to the best patches out of those selected patches. In its simplest form, the main steps of the Bees Algorithm can be summarized as follows [2]:

1.   Initialize population with (n) random solutions.
2.   Evaluate fitness of the population.
3.   While (stopping criterion not met)
        // Forming new population
4.   Select (m) sites for neighborhood search.
5.   Determine the patch of each selected site to have an initial size of (ngh).

6.   Recruit bees for selected sites or patches (more bees (*nep*) for best (*e*) sites out of the (*m*) selected sites and less bees (*nsp*) for the remaining (*m-e*) selected sites).
7.   Evaluate the fitness of the bees of the selected patches.
8.   Select the fittest bee from each patch.
9.   Assign remaining bees (*n-m*) to search randomly and evaluate their fitness.
10.   End while.

3   Levy Flight

In nature, resources such as flowers, fish, krill, etc. are usually distributed in patches which regenerate and are rarely completely depleted. In the absence of prior knowledge about the locations of patches of food sources, stochastic search strategies are considered as more efficient strategies than systematic deterministic strategies [15]. So the random walks can be used as models of animal movement. Among random walks, it was demonstrated that Levy flight motion is the optimal search strategy for the forager or searcher to search for food patches [11], [13] and [16].

Levy Flights were demonstrated theoretically and empirically to characterize foraging patterns of various animals and species such as the wandering albatross, reindeer, jackals, dinoflagellates, spider monkeys, sharks, bony fish, sea turtles, penguins, drosophila fruit flies, bumblebees and honeybees [17]. Even foraging patterns in humans such as the Dobe Ju/'hoansi hunter-gatherers presented evidence of using human foragers of Levy flights as search patterns [18]. The evidence of Levy flight foraging patterns in honeybees is specifically strong [17]. This is because compared with other studies, Reynolds et al. [14] and [16] use techniques where few of which have been criticized by recent studies [19] and [20].

Levy flights are random walks named after Paul Levy, the French mathematician. Levy flights comprise sequences of independent, randomly oriented steps with lengths l, drawn at random from an inverse power-law distribution with heavy and long tail, $p(l) \sim l^{-\mu}$ where $1 < \mu \le 3$. Levy flights are scale-free since they do not have any characteristic scale because of the divergent variance of $p(l)$ and they present the same fractal patterns regardless of the range over which they are viewed. The pattern in Levy flights can be described by many relatively short steps (corresponding to the detection range of the searcher) that are separated by occasional longer jumps.

It can be easily noticed from the definition of Levy flights that two steps are required to mimic the Levy movements in nature and to implement these flights. The first step is generating a random direction d to mimic the random choice of direction by drawing it from a uniform random distribution. In the proposed initialization algorithm, the direction is drawn from a uniform distribution between -1 and 1. The second step is generating the step length that obeys a Levy distribution. One of the most efficient and at the same time straightforward algorithms to generate Levy stable variable is the Mantegna's Algorithm for symmetrical Levy stable distribution [21]. A stochastic variable z can be called stable if it satisfies the flowing property: a linear combination of several independent stochastic copies x_i of the stochastic variable z has the same distribution of x_i variables [21].

The symmetrical Levy stable distribution can be defined as

$$L_{\alpha,\gamma}(z) = \frac{1}{\pi} \int_0^\infty \exp(-\gamma q^\alpha) \cos(qz) \, dq, \tag{1}$$

Where $0 < \alpha \le 2$ define the index of the distribution and determine the shape of the distribution and $\gamma > 0$ is a scale parameter selects the scale unit of the distribution. The Mantegna's algorithm is divided into three steps. The first one is to generate a stochastic variable v as follows

$$v = \frac{x}{|y|^{1/\alpha}}, \tag{2}$$

where x and y are normally distributed. That is

$$x \sim N(0, \sigma_x^2), \qquad y \sim N(0, \sigma_y^2), \tag{3}$$

where

$$\sigma_x(\alpha) = \left[\frac{\Gamma(1+\alpha)\sin(\pi\alpha/2)}{\Gamma((1+\alpha)/2)\alpha 2^{(\alpha-1)/2}} \right]^{1/\alpha}, \quad \sigma_y = 1, \tag{4}$$

where Γ is the gamma function. The distribution of the stochastic variable v has the same behavior as the Levy stable distribution for large values of this variable.

To obtain a random variable with a distribution that converges rapidly to the Levy stable distribution all over the range of this stochastic variable, a nonlinear transformation is achieved as follows

$$w = \left\{ [K(\alpha) - 1] \left[\exp(-|v|/C(\alpha)) \right] + 1 \right\} v, \tag{5}$$

where the values of the parameters $K(\alpha)$ and $C(\alpha)$ are tabulated by Mantegna [21] numerically for specific values of α ($0.75 \le \alpha \le 1.95$), where it was found that the algorithm is very fast and efficient for these values of α. In the proposed initialization algorithm, α is considered to be 1.5 and thus the corresponding $K(\alpha)$ and $C(\alpha)$ are set to 1.59922 and 2.737 respectively as tabulated by Mantegna [21].

Then, n independent copies of w are generated and the central limit theorem is used to generate a random variable z that converges quickly to the Levy stable distribution as follows

$$z = \frac{1}{n^{1/\alpha}} \sum_{k=1}^{n} w_k. \tag{6}$$

The distribution for z obeys the Levy distribution all over rang values of z, not for the large values only. So the random variable s which represents the step length is given by

$$s = \gamma^{1/\alpha} z \qquad\qquad (7)$$

where γ is related to the scale of the problem under study.

4   Proposed Bees Algorithm

In this paper, the initialization part of the original Bees Algorithm is improved and made to be closer to nature. The food sources in the improved Bees Algorithm are patchily distributed and the bees are randomly distributed according to the Levy flight distribution. In this section, the proposed initialization algorithm is presented, and then improved Bees Algorithm is devised based on the new initialization procedure.

4.1   Proposed Initialization Algorithm

Since starting forging or searching from good points can significantly enhance the quality of the last solution and accelerate getting to the optimal solution, population initialization is considered to be a critical part in the Bees Algorithm and other population-based metaheuristics. Therefore, a new population initialization inspired by nature is adopted and an improved Bees Algorithm is proposed.

Two aspects in nature are modeled in the proposed initialization algorithm. The first aspect modeled is the existence of food sources (flowers) in patches. Even though the patch concept is considered in the neighborhood search of the standard Bees Algorithm, this concept is not used clearly in the initialization step. The improved Bees Algorithm, in contrast, uses the patches clearly and the search space is divided into a number of equal segments (P), each of which is called a patch. This number of patches P is determined by the user using a small number of trials, where $1 \leq P \leq n$, n is the number of scout bees.

The second natural aspect which is imitated by the new initialization algorithm is the Levy flight patterns characterizing the forging patterns of the honeybees. Therefore, the initial distribution of scout bees into the search space in the improved Bees Algorithm is achieved according to Levy flight distribution. In nature, bees occupy different spaces or areas inside their hives from which bees start to fly away from the hive. As a result, in order to model this difference of areas occupied by bees, one point out of the points in a patch is selected to represent a space or an area in the beehive. One bee or more can fly from this area foraging for food. Hence, any patch will be near to the bees which will fly from the space represented by the point chosen from this patch. Conversely, the other patches will be far from those bees. This difference in distance of fly is modeled properly using the Levy flights since they consist of frequent short steps separated by occasional longer jumps. In the proposed initialization algorithm, the point representing an area of hive is chosen to be the center of a patch because it represents the mean.

To summarize, in the proposed algorithm, the food sources are patchily distributed in a clear way. In addition, the scout bees fly initially according to the Levy flight distribution demonstrated to characterize the forging pattern of honeybees.

Based on these operations, a new initialization algorithm is proposed to generate the initial population in the Bees Algorithm. The steps of this algorithm are displayed in the flowchart depicted in Fig. 1.

4.2   The Improved Bees Algorithm

Having presented the proposed initialization procedure, the main steps of the improved Bees Algorithm can be summarized as follows.

1. Initialize population using the proposed initialization algorithm.
2. Evaluate fitness of the population.
3. While (stopping criterion not met)
   // Forming new population
4. Select sites for neighborhood search.
5. Recruit bees for selected sites (more bees for best e sites) and evaluate their fitness.
6. Select the fittest bee from each patch.
7. Assign remaining bees to search randomly and evaluate their fitness.
8. End while.

5   Experimental Results and Discussion

The proposed Bees Algorithm has been tested with six benchmark functions [22] and [23]. Table 1 shows the benchmark functions and their global optima. As the Bees Algorithm and, thus the improved Bees Algorithm were designed to search for the maximum, the functions to be minimized were inverted by multiplying them by (-1) before applying the algorithms.

To demonstrate the effectiveness of the proposed initialization algorithm, the performance of the improved Bees Algorithm employing this initialization procedure is evaluated and compared with the performance of the original Bees Algorithm (BA). To evaluate the performance of the proposed Bees Algorithm, two performance metrics were used, namely, the success rate and the mean number of evaluations. Moreover, the convergence behavior of the proposed algorithm was analyzed.

The success rate is one of the performance criteria used in the literature for evaluating the performance of the algorithms. The success rate for an algorithm is the percentage of runs in which an algorithm succeeded to converge to the optimal solution among a number of runs for the algorithm. With respect to the second criterion, the number of times any benchmark function had to be evaluated to reach the optimal solution was taken for the proposed and original algorithms to get an indication of the convergence speed of the algorithms.

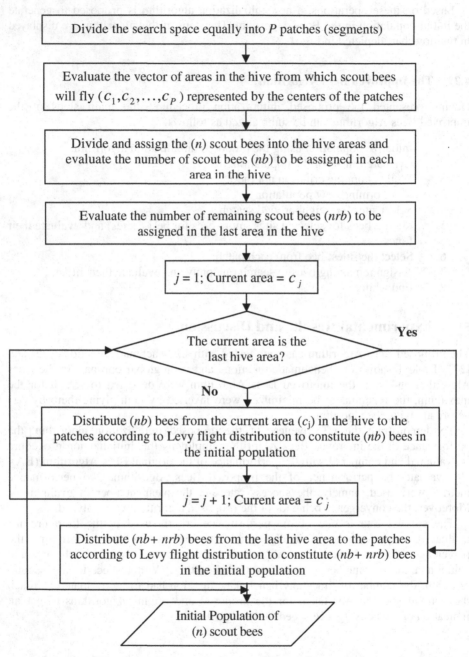

Fig. 1. Flowchart of the proposed initialization algorithm

The parameter setting of the algorithms which used for the different test functions is shown in Table 2. The same parameters and values were used for the original and improved versions of the Bees Algorithm except for the last parameter (*P*) which was only set for the proposed Bees Algorithm. In addition, to take into account the stochastic nature of the original and proposed algorithms, both algorithms were executed and the average results for 100 independent runs were considered.

5.1   Success Rate and Mean Number of Evaluations

Table 3 shows the average number of evaluations and success rates resulted from applying both the original and proposed Bees Algorithm to the six benchmarks. The first test function was Schaffer's function, for which the proposed Bees Algorithm showed an improvement of approximately 95% compared with the basic Bees Algorithm (BA) in terms of convergence speed, with 100% success for both algorithms. The second function was Easom's function, for which the Improved Bees Algorithm also satisfied 100% success and an improvement of roughly 53% over the basic algorithm. For test function 3, it was the fifth function of De Jong's test functions which is called the two-dimensional Shekel's Foxholes function. Shekel's Foxholes function is a multimodal function with many local optima. With regard to this function, the function evaluations required for the improved Bees Algorithm to converge to the optimum is 39 times less than that required by the standard Bees Algorithm, also with 100% success. With Rastrigin's function, the proposed Bees Algorithm delivered a success rate of 100% and was able to locate the global optimum 184 times faster than the basic Bees Algorithm.

The fifth function was a multimodal test function called Ackley's function with ten dimensions. For this function, the improvement delivered by the improved Bees Algorithm in terms of the number of function evaluations was very highly significant. It was found that only 167 function evaluations were necessary for the proposed Bees Algorithm to converge to the global optimum with 100% success. On the other hand, the standard Bees Algorithm required a number of evaluations of 901,611 evaluations which was approximately 5,398 times higher than that required by the improved Bees Algorithm. The last test function was shifted Griewank's function with fifty dimensions. This function is highly multimodal and has many local optima which make this function difficult to handle. For this fifty-dimensional problem, the improved Bees Algorithm performed much better and satisfied a highly significant improvement over the basic Bees Algorithm. While the basic Bees Algorithm was able to locate the global optimal solution with a 96% success rate, the improved Bees Algorithm attained convergence to the global optimum with a success rate of 100%. Furthermore, the improved Bees Algorithm could converge to the optimal solution in just 165 function evaluations, whereas the original Bees Algorithm required 560,905 function evaluations, which was about 3,398 times more than that required by the Improved Bees Algorithm.

5.2   Convergence of the Proposed Bees Algorithm

To further analyze the convergence of the Bees Algorithm based on the proposed initialization procedure, an additional test function was used to compare the proposed Bees Algorithm with the basic Bees Algorithm. This test function was the Schwefel's function with six dimensions. As mentioned before, as the proposed and basic Bees Algorithms search for the maximum and this test function is a minimization function, the function was inverted by multiplying it by (-1) before applying the algorithms. Fig. 3 shows the convergence progress of the proposed and basic Bees algorithms to the best fitness with the number of evaluations of the inverted test function.

The Schwefel's function is a highly multimodal test function with great number of peaks and valleys. In addition, this test function is deceptive in that global minimum is geometrically far from the second best local minimum where a search algorithm can be trapped. Consequently, a search algorithm is likely to converge in the wrong direction [22]. The definition of this function is as the following

$$f(x) = \sum_{i=1}^{6} \left[-x_i \sin\left(\sqrt{|x_i|}\right) \right],$$

(8)

where $-500 \le x_i \le 500, i = 1,...6$. The global minimum is $f\left(\vec{x}_{min}\right) = -2513.8974$ and $\vec{x}_{min,i} = 420.9687$, $i = 1,...6$. An overview of the Schwefel's function in 2D is shown in Fig. 2. With respect to the inverted six-dimensional Schwefel's function, the improved Bees Algorithm converged to the global maximum (2513.8974) much faster than the basic Bees Algorithm. It can be seen clearly from Fig. 3 that the improved algorithm based on the proposed initialization algorithm could converge to the global optimum after exactly 1,535 function evaluations which correspond exactly to only one iteration. On the other hand, the standard Bees Algorithm required a large number of evaluations to locate the optimum. The average number of evaluations in 100 independent runs was plotted against the best fitness value as shown in Fig. 3. It is clear that the basic Bees Algorithm was able to find solutions close to the optimum after approximately 3,500,000 evaluations. For this test, the parameter values used were $n = 500$, $m = 15$, $e = 5$, $nsp = 30$, $nep = 50$ and initial patch size $ngh = 20$. The parameter related to the number of patches in the proposed algorithm was set to 500 ($P = 500$).

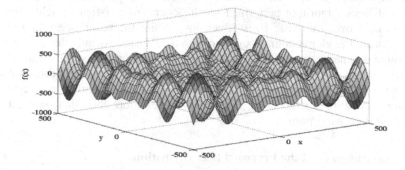

Fig. 2. An overview of Schwefel's function in 2D

In general, it can be observed that the proposed Bees Algorithm employing the proposed initialization algorithm needs less number of function evaluations than the original Bees Algorithm. Moreover, the proposed Bees Algorithm succeeds to converge to the global optimum in all runs of the algorithm. As a result, it can be concluded that the proposed initialization procedure accelerates the convergence of the basic Bees Algorithm to the global optimum. In addition, the Bees Algorithm based on this initialization algorithm performs better than the standard Bees Algorithm in terms of the convergence speed and success rate.

a

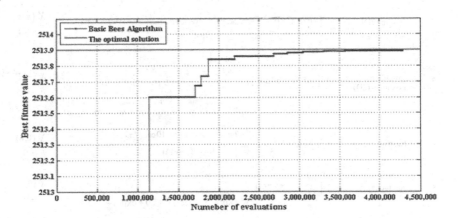

b

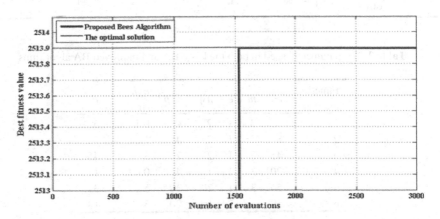

Fig. 3. Convergence behavior of: (a) basic Bees Algorithm on Inverted Schwefel's function, (b) proposed Bees Algorithm on Inverted Schwefel's function

Table 1. Test functions and their global optima

No	Function Name	Search Space	Function	Global Optimum
1	Schaffer (2D)	$X[-100,100]$	$\min F = 0.5 + \dfrac{\left(\sin\sqrt{x_1^2+x_2^2}\right)^2 - 0.5}{\left[1.0+0.001\left(x_1^2+x_2^2\right)\right]^2}$	$X(0,0)$ $F=0$
2	Easom (2D)	$X[-100, 100]$	$\min F = -\cos(x_1)\cos(x_2)\exp\left(-(x_1-\pi)^2-(x_2-\pi)^2\right)$	$X(\pi,\pi)$ $F=-1$
3	Shekel's Foxholes (2D)	$X[-65.536, 65.536]$	$\min F = \left[\dfrac{1}{500} + \sum_{j=1}^{25}\dfrac{1}{j + \sum_{i=1}^{2}(x_i - a_{ij})^6}\right]^{-1}$ $a_{ij} = \begin{pmatrix} -32 & -16 & 0 & 16 & 32 & -32 & \dots & 0 & 16 & 32 \\ -32 & -32 & -32 & -32 & -32 & -16 & \dots & 32 & 32 & 32 \end{pmatrix}$	$X(-32,-32)$ $F=1$
4	Rastrigin (2D)	$X[-5.12, 5.12]$	$\min F = 20 + \left(x_1^2 - 10\cos(2\pi x_1)\right) + \left(x_2^2 - 10\cos(2\pi x_2)\right)$	$X(0,0)$ $F=0$
5	Ackley (10D)	$X[-32.768, 32.768]$	$\min F = -20\exp\left(-0.2\sqrt{0.1\sum_{i=1}^{10} x_i^2}\right) - \exp\left(0.1\sum_{i=1}^{10}\cos(2\pi x_i)\right) + 20 + \exp(1)$	$X(\bar{0})$ $F=0$
6	Griewank (shifted) (50D)	$X[-600,600]$	$\min F = \sum_{i=1}^{50}\dfrac{(x_i-100)^2}{4000} - \prod_{i=1}^{50}\cos\left(\dfrac{x_i-100}{\sqrt{i}}\right) + 1$	$X(\overline{100})$ $F=0$

Table 2. The parameters used in both the basic BA and improved BA algorithms

Function number	n	m	e	nsp	nep	ngh (initial)	P
1	8	3	1	2	4	0.1	5
2	8	3	1	2	4	1	1
3	45	3	1	2	7	3	45
4	20	3	1	1	4	0.09	1
5	50	3	1	20	30	5	31
6	50	5	2	10	20	5	18

Table 3. Success rate, mean number of evaluations test functions using basic BA and improved BA algorithms

No	Function Name	Bees Algorithm			Improved Bees Algorithm		
		Mean no. of evaluations	Std of no. of evaluations	Success rate	Mean no. of evaluations	Std of no. of evaluations	Success rate
1	Schaffer (2D)	400	197	100%	21	0	100%
2	Easom (2D)	2,341	1,850	100%	1,091	832	100%
3	Foxholes (2D)	3,897	4,139	100%	98	0	100%
4	Rastrigin (2D)	7,958	6,678	100%	43	0	100%
5	Ackley (10D)	901,611	16,750	100%	167	0	100%
6	Griewank (shifted) (50D)	560,905	250,339	96%	165	0	100%

6   Conclusion

The Bees Algorithm (BA) is a population-based metaheuristic algorithm inspired by the foraging behavior of honeybees. This algorithm has been proven to be robust and efficient in the optimization field. In addition, the algorithm is very simple, more intuitive and more natural. However, the convergence speed of the Bees Algorithm to the optimal solution is still in need to be improved more without being trapped in the local optima. In addition, there are some biological aspects in nature related to the honeybee foragers and food sources that can be modeled and incorporated in the Bees Algorithm. Among these aspects which can be modeled are the distribution of food sources and the distribution of honeybees when they fly away from the hive foraging for food. It has been noted that food sources, in nature, are usually found in patches. Furthermore, it has been found that the Levy flights characterize the foraging flight patterns of honeybees. In this paper, a novel initialization algorithm based on the patch concept and the Levy motion has been proposed. Thus, an improved Bees Algorithm employing this initialization algorithm has been adopted. The experiments confirmed that the improved Bees Algorithm converged rapidly to the optimal solution without getting stuck at local optima. In addition, the experimental studies proved that the proposed algorithm outperformed the original Bees Algorithm in terms of the convergence speed to the optimal solution and the rate of succeeding to converge to the optimal solution.

Acknowledgments. The authors would like to thank the Faculty of Information Science and Technology, Universiti Kebangsaan Malaysia, for providing facilities and financial support under Fundamental Research Grant Scheme No. AP-2102-019 entitled "Automated Medical Imaging Diagnostic Based on Four Critical Diseases: Brain, Breast, Prostate and Lung Cancer".

References

1. Yang, X.-S.: Review of metaheuristics and generalised evolutionary walk algorithm. Int. J. Bio-Inspired Comput. 3(2), 77–84 (2011)
2. Pham, D., et al.: The bees algorithm–a novel tool for complex optimisation problems. In: Proceedings of IPROMS 2006 Conference (2006)
3. Pham, D., et al.: Using the bees algorithm to schedule jobs for a machine. In: Proceedings of Eighth International Conference on Laser Metrology, CMM and Machine Tool Performance (2007)
4. Pham, D., Otri, S., Darwish, A.H.: Application of the Bees Algorithm to PCB assembly optimisation. In: IPROMS, 3rd International Virtual Conference on Intelligent Production Machines and Systems (2007)
5. Lara, C., Flores, J.J., Calderón, F.: Solving a School Timetabling Problem Using a Bee Algorithm. In: Gelbukh, A., Morales, E.F. (eds.) MICAI 2008. LNCS (LNAI), vol. 5317, pp. 664–674. Springer, Heidelberg (2008)
6. Pham, D., Darwish, A.H.: Fuzzy selection of local search sites in the Bees Algorithm. In: 4th International Virtual Conference on Intelligent Production Machines and Systems, IPROMS 2007 (2008)
7. Ahmad, S.: A study of search neighbourhood in the bees algorithm. Cardiff University (2012)
8. Ghanbarzadeh, A.: The Bees algorithm. A novel optimisation tool. Cardiff University
9. Otri, S.: Improving the bees algorithm for complex optimisation problems. Cardiff University (2011)
10. Packianather, M., Landy, M., Pham, D.: Enhancing the speed of the Bees Algorithm using Pheromone-based Recruitment. In: 7th IEEE International Conference on Industrial Informatics. IEEE (2009)
11. Viswanathan, G., et al.: Optimizing the success of random searches. Nature 401(6756), 911–914 (1999)
12. Bailis, P., Nagpal, R., Werfel, J.: Positional communication and private information in honeybee foraging models. Swarm Intelligence, 263–274 (2010)
13. Reynolds, A.M.: Cooperative random Lévy flight searches and the flight patterns of honeybees. Physics Letters A 354(5), 384–388 (2006)
14. Reynolds, A.M., et al.: Honeybees perform optimal scale-free searching flights when attempting to locate a food source. Journal of Experimental Biology 210(21), 3763–3770 (2007)
15. Bartumeus, F., Catalan, J.: Optimal search behavior and classic foraging theory. Journal of Physics A: Mathematical and Theoretical 42(43) (2009)
16. Reynolds, A.M., et al.: Displaced honey bees perform optimal scale-free search flights. Ecology 88(8), 1955–1961 (2007)
17. Viswanathan, G., Raposo, E., Da Luz, M.: Lévy flights and superdiffusion in the context of biological encounters and random searches. Physics of Life Review 5(3), 133–150 (2008)
18. Brown, C.T., Liebovitch, L.S., Glendon, R.: Lévy flights in Dobe Ju/'hoansi foraging patterns. Human Ecology 35(1), 129–138 (2007)
19. Edwards, A.M.: Overturning conclusions of Lévy flight movement patterns by fishing boats and foraging animals. Ecology 92(6), 1247–1257 (2011)
20. Edwards, A.M.: Using likelihood to test for Lévy flight search patterns and for general power-law distributions in nature. Journal of Animal Ecology 77(6), 1212–1222 (2008)
21. Mantegna, R.N.: Fast, accurate algorithm for numerical simulation of Levy stable stochastic processes. Physical Review E 49, 4677–4683 (1994)
22. Molga, M., Smutnicki, C.: Test functions for optimization needs, http://www.zsd.ict.pwr.wroc.pl/files/docs/functions.pdf
23. Adorio, E.P., Diliman, U.: MVF-Multivariate Test Functions Library in C for Unconstrained Global Optimization, http://www.geocities.ws/eadorio/mvf.pdf

Part-of-Speech for Old Malay Manuscript Corpus: A Review

Juhaida Abu Bakar[1], Khairuddin Omar[2],
Mohammad Faidzul Nasrudin[2], and Mohd Zamri Murah[2]

[1] School of Computing, Universiti Utara Malaysia,
06010 Sintok, Kedah, Malaysia
[2] Center for Artificial Intelligence Technology, Faculty of Information Science and
Technology, Universiti Kebangsaan Malaysia,
43600 Bangi, Selangor, Malaysia
juhaida.ab@uum.edu.my, {ko,mfn,zamri}@ftsm.ukm.my

Abstract. Research in Malay Part-of-Speech (POS) has increased considerably in the past few years. From the literature, POS are known as the first stage in automated text analysis and the development of language technologies can scarcely begun without this initial phase. Malay language can be written in Roman or Jawi. Three different spelling between Roman and Jawi make this study essential. In this paper, we highlighted the problem and issues related to Malay language, POS general framework, POS approaches and techniques. POS at basis was introduced to get information from Old Malay Manuscripts that contain important information in various spheres of knowledge. Promising result for the auto-tagging of Malay written in Jawi is expected.

Keywords: Part-of-speech tagging, tagging framework, malay language, Jawi.

1   Introduction

Assigning lexical categories to words is an important first step in the automated analysis of a text [1]. The task of part-of-speech assignment consists of assigning a word to its appropriate word class. The significance of part-of-speech (also known as POS, word classes, morphological classes, or lexical tags) for language processing is the large amount of information they give about a word and its neighbors [2], hence can reduce ambiguous, homonym, and homograph [1].

1.1   Scope of the Problem

Jawi script is an Arabic script that adopts Malay language writing. Jawi spelling system has existed since the 17th century, however, the system (old Jawi system) is still not fixed (non-systematic) and neat [3] and [4]. In addition, [4] added there are many homographs (same word but different meaning), and dialect in old Jawi writing. Recent work done by [5] also stated the same issue raised by the linguist. But, with

S.A. Noah et al. (Eds.): M-CAIT 2013, CCIS 378, pp. 53–66, 2013.
© Springer-Verlag Berlin Heidelberg 2013

the existing recording and writing, the old Jawi writing considered has adopted the basic standardization system to guide pronunciation and writing at that time.

Malay language can be written either in Roman or Jawi. There are differences between Roman and Jawi. Roman should be written from left to right and it is more like English character but Jawi should be written from right to left and it is similar to Arabic character. Spelling of Malay word in Jawi is different compare to spelling Malay word in Roman [5], [6] and [7].

Based on [8], three different spelling between Roman/Jawi and related to the preparation of POS are, acronyms/abbreviations, reduplicative and spelling preposition.

First, there are no distinct upper and lower case letter forms in writing acronyms/abbreviations in Jawi. The distinctions are follows the Arabic character. In difference, Roman that follows English character can differentiate the acronyms such as Proper Noun (PN) with capital letter. State-of-the-art technique for English language, TnT tagger [9], highlighted that the additional information that turned out to be useful for disambiguation process for several corpora and tagsets is capitalization information. Hence, we can detected the acronyms and abbreviations with much easier using Roman than Jawi. Table 1 shows detail on writing acronyms and abbreviation for Roman and Jawi by [8], [10].

Table 1. Writing acronyms and abbreviations for Jawi and Roman in Malay language

No	Form	Roman	Jawi
1.	Special name; name of the position, rank, title, and proper name	Profesor	ڤروفيسور
2.	Initialism abbreviation for the name of department, and the position of the Malay language	TUDM	ت.او.د.م
3.	Initialism abbreviation for the name of department, and the position of the English language	MCA	عيم.سي.أي
4.	Acronyms special name / common in Malay / English	Pas	ڤاس

Second, reduplicative word in Jawi spelled by Arabic numerals digit two (٢) and differ to Roman which spelled with hypens. For example, sheeps spelled as (بيري٢) in Jawi, but (*biri-biri*) in Roman. Third, spelling system is differ for both Roman and Jawi. For example, in writing preposition such as *di-*, *ke-* in Jawi, we must lifting or attached the word to the base word but in Roman, there is a cases where we separated the preposition with the base word. Here, proves the difference in the formation of Jawi and Roman character and make it the motivation of this study.

Other than that, there are two main things important in this study; the methods of learning and techniques as the consequence. Transition from supervised to unsupervised learning becomes necessity now as some of the problems encountered in supervised learning. The problems rise in supervised learning from the needs of annotated data. Those high-quality resources are typically unavailable for many languages and their creation is labour-intensive [11], [12], [13], [14] and [15]. At least two possible solutions can be used to overcome the problem either to improve prediction algorithm by adding linguistic data or updating lexicon model [16]. Both solutions are not easy because of the continuous updating task from time to time. Even for languages with rich resources language like English, tagger performance breaks down on noisy input. Texts of a different genre than the training material may also create problems, e.g. e-mails as opposed to newswire or literature [12]. Table 2 shows the direction from supervised to unsupervised for several languages.

Table 2. Direction of learning approaches in POS tags

Methods of learning	Malay language	Indonesian language	Arabic language	English language	References
Supervised	✓	✓	✓	✓	[17–22]
Semi-supervised /Unsupervised			✓	✓	[11], [12], [14], [23], [24]

In Malay POS, the prior research [17], [25], [26] and [27] are only done for Roman character. Minority languages including Malay facing problem in preparing annotated data. [17] stated the problem arise when building a statistical tagger for Malay because there is no readily Malay tagged corpus publicly accessible. [27] also stated the same issue. To date, however, there is no current research for Jawi POS.

2   Related Work

Based on the issues raised above, the previous study related to Jawi script is discussing here. Table 3 and 4 shows the previous study in script Jawi and relation to natural language processing (NLP).

Table 3. Script Jawi study related to NLP

No	Researcher	Sub field Related to NLP	Subject
1.	Sulaiman [6], [28]	Morphology analysis	Stemmer
2.	Razak, Othman [29], [30]	Morpheme analysis	Speech recognition
3.	Ghani, Bakar, Yonhendri, Ahmad [31–34]	NLP application	Transliteration

Table 4. Script Jawi study unrelated to NLP

No	Researcher	Sub field Unrelated to NLP	Subject
1.	Nasrudin, Azmi, Heryanto, Zulcaffle, Redika [35–40]	Pattern recognition	Character recognition
2.	Shittiq [41]	Multimedia	Teaching & Learning
3.	Diah, Abdullah [42–44]	Multimedia	Mobile games (apps)
4.	Ismail [45]	Engineering	Hardware

Computer processing of natural language normally follows a sequence of steps, beginning with a phoneme and morpheme-based and stepping toward semantics and discourse analyses [13]. There are six different phases of knowledge to understand a natural language; phonology, morphology, syntax, semantic, pragmatic and discourse [46]. Computer processing is an important pre-processing step for most NLP applications. Computer processing are used for speech recognition, text processing, information retrieval, information extraction, machine translation. The previous study related to script Jawi and machine translation done by [31], [32], [33] an d [34].

Phonology is a field regarding how words are related to the sound. The simplest abstract class is called the phoneme [2]. More precisely, phoneme is speech sound that have differential meaning components, such as /p/ and /b/ in the pagi (morning) and bagi (for). Previous study done by [29] and [30] related to Jawi script and phonology. Since this study started from the word, phoneme studies excluded from this study discussion.

Morphology is the study of the way words are built up from smaller meaning-bearing units, morphemes [2]. A morpheme is often defined as the minimal meaning-bearing unit in a language. For example, the word fox consists of one morpheme (the morpheme fox) and the word cats consists of two: the morpheme cat and the morpheme –s. On other case, for example, the word 'recently' comes from the base word 'recent' (adjective) and coupled with the suffix –ly. When both are joined together, the word 'recently' becomes an adverb [46]. Here we can see some cases when morphology word can determine parts-of-speech of the word. Malay linguist stated that [47] morphological field focused on the formation of the Malay words, the structures and parts-of-speech. Parts-of-speech can be used in a stems as knowing parts-of-speech can help in identifying the appropriate affixation word in a word. Previous study done by [6] and [28] related to Jawi script and stemming. This study will revolve around this part.

The next phase in the computer processing is syntactic. This phase concerns how words can be put together to form a correct sentence. Syntax can be defined as areas of linguistics that studies the shape, structure, and building or construct sentences [48]. For example, the structure of the sentence 'she reads a book' is (S (NP (N she)) (VP (V reads) (VP (ART a) (N book)))). The fourth phase is semantic knowledge. The semantic knowledge concerns about the meaning of the words and how these meanings are combined in sentences to form sentence meanings [2] and [47]. For instance, the sentence "Colorless green ideas sleep furiously" [49] would be rejected as semantically anomalous.

Next is the pragmatic phase. It deals with the use of a sentence in different situations. For instance, the sentence "you have a green light" can give two interpretations. The first interpretation is "you are holding a green light bulb" and the second interpretation is "you have a green light to drive your car". The last phase is the discourse phase. Discourse phase is a phase which concerns how the previous sentence affects the next sentence. For example, "Ahmad wanted it". "It" is referred to previous sentence.

3   Methods of Learning

Methods of learning can be divided to four different systems; supervised systems, semi-supervised systems, bootstrapping, and unsupervised systems [50].

3.1   Supervised Systems

In supervised systems, the data as presented to a machine learning algorithm is fully labelled either manually hand tagged or annotated by a lexicon. That means all examples are presented with a classification that the machine is meant to reproduce. For this, a classifier is learned from the data, the process of assigning labels to yet unseen instances is called classification [50].

In supervised learning, there are several approaches to automatically classify words that can be used like rule-based approach, probability-based approach and transformation-based rule. Rule-based classification of words [20] defines the word based on rules given by linguists manually, for example, nouns are classified word if it is after the adjective. Probability approach [17] , [18], [19], [22] and [51] determine the word with the highest possibility occur, and the context in which nearby. Probability percent given by the annotated corpus using different models such as Conditional Random Fields (CRF), Maximum entropy (ME), Hidden Markov Models (HMM), and Decision Trees (DT) distinguishes performance evaluation of the lexicon. Third, transformation-based approach combined rule-based approach and the possibility of automating the rules directly from corpus linguistics. One of the transformation-based approaches is well known [52]. Assessment in terms of classification accuracy measured words. For English, the classification accuracy is between 97% up to 100% depending on the size of the set of words [53], [54], [55], [56], [57] and [58].

For Malay language, studies by [17] and [25] are the starting point in word classification of Malay. In [17] study, a modified HMM model with Malay morphological information is used to learn the word classification. 18,135 corpus data were used with 1,381 ambiguous words, the classification accuracy up to 67.9% for 21 tagset.

Other work improved on previous techniques such as that of Zamin in [27] she proposed an unsupervised technique using bitext mapping to tag a Malay text. This approach involves translation of texts into a resource-rich language, i.e. English, and a dictionary lookup. The results demonstrate that the system achieved 76% precision and 67% recall. This study is a starting point for a development of an unsupervised automated tagger for Malay. This implementation of this technique shows well when the sentence pair is good, i.e when there is no data sparsity problem. Additionally, there is a problem to generate many-to-one alignments, e.g. to map *melarikan diri* with fled. Since this is ongoing research, several new features will be added to increase the performance. In my opinion, the tagset also need to be considered in ensuring the consistency for both languages. There is also a study of similar language are summarized in Table 5.

Table 5. POS Tagging from different language

Language	Researcher
Indonesian	Pisceldo, Wicaksono and Syandra [18–20]
Arabic	Albared, AlQrainy, Diab and Khoja [22], [23], [59], [60]
English	Lluis, Schroder, Brants, Nugues, Dickinson, Mohd Fadzil, and MacCartney [53–58], [61]

For the same root language, Indonesian language, the highest percentage of word classification is [18] with 99.4%. [18] used HMM model where a word not in the dictionary (*Out-of-Vocabulary* (OOV)) is not included, if OOV included, the accuracy dropped to 80.4%. This study shows that the OOV words are the problem that should be considered when supervised learning is used. Other studies [19] from Indonesian language using two different models, namely Conditional Random Fields (CRF) and Max entropy with learning 10 fold has an accuracy of 91.5% for CRF and 97.57% for Max entropy. Study by [20] where adopted the approach by [52] obtain 88% percent accuracy.

In Arabic language, [22] shows highest percentage in classifying word using smoothing algorithms with Hidden Markov Model (HMM). Several lexical model are also have been defined and implemented to handle unknown word POS guessing based on word substring i.e. prefix probability, suffix probability or the linear interpolation for both of them. The average overall accuracy is 95.8%. Study by [23] with 91% accuracy were followed using pattern of diacritics (wazn) to build a generic POS tagger system without a lexicon (dictionary) and depends on the language and the characteristics of its grammar, both the morphological and the syntactical systems of the language.

Two difficulties need to be handle in POS tagging which are ambiguous word and unknown word [16], [17], [62] and [63]. More than one annotated tag given for one word called ambiguous word, and if null annotated tag given for a word called unknown word. Unknown word basically happen when word is unseen in training corpus [13] and [17] and word not in the lexicon (dictionary) [18]. As mention in Section 1.1, Roman writing used upper letter to differentiate acronyms and abbreviation. Here, unknown word always pre-tagged with the assumption it is Noun or Proper Noun if the word not in the lexicon such as Person Name, Place Name, etc. This problem named Out-of-Vocabulary (OOV) problem and the task to identify the OOV problem is called Named Entity Recognition (NER). It is easier to cater when we using Roman writing because of the upper letter used but not easier for Jawi writing. Majority language like European languages such as English, German, and Spanish mainly used POS tagging general framework were shown in Fig. 1.

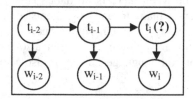

Fig. 1. POS tagging general framework

In determining the correct tag sequence, we make use of the morphological and syntactic (and maybe semantic) relationships within the sentence (the context) [13]. Tagger basically simply encodes and uses the constraints enforced by these relationships by restricting the context to the some of the words that close to the target words (ambiguous word), and use the information provided by w_{i-2}, t_{i-2}, w_{i-1}, t_{i-1}, and w_i. Most studies scan the word from left to right and use context information in the left. In the case of unknown word, two methods used, i.e., using context information and used the information of target words (unknown word). It seems useful to describe how POS taggers work. Fig. 2 shows a schematic design for a part-of-speech tagger by McEnery [1].

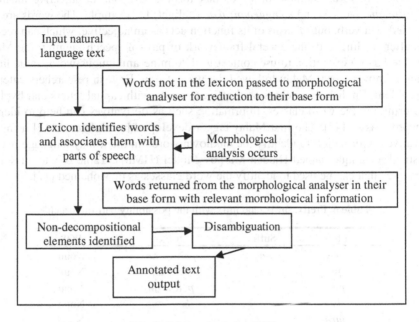

Fig. 2. Schematic design POS

The system first tries to see if each word is present in a machine-readable lexicon it has available. These lexicons are typically of the form <word><part of speech 1, ... part of speech n>. If the word is present in the lexicon, then the system assigns to the word. If the word is not in the lexicon, morphological analyzer will comes to play by guessing word by reduce word to their base form. If the interaction between lexicon and morphological component cannot identify the word, the system assumes that the word may have any of the accepted parts of speech (tag to the likeliest given the known surrounding context) and leave the assignment of a unique part of speech to the disambiguation phase.

If word does not decompose to the word level, i.e., where two or more words together combine to have one syntactic function, for example, idiomatic word groups such as *cock and bull*, or compound word such as *red tape, ice cream* in English and *kapal terbang, air mata* in Malay language. In such cases, POS analysis systems usually have a specialized lexicon to identify such syntactic idioms and to assign a special set of tags to the words involved in these complex sequences. After a text has been analyzed by the lexicon/morphological processor and syntactic idioms, the task of assigning unique POS codes to words is far from complete. Stochastic process been used to help in probability decision making when choose the tag that is the likeliest given the known surrounding context.

Morphological data traditionally have a prefix and suffix words, whether the word starts with capital letters and whether the word has a hyphen or not. For example, Brill [52] classified the capital letters as a proper noun and the other as a noun [13]. Malay were similar to English in determining word classes as described by Karim [47], where there are number of words that form a verb but an adjective meaning, *menyedihkan* (sadden) and *menggembirakan* (delight) for example. The words are still classified as a verb, but in terms of its function acts as an adjective, which can receive intensifier. In line with the general framework of part-of-speech tagging, the Malay language has no exception to use context to determine ambiguous word. In addition, for an unknown word, Malay followed the way of how English researchers cater for the problem. English researchers used affixes, words with capital letters and hyphens in determining the word classes. Information such as derivatives can help to identify the word classes [47]. The first Malay tagger developed by Mohamed [17] using the derivatives information to identify the unknown words in the Malay language using a statistical technique named Hidden Markov Model (HMM). Table 6 below refers to derivatives that can be used to identify the word classes as in Mohamed [17].

Table 6. Prefix, suffix, and infix with the possibility part-of-speech

Prefix	Suffix	Infix	POS
peN-	*-an*	*peN-....-an*	Noun
pe-		*pe-..-an*	Noun
peR-		*peR-..-an*	Noun
ke-		*ke-..-an*	Noun
juru-			Noun
meN-	*-kan*	*meN-..-kan*	Verb
beR-	*-i*	*ber-..-kan*	Verb
teR-		*beR-..-an*	Verb
di-		*di-..-kan*	Verb
mempeR-		*meN-..-i*	Verb
dipeR-		*di-....-i*	Verb
		mempeR-....-kan	Verb
		mempeR-..-i	Verb
		ke-....-an	Verb
		dipeR-..-kan	Verb
		dipeR-....-i	Verb
ter-		*ke-....-an*	Adjective
se-			Adjective

3.2   Semi-supervised Systems

In semi-supervised systems, the machine is allowed to additionally take unlabelled data into account [50].

The reason for this improvement is that more unlabelled data enables the system to model the inherent structure of the data more accurately. Semi-supervised learning suitable to solve the drawbacks in supervised learning. [64] expressed that semi-supervised is a good idea to reduce human labor and improve accuracy. Even though you (or your domain expert) do not spend as much time in labeling the training data, you need to spend reasonable amount of effort to design good models / features / kernels / similarity functions for semi-supervised learning. Semi-supervised seems suitable for minority languages like Malay language because the difficulties in obtaining the annotated corpus.

3.3   Bootstrapping

Bootstrapping used less training examples compared to supervised and semi-supervised learning. [50] state that bootstrapping starts with a few training examples, trains a classifier, and uses thought-to-be positive examples as yielded by this classifier for retraining. As the set of training examples grows, the classifier improves, provided that not too many negative examples are misclassified as positive, which could lead to deterioration of performance.

3.4   Unsupervised Systems

Unsupervised systems are not provided any training examples at all and conduct clustering [50]. Data corpus will be divided into several groups based on their semantic similarities. Several approaches have been used for English language to derive syntactic categories. [12] state that all of them employ a syntactic version of Harris' distributional hypothesis [65]: words of similar parts of speech can be observed in the same syntactic contexts. Measuring to what extent two words appear in similar contexts measures their similarity. The general methodology for inducing word class information can be outlined as follows [12] :-

1. Collect global context vectors of target words by counting how often feature words appear in neighboring positions.
2. Apply a clustering algorithm on these vectors to obtain word classes.

[13] stated that the task of POS tagging is based on a predetermined tagset and therefore adopts the assumptions as word classes. However, this may not be appropriate always, especially when we are using texts from different genres or from different languages. So, labeling the words with tags that reflect the characteristics of the text in question may be better than trying to label with an inappropriate set of tags.

Unsupervised learning is not follow the general framework of POS tagging as shown in Fig. 1. In this situation, the decision is not depend on the tag that is the likeliest given the known surrounding context. Arabic study [23], using diacritical pattern in determining the part-of-speech of Arabic language. As Jawi has no diacritical signs in his writings, this method cannot be applied in determining part-of-speech of Malay language. Other than that, English study [11] used part-of-speech of other language to tag word in minority language. This method were followed by Malay researcher [27]. Majority of the researcher in English language [12], [14], [15], [24], [66], [67] and [68] used context vectors of clustering techniques than look in the context of supervised learning.

4   Future Direction and Conclusion

In this literature, the learning systems used in POS were discussed in details. The most obvious constraint of previous study with this study is in the use of Malay character and writing style. Jawi character and writing style is different from Roman character and writing style. Although the POS of Malay word written in Jawi or Roman is the same, but if the data corpus is in Jawi script such as Old Malay Manuscipt, the tagging cannot be done. As mention above, the previous study in Malay POS tagging are leads to Roman character, e.g. [17], [25] and [27]. To date based on author's knowledge, there is no current research for POS tagging for Jawi. Three highlighted issue arising in the scope of the problem indicates the need of this research.

In addition, morphology used in Jawi script has its own system. With this, a proposing POS for Jawi script will be conducted which will take into account the criteria of syntactic and semantic of Malay language. For the first experiment, Mohamed [17] pattern of word and tagset applied has been studied. Word forms are often ambiguous in their POS. Simple testing has been done to a corpus given by [17] with 154,351 token. The result shows 17,003 token or 11.02% words have ambiguous tags. The algorithm for detected the ambiguous is described as follows:

```
READ corpus;
LOAD word, tag;
   IF tag for word more than 3;
   THEN convert list of tags to strings;
   WRITE the word and tags;
END IF
END
```

From the ambiguous tags in Malay corpus above, it can be concluded here, there are a lot more space in automatic POS tagging research in Malay language. Some of that can be studied, is to answer questions such as 1) How to determine the correct tag for words that has more than one tag? 2) What is the morphological and syntactic information that can be used in determining the exact tag? 3) Do corpus used is reliable? and 4) Is there a semantic similarity information for Malay language text as methods in the study of the English language?, just to name a few.

In addition to the above data sets, two more corpus were collected. First corpus written in Jawi with 114 chapters of the Quran used in the Sulaiman [6] research. Second corpus written in Roman, developed by Dewan Bahasa dan Pustaka is a collection of Old Malay Text Corpus that can be accessed through the website http://prpm.dbp.gov.my/. Both two mentioned corpus are not annotated corpus. Due to the lack of annotated data sets, it gives space to the study of semi-supervised learning method.

This paper has discussed research on the automatic POS tags for Malay language. Four methods of learning used in automating the POS tags. All the methods have been explained according to a general POS tags framework. Past research in manuscript only focused on the translation [31], [32] and [33] of the manuscript. In order to reduce ambiguous, homonym, and homograph [5], POS tagging is one of the solution.

POS tagging is the first step in the process of natural language processing research involving morphology and syntax of the manuscript. Furthermore, the study of morphology and syntax can be used to understand the content (content extraction) and manuscript classification. This allows the study of Malay manuscripts can be done with more extensive and comprehensive. Results of the study will form the basis of morphology and syntax of Malay manuscripts, which in turn can be used to examine the contents of the manuscript. This research will benefit the community, pattern recognition group, and manuscript researcher generally. Promising result for the auto-tagging of Malay written in Jawi is expected to catch up with the advancement in Malay language written in Roman.

References

1. McEnery, T., Wilson, A.: Corpus Linguistics: An Introduction, 2nd edn. Edinburgh University Press, Edinburgh (2004)
2. Jurafsky, D., Martin, J.H.: Speech and Language Processing An Introduction to Natural Language Processing, Computational Linguistics, and Speech Recognition, 2nd edn. Pearson Education, Inc., New Jersey (2009)
3. Mohammed, N.: Sejarah sosiolinguistik Bahasa Melayu lama. Universiti Sains Malaysia, Pulau Pinang (1999)
4. Abdullah, W.M.S.: Tulisan Melayu/Jawi dalam manuskrip dan kitab bercetak: Suatu analisis perbandingan. In: Tradisi Penulisan Manusrip Melayu, pp. 87–105. Perpustakaan Negara Malaysia, Kuala Lumpur (1997)
5. Shamsul, C.W., Omar, K., Nasrudin, M.F., Murah, M.Z.: Machine Transliteration for Old Malay Manuscript. In: The 2nd National Doctoral Seminar on Artificial Intelligence Technology, pp. 19–25. Selangor (2012)
6. Sulaiman, S., Omar, K., Omar, N., Murah, M.Z., Rahman, H.A.: A Malay Stemmers for Jawi Characters. In: Wang, D., Reynolds, M. (eds.) AI 2011. LNCS, vol. 7106, pp. 668–676. Springer, Heidelberg (2011)
7. Nasrudin, M.F., Omar, K., Zakaria, M.F., Yeun, L.C.: Handwritten Cursive Jawi Character Recognition: A Survey. In: 2008 Fifth International Conference on Computer Graphics, Imaging and Visualisation, pp. 247–256. IEEE, Penang (2008)
8. Rahman, H.A.: Panduan menulis dan mengeja Jawi. Dewan Bahasa dan Pustaka, Kuala Lumpur (1999)
9. Brants, T.: TnT – A Statistical Part-of-Speech Tagger. In: 6th Conference on Applied Natural Language Processing, pp. 224–231. ACL, USA (2000)
10. Pustaka, D.B.: Daftar Kata Bahasa Melayu Rumi-Sebutan-Jawi, Edisi Kedua. Dawama Sdn. Bhd., Kuala Lumpur (2008)
11. Das, D., Petrov, S.: Unsupervised Part-of-Speech Tagging with Bilingual Graph-Based Projections. In: Proceedings of the 49th Annual Meeting of the Association for Computational Linguistics (ACL 2011), pp. 600–609. ACL, Stroudsburg (2011)
12. Biemann, C.: Unsupervised Part-of-Speech Tagging in the Large. Research on Language and Computation 7(2-4), 101–135 (2010)
13. Gungor, T.: Part-of-Speech Tagging. In: Indurkhya, N., Damerau, F.J. (eds.) Handbook of Natural Language Processing, 2nd edn., pp. 205–235. Chapman & Hall/CRC (2010)
14. Haghighi, A., Klein, D.: Prototype-Driven Learning for Sequence Models. In: Proceedings of the Human Language Technology Conference of the North American Chapter of the ACL, pp. 320–327. ACL, Stroudsburg (2006)

15. Ninomiya, D., Mozgovoy, M.: Improving POS tagging for ungrammatical phrases. In: Proceedings of the 2012 Joint International Conference on Human-Centered Computer Environments (HCCE 2012), pp. 28–31. ACM, New York (2012)
16. Teodorescu, L.R., Boldizsar, R., Ordean, M., Duma, M., Detesan, L., Ordean, M.: Part of Speech Tagging for Romanian Text-to-Speech System. In: 2011 13th International Symposium on Symbolic and Numeric Algorithms for Scientific Computing, pp. 153–159. IEEE, Timisoara (2011)
17. Mohamed, H., Omar, N., Aziz, M.J.A.: Statistical Malay Part-of-Speech (POS) Tagger using Hidden Markov Approach. In: 2011 International Conference on Semantic Technology and Information Retrieval, pp. 231–236. IEEE, Putrajaya (2011)
18. Wicaksono, A., Purwarianti, A.F.: HMM Based Part-of-speech Tagger for Bahasa Indonesia. In: The 4th International MALINDO (Malay and Indonesian Language) Workshop (2010)
19. Pisceldo, F., Adriani, M., Manurung, R.: Probabilistic Part of Speech Tagging for Bahasa Indonesia. In: Third International MALINDO Workshop, Colocated Event ACL-IJCNLP (2009)
20. Syandra, S., Hayurani, H., Adriani, M., Bressan, S.: Developing Part of Speech Tagger for Bahasa Indonesia Using Brill Tagger. In: Second International MALINDO Workshop (2008)
21. Kubler, S., Mohamed, E.: Part of speech tagging for Arabic. Natural Language Engineering 18(4), 521–548 (2012), doi:10.1017/S1351324911000325
22. Albared, M., Omar, N., Aziz, M.J.A., Ahmad Nazri, M.Z.: Automatic Part of Speech Tagging for Arabic: An Experiment Using Bigram Hidden Markov Model. In: Yu, J., Greco, S., Lingras, P., Wang, G., Skowron, A. (eds.) RSKT 2010. LNCS, vol. 6401, pp. 361–370. Springer, Heidelberg (2010)
23. Alqrainy, S., AlSerhan, H.M., Ayesh, A.: Pattern-based algorithm for Part-of-Speech tagging Arabic text. In: 2008 International Conference on Computer Engineering & Systems, pp. 119–124. IEEE, Cairo (2008)
24. Teichert, A.R., Daume III, H.: Unsupervised Part of Speech Tagging Without a Lexicon. In: NIPS Workshop on Grammar Induction, Representation of Language and Language Learning, pp. 1–6 (2009)
25. Don, Z.M.: Processing Natural Malay Texts: a Data-Driven Approach. Trames. Journal of the Humanities and Social Sciences 14(1), 90–103 (2010)
26. Ranaivo-Malancon, B.: Malay lexical analysis through corpus-based approach. In: International Conference of Malay Lexicology and Lexicography (PALMA). Universiti Sains Malaysia (2005)
27. Zamin, N., Oxley, A., Bakar, Z.A., Farhan, S.A.: A Statistical Dictionary-based Word Alignment Algorithm: An Unsupervised Approach. In: 2012 International Conference on Computer & Information Science (ICCIS), pp. 396–402. IEEE, Kuala Lumpur (2012)
28. Sulaiman, S., Omar, K., Omar, N., Murah, M.Z., Rahman, H.A.: Spelling Error Detector Rule for Jawi Stemmer. In: 2011 International Conference on Pattern Analysis and Intelligent Robotics, pp. 78–82. IEEE, Putrajaya (2011)
29. Razak, Z., Sumali, S.R., Idris, M.Y.I., Ahmedy, I., Yusoff, M.Y.Z.B.M.: Review of Hardware Implementation of Speech-To-Text Engine for Jawi Character. In: 2011 International Conference on Science and Social Research (CSSR 2010), pp. 565–568. IEEE, Kuala Lumpur (2010)
30. Othman, Z.A., Razak, Z., Abdullah, N.A., Yusoff, M.Y.Z.B.M.: Jawi Character Speech-to-Text Engine Using Linear Predictive and Neural Network for Effective Reading. In: 2009 Third Asia International Conference on Modelling & Simulation, pp. 348–352. IEEE, Bali (2009)

31. Ghani, R.A.A., Zakaria, M.S., Omar, K.: Jawi-Malay Transliteration. In: 2009 International Conference on Electrical Engineering and Informatics, pp. 154–157. IEEE, Selangor (2009)
32. Bakar, J.A.: Transliterasi Jawi Lama-Jawi Baru berasaskan Grafem (Kajian Kes Pada Hikayat Merong Mahawangsa). Universiti Kebangsaan Malaysia (2008)
33. Yonhendri: Enjin Transliterasi Rumi Jawi. Universiti Kebangsaan Malaysia (2008)
34. Ahmad, C.W.S.C.W.: Penterjemah Jawi lama kepada Jawi baru. Universiti Kebangsaan Malaysia (2007)
35. Nasrudin, M.F., Petrou, M.: Offline Handwritten Jawi Recognition using the Trace Transform. In: 2011 International Conference on Pattern Analysis and Intelligent Robotics, pp. 87–91. IEEE, Putrajaya (2011)
36. Nasrudin, M.F., Petrou, M., Kotoulas, L.: Jawi Character Recognition Using the Trace Transform. In: 2010 Seventh International Conference on Computer Graphics, Imaging and Visualization, Sydney, pp. 151–156 (2010)
37. Azmi, M.S., Omar, K., Faidzul, M., Khadijah, N., Mohd, W.: Arabic Calligraphy Identification for Digital Jawi Paleography using Triangle Blocks. In: Proceeding of the International Conference on Electrical Engineering and Informatics (ICEEI), pp. 1–5. IEEE, Bandung (2011)
38. Heryanto, A., Nasrudin, M.F., Omar, K.: Offline Jawi Handwritten Recognizer Using Hybrid Artificial Neural Networks and Dynamic Programming. In: International Symposium on Information Technology (ITSim 2008), pp. 1–6. IEEE, Kuala Lumpur (2008)
39. Zulcaffle, T.M.A., Othman, A.K., Abidin, W.A.W.Z., Mohammaddan, S., Marzuki, A.S.W.: A Thresholding Algorithm for Text/Background Segmentation in Degraded Handwritten Jawi Documents. In: 2010 Second International Conference on Advances in Computing, Control, and Telecommunication Technologies, pp. 80–84. IEEE, Jakarta (2010)
40. Redika, R., Omar, K., Nasrudin, M.F.: Handwritten Jawi Words Recognition Using Hidden Markov Models. In: International Symposium on Information Technology (ITSim 2008), pp. 1–5. IEEE, Kuala Lumpur (2008)
41. Shitiq, H.A.A.H., Mahmud, R.: Using an Edutainment Approach of a Snake and Ladder game for teaching Jawi Script. In: 2010 International Conference on Education and Management Technology (ICEMT 2010), pp. 228–232. IEEE, Cairo (2010)
42. Diah, N.M., Ismail, M., Hami, P.M.A., Ahmad, S.: Assisted Jawi-writing (AJaW) software for children. In: 2011 IEEE Conference on Open Systems, pp. 322–326. IEEE, Langkawi (2011)
43. Diah, N.M., Ismail, M., Ahmad, S., Abdullah, S.A.S.S.: Jawi on Mobile Devices with Jawi WordSearch Game Application. In: 2010 International Conference on Science and Social Research (CSSR 2010), pp. 326–329. IEEE, Kuala Lumpur (2010)
44. Abdullah, N.A., Raja, R.H., Kamaruddin, A., Razak, Z., Yusoff, M.Y.Z.B.M.: An authoring toolkit design for educational game content. In: 2008 International Symposium on Information Technology, pp. 1–6. IEEE, Kuala Lumpur (2008)
45. Ismail, K., Yusof, R.J.R., Jomhari, N.: A case study of Jawi Editor in the XO-laptop simulated environment. In: 2010 International Conference on User Science and Engineering (i-USEr), pp. 21–25. IEEE, Shah Alam (2010)
46. Rahim, N.H.A.: A Statistical Parser To Reduce Structural Ambiguity in Malay Grammar Rules. Universiti Malaya (2011)
47. Karim, N.S., Onn, F.M., Musa, H., Mahmood, A.H.: Pembentukan Kata. In: Tatabahasa Dewan Edisi Ketiga, p. 57 (2010)

48. Karim, N.S., Onn, F.M., Musa, H., Mahmood, A.H.: Sintaksis, Satu Pengenalan. In: Tatabahasa Dewan Edisi Ketiga, p. 339 (2010)
49. Chomsky, H.: Syntactic Structure. The Hague, The Netherlands (1957)
50. Biemann, C.: Unsupervised and Knowledge-free Natural Language Processing in the Structure Discovery Paradigm. University of Leipzig (2007)
51. van der Maaten, L., Welling, M., Saul, L.K.: Hidden-Unit Conditional Random Fields. In: The 14th International Conference on Artificial Intelligence and Statistics (AISTATS 2011), USA, vol. 15, pp. 479–488 (2011)
52. Brill, E.: Transformation-Based Error-Driven Learning and Natural Language Processing: A Case Study in Part-of-Speech Tagging. Computational Linguistics 21(4), 543–565 (1995)
53. Lluis, M., Lluis, P., Horacio, R.: A Machine Approach to POS Tagging. Machine Learning 39, 59–91 (2000)
54. Brants, T.: Part-of-Speech Tagging. In: Encyclopedia of Language & Linguistics, 2nd edn., pp. 221–230 (2006)
55. Dickinson, M.: Determining Ambiguity Classes for Part-of-Speech Tagging. In: The Recent Advances in Natural Language Processing (RANLP 2007), Bulgaria (2007)
56. Stanford Log-linear Part-of-Speech Tagger,
   http://nlp.stanford.edu/software/tagger.shtml
57. Nugues, P.M.: An introduction to language processing with Perl and Prolog: an outline of theories, implementation, and application with special consideration of English, French, and German. Springer, New York (2006)
58. Schroder, I.: Case Study in Part-of Speech Tagging Using the ICOPOST Toolkit. Univ. Bibliothek des Fachbereichs Informatik (2002)
59. Diab, M., Hacioglu, K., Jurafsky, D.: Automatic Tagging of Arabic Text: From Raw Text to Base Phrase Chunks. In: The Human Language Technology Conference/North American (2004)
60. Khoja, S.: APT: Arabic Part-of-speech Tagger. In: The Proceedings of the Student Workshop at the Second Meeting of NAACL 2001, pp. 20–25 (2001)
61. Hassan, Y.S.M.F., Zamin, N.: Creating Extraction Pattern by Combining Part of Speech Tagger and Grammatical Parser. In: Proceeding of the International Conference on Computer Technology and Development, pp. 515–519. IEEE, Kota Kinabalu (2009)
62. Jahangiri, N., Kahani, M., Ahamdi, R., Sazvar, M.: A study on part of speech tagging. Review Literature and Arts of the Americas (2011)
63. Zin, K.K., Thein, N.L.: Part of Speech Tagging for Myanmar Using Hidden Markov Model. In: Proceedings of 3rd International Conference on Communications and Information, pp. 123–128 (2009)
64. Zhu, X.: Semi-Supervised Learning Literature Survey Contents. Univ. of Winconsin, Madison (2008)
65. Harris, Z.S.: Mathematical structures of language. Interscience Publishers, New York (1968)
66. Berg-Kirkpatrick, T., Bouchard-Cote, A., DeNero, J., Klein, D.: Painless Unsupervised Learning with Features. In: Proceedings of NAACL 2010, California, pp. 582–590 (2010)
67. Naseem, T., Snyder, B., Eisenstein, J., Barzilay, R.: Multilingual Part-of-Speech Tagging: Two Unsupervised Approaches. Journal of Artificial Intelligence Research 36, 341–385 (2009)
68. Biemann, C.: Chinese Whispers - an Efficient Graph Clustering Algorithm and its Application to Natural Language Processing Problems. In: Proceedings of the First Workshop on Graph Based Methods for Natural Language Processing, pp. 73–80. ACL, USA (2006)

A Novel Baseline Detection Method of Handwritten Arabic-Script Documents Based on Sub-Words

Tarik Abu-Ain[1], Siti Norul Huda Sheikh Abdullah[1], Bilal Bataineh[1],
Khairuddin Omar[1], and Ashraf Abu-Ein[2]

[1] Pattern Recognition Research Group, Center for Artificial Intelligence Technology,
Faculty of Information Science and Technology, Universiti Kebangsaan Malaysia, 43600,
Bangi, Selangor, Malaysia
[2] Computer Engineering Department, Al-Balqa' Applied University, Faculty of Engineering
Technology, Amman, Jordan
tabuain@siswa.ukm.edu.my, {mimi,ko}@ftsm.ukm.my,
b.btnh@yahoo.com, ashraf.abuain@bau.edu.jo

Abstract. Baseline detection is an important process in document image analysis and recognition systems. It is extensively used to many various preprocessing stages such as text normalization, skew correction, characters segmentation, slant and slop correction as well as in feature extraction. in this work, we proposed a new method for baseline detection based on horizontal projection histogram and directions features of subwords skeleton for Arabic script; which form the main component of the text that may consist of at least one letter, in addition of diacritic and dots. The efficiency of the proposed method is has been proven by the experiment's results on an IFN/ENIT Arabic benchmark dataset.

Keywords: Preprocessing, Text normalization, Arabic handwriting, Baseline detection, Sub-word extraction.

1   Introduction

Arabic language is one of six international languages recognized in the United Nations and it has been widely adopted by many other languages such as Jawi, Persian, Kurdish, Pashto, Urdu, and Hausa [1] and [2]. In Arabic script, there are three types of written forms: the printed, handwritten and calligraphy [3].

Baseline in Arabic script defined as a virtual straight line whereas all characters aligns and connected over it in a specific part of each character [4]. Baseline leads to important information about the orientation of the text and the location of connection points between characters; the ascenders, descenders, dots and the diacritics location.

Text line in Arabic scripts is splits into three imaginary regions: upper, middle and lower regions. The upper region contains ascenders dots, and upper diacritic; while the lower region contains descenders dots and lower diacritic points whereas the main contents of the text and loops are lies in the middle region (baseline region).

S.A. Noah et al. (Eds.): M-CAIT 2013, CCIS 378, pp. 67–77, 2013.

For the printed scripts, the baseline can be detected ideally using the general horizontal projection histogram as shown in Fig. 1(b-d). While in handwritten scripts, this method is not suitable due to extensive variety of writing styles and variation characteristics such as cursive writing and large number of dots and diacritics as in Arabic scripts as shown in Fig. 1(e-g).

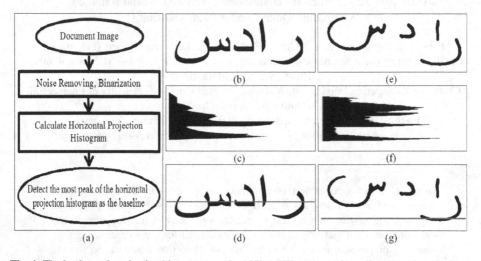

Fig. 1. The horizontal projection histogram method [5], (a) Flowchart of baseline detection, (b-d) an example of success in baseline detection process for a printed text, (e-g) an example of failure in baseline detection process for a handwritten text

In document image analysis process, the baseline detection process is a substantial step, leading to more accurate result especially if a set of logical and language topology dependent rules are used simultaneously.

The aim of the presented work is the baseline detection of off-line handwriting words of Arabic scripts, which may consist of one or more of Part of Arabic Word (PAW), which are one of the main distinguishing properties of Arabic script that differs from the other scripts. It is appearing in case of any one of these letters: د،ذ،ر،ز،و،أ appear in the middle of the word which causes the division of the word into PAWs (Table 1).

These PAWs are distributed irregularly because of the free style of writing for every person. That leads to inaccurate in detecting a straight baseline for each word/line of the text. These facts conflict with the definition of baseline as introduced previously.

This paper consists of four main sections; first we introduce the work and its importance. Then, an overview of related works is discussed. The proposed method is explored and the experiment's results are reported in the next section. Finally, the conclusion and the future directions are noted subsequently.

Table 1. Arabic letters and their shapes depending on the position in the text

Beginning	Middle	End	Isolated	Beginning	Middle	End	Isolated
-	-	ﻟ	ﺍ	ﺿ	ﻀ	ﺽ	ض
ﺑ	ﺒ	ﺐ	ب	ﻃ	ﻄ	ﻂ	ط
ﺗ	ﺘ	ﺖ	ت	ﻇ	ﻈ	ﻆ	ظ
ﺛ	ﺜ	ﺚ	ث	ﻋ	ﻌ	ﻊ	ع
ﺟ	ﺠ	ﺞ	ج	ﻏ	ﻐ	ﻎ	غ
ﺣ	ﺤ	ﺢ	ح	ﻓ	ﻔ	ﻒ	ف
ﺧ	ﺨ	ﺦ	خ	ﻗ	ﻘ	ﻖ	ق
-	-	ﺪ	د	ﻛ	ﻜ	ﻚ	ك
-	-	ﺬ	ذ	ﻟ	ﻠ	ﻞ	ل
-	-	ﺮ	ر	ﻣ	ﻤ	ﻢ	م
-	-	ﺰ	ز	ﻧ	ﻨ	ﻦ	ن
ﺳ	ﺴ	ﺲ	س	ﻫ	ﻬ	ﻪ	ه
ﺷ	ﺸ	ﺶ	ش	-	-	ﻮ	و
ﺻ	ﺼ	ﺺ	ص	ﻳ	ﻴ	ﻲ	ي

2   State of the Art

This section provides an overview of the related methods of baseline detection for handwritten Arabic scripts and a review of previous work on this topic.

2.1   Baseline Properties

Scripts are divided into two main categories based on the text generator either via machine or human. The script lay on straight line in machine printed text due to non-intervention from humans and preformatting rules dictated by text editor programs which is absolutely confirmed with the definition of baseline. However, the baseline detection challenges appear in human handwritten scripts due to free style of writing, writing habitats, circumstances environment, writer psychology and physically, writing tools. All above factors caused problem in perfection of baseline detection process.

Each script has unique baseline properties depending on the nature of the way of writing the characters. Since the Arabic script is cursive, the baseline defined as a horizontal straight line that all binding points between the characters as well as certain position of these characters should be laid over it.

2.2 Arabic Script Baseline Detection Methods

Pechwitz et. al. (2002) proposed a baseline detection method based on polygonally approximated skeleton processing [6]. However, the method is conflict of baseline definition. Since, it is using a linear regression algorithm which is not working well with unaligned text (Fig. 4 (k-l)). A little enhancement of Pechwitz baseline detection method by Farooq et. Al. (2005), which uses a two-steps linear regression after locating the local minima points of word contour [7]. However, it is still suffer same problems (Fig. 4 (m-n)).

Ziaratban et. al. (2008) proposes a baseline estimating algorithm using a template matching and a polynomial fitting algorithm [8]. However, the method is less effective in presents of short words, dots and diacritics.

A method of baseline detection for thinned text then find the relation between the text point's alignment and their trajectory neighbor directions is proposed by Bobaker [9]. However, the algorithm is not efficient when it is deal with short words that consist of isolated characters only.

Boukerma et. al. (2009) proposed an algorithm based on subwords skeleton where some feature points use to estimate a horizontal band of the text using a linear interpolation algorithm [10]. However, the method fails in case of ligatures and small diacritics. In addition, the final result is not a straight line which is conflict with the definition on baseline.

Nagabhushan et. al. (2010) proposed an algorithm that uses a piece-wise painting scheme to identify points that will be used to estimate the baseline [11]. However, the algorithm is less effective if large diacritics and small characters are exists (Fig. 4 (o-p)).

From literature [6 - 11], it clearly most methods are defected by many factors such as diacritics, isolated characters, long words. In addition, some binding points between characters are not laying over the baseline as well as some of subwords are not intersecting the straight baseline in the right points.

3 The Proposed Method

The accurate detection of baseline location is help in extracting more accurate meaningful information such as writing directions, ascenders, descenders, dots and diacritics. Irregularity in Arabic script handwriting style is leading to irregularity in word/line components straightness. Baseline detection and straightness is crucial step in preprocessing stage as a text normalization process. From literature, most of the methods did not detect the correct baseline in case of short characters and when large diacritics exist. To overcome the problem, a new method is proposed to estimate the local baseline for each PAW which will be used later to detect the global baseline of the whole word/line. Fig. 2, shows the proposed method framework.

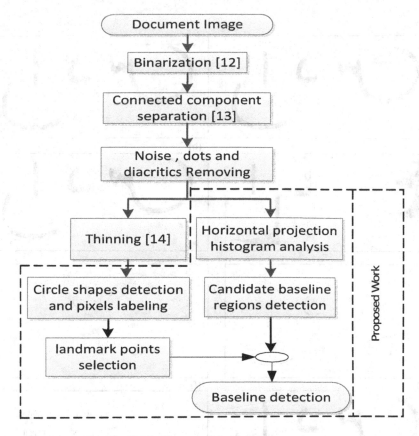

Fig. 2. The Proposed Method Framework

3.1 Binarization

An adaptive threshold technique is used as shown in the equation 1 [12]. From literature, it is considered as one the best method that implements with fine and degraded document images [3], as shown in Fig. 3(a).

$$T_w = m_w - \frac{m_w{}^2 * \sigma_w}{(m_g + \sigma_w)(\sigma_{Adaptive} + \sigma_w)}. \tag{1}$$

where T_W is the thresholding value of the binarization window, m_w is the mean value of the pixels in the window; m_g is the mean value of the global image pixels. $\sigma_{Adaptive}$ is the adaptive standard deviation for the window, σ_w is the standard deviation of the window.

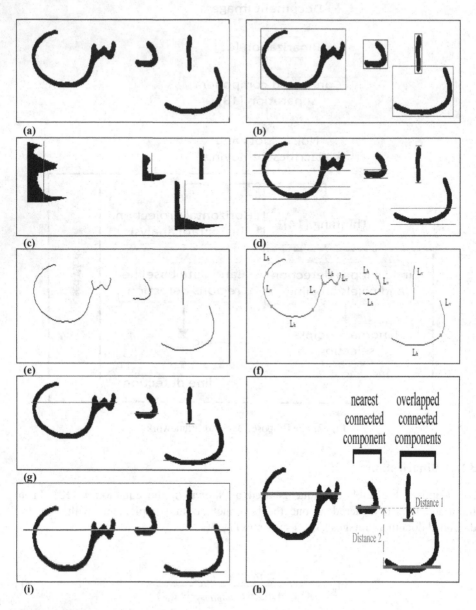

Fig. 3. An example of a handwritten text image (رادس) results after complete every step of the proposed method (a) text after binarization and removing the noise, dots and diacritics, (b) subwords detection, (c) horizontal projection histogram for each of subwords and calculate the T2, (d) candidate baseline regions detection, (e) thinned text, (f) circle shapes detection, pixels labeling and landmark spots selection, (g) local baseline location, (h) process of the baseline straightness, (i) final baseline location.

3.2   Connected Component Detection

This step uses two-pass scanning algorithm for connected-region detection [13] which works as follows: on the first pass, for all foreground pixels do the following: (i) get the 8-neighboring pixels of the current pixel (ii) if there are no neighbors, uniquely label the current pixel and continue, otherwise, find the neighbor with the smallest label and assign it to the current pixel (iii) store the equivalence between neighboring labels. While on the second pass, all the foreground pixels are re-labeled to the lowest equivalent label as shown in Fig. 3(b).

3.3   Noise, Dots and Diacritics Removing

From the definition of baseline, only the main parts of the characters are align over it not the dots or diacritics. For those components that their size is less than a threshold value (*T1*) calculated by the equation 2, they will be removed.

$$T_1 = (\frac{\sum black\ pixels}{number\ of\ connected\ components})/\ V_e\ . \tag{2}$$

where $3 < V_e < 4$. V_e *is a non-constant value that depends on the size of noise, dots and diacritics sizes.*

3.4   Horizontal Projection Histogram and Candidate Baseline Regions Detection

The method calculates the intensity of horizontal histogram for each subword, and then a threshold value (T_2) performed to identify the candidate regions that may contain the actual local baseline using equation 3 as shown in Fig. 3(c) [5].

$$T_2 = \frac{\sum black\ pixels}{number\ of\ rows}\ . \tag{3}$$

For each set of continues rows that are equal or greater than the T_2, it will be considered as a candidate baseline region as shown in Fig. 3(d).

3.5   Thinning

One of the most known thinning techniques from literature is applied [14]. The technique keeps the topology and connectivity of the text with a one-pixel width efficiently as shown in Fig. 3(e).

3.6   Circle Shapes Detection, Pixels Labeling and Landmark Spots Selection

A set of sequence sub-processes are performed as following:

a.   Circle shapes are detected using connected component technique [13] and labeled by L_c.

b. Then, the rest of pixels are labeled based on adjacent pixels relationship.
    (i)   All pixels that are adjacent horizontally are labeled by L_h.
    (ii)   All pixels that are adjacent vertically are labeled by L_v.
    (iii)   Unlabeled pixels are labeled by L_h if they lay between two L_h labeled pixels, otherwise, labeled by L_v.
c. Finally, mark the pixels that connect between the different labeled sets as a landmark points as shown in Fig. 3(f).

3.7 Baseline Detection and Straightness

For each subword, baseline is highlighted into the candidate reign that has the most number of these landmark points, which will be the highest pixel Intensity from the horizontal projection histogram as shown (Fig. 3(g)). As a result of existence many subwords in the same text line, there will be a baseline for each of them which called local baselines. All these local baselines are aligned into one straight line called a global baseline except in case of overlapping subwords which are treated as a special case as following; measure the vertical distance between each of the overlapped subwords baseline and the nearest un-overlapped subwords baseline as shown in Fig. 3(h), the subword that have the shorter distance is aligned onto the global baseline (Fig. 3(i)).

4   Experiments and Results

In this section, several experiments are conducted on 2684 images from the set_a of IFN/ENIT dataset of Tunisian town names by 42 writers to demonstrate the capabilities of the proposed method on various binary images [15]. Only subsets of the results are presented owing to space constraints. To evaluate the performance of the proposed method, it is compared with that of other widely used methods, including the algorithms proposed by horizontal projection histogram [5], Pechwitz [6], Farooq [7] and Boukerma [10] (Fig. 4), and it is clear that the proposed algorithm shows better results and setting right the failure for their methods. The proposed method is able to process machine printed texts as well handwritten without any modification. The experiments visually illustrate various important problems that arise in baseline detection and their solutions.

The baseline detection for handwritten text images problems, have largely been solved in the proposed method such as in the case of diacritics, isolated characters and short words as well as the binding points between characters are laying over the baseline accurately, also all subwords are intersecting the straight baseline in the right points Fig. 4(a-e).

Fig. 4. Results of (a-e) the proposed method, (f-j) horizontal projection histogram method [5], (k-l) Pechwitz method [6], (m-n) Farooq method [7], (o-p) Boukerma method [10]

5   Conclusion

In this paper, a new baseline detection method for Arabic script is proposed. The method consists of four main stages: connected component separation, calculation the average of the horizontal projection histogram for each component, circle shapes detection, pixels labeling and landmark spots selection and finally baseline detection and straightness. The visual experiments demonstrate the high-quality performance of the proposed method on textual binary images. IFN/ENIT dataset is used in the experiments, and the results of the proposed method are compared with some other methods and it achieves a superior performance compared to them. Currently we are looking for further robust baseline relevant features to be used in both preprocessing and feature extraction stages to be tested in a complete recognition system.

Acknowledgments. The authors would like to thank the Faculty of Information Science and Technology and Center for Research and Instrumentation Management of the Universiti Kebangsaan Malaysia for providing facilities and financial support under Exploration Research Grant Scheme Project No. ERGS/1/2011/STG/UKM/01/ 18 entitled "Calligraphy Recognition in Jawi Manuscripts using Paleography Concepts Based on Perception Based Model" and Fundamental Research Grant Scheme No. FRGS/1/2012/SG05/UKM/02/8 entitled "Generic Object Localization Algorithm for Image Segmentation".

References

1. U. Nations, http://www.un.org (March 13, 2013)
2. Abu-Ain, T.A.H., Abu-Ain, W.A.H., Sheikh Abdullah, S.N.H., Omar, K.: Off-line Arabic Character-Based Writer Identification – a Survey. In: International Journal on Advanced Science, Engineering and Information Technology, pp. 161–166 (2011); Proceeding of the International Conference on Advanced Science, Engineering and Information Technology Bangi, Malaysia
3. Bataineh, B., Abdullah, S.N.H.S., Omar, K.: Arabic calligraphy recognition based on binarization methods and degraded images. In: International Conference in Pattern Analysis and Intelligent Robotics (ICPAIR 2011), pp. 65–70 (2011)
4. Gacek, A.: Arabic Manuscripts: A Vademecum for Readers, BRILL (2009)
5. Parhami, B., Taraghi, M.: Automatic Recognition of Printed Farsi Texts. Presented at the Proc. Conf. Pattern Recognition, England (1980)
6. Pechwitz, M., Margner, V.: Baseline estimation for Arabic handwritten words. In: Proceeding in Eighth International Workshop on Frontiers and Handwriting Recognition, pp. 479–484 (2002)
7. Farooq, F., Govindaraju, V., Perrone, M.: Pre-processing methods for handwritten Arabic documents. In: Proceedings in Eighth International Conference on Document Analysis and Recognition, vol. 1, pp. 267–271 (2005)
8. Ziaratban, M., Faez, K.: A novel two-stage algorithm for baseline estimation and correction in Farsi and Arabic handwritten text line. In: 19th International Conference on Pattern Recognition, ICPR 2008, pp. 1–5 (2008)

9. Boubaker, H., Kherallah, M., Alimi, A.M.: New Algorithm of Straight or Curved Baseline Detection for Short Arabic Handwritten Writing. In: 10th International Conference on Document Analysis and Recognition, ICDAR 2009, pp. 778–782 (2009)
10. Boukerma, H., Farah, N.: A Novel Arabic Baseline Estimation Algorithm Based on Sub-Words Treatment. In: International Conference on Frontiers in Handwriting Recognition (ICFHR 2010), pp. 335–338 (2010)
11. Nagabhushan, P., Alaei, A.: Tracing and Straightening the Baseline in Handwritten Persian/Arabic Text-line: A New Approach Based on Painting-technique. International Journal on Computer Science and Engineering 2, 907–916 (2010)
12. Bataineh, B., Abdullah, S.N.H.S., Omar, K.: An adaptive local binarization method for document images based on a novel thresholding method and dynamic windows. Pattern Recognition Letters 32, 1805–1813 (2011)
13. Linda, G.C.S., Shapiro, G.: Computer Vision. Prentice Hall (2002)
14. Abu-Ain, W., Abdullah, S.N.H.S., Bataineh, B., Abu-Ain, T., Omar, K.: Skeletonization Algorithm for Binary Images. In: International Conference on Electrical Engineering and Informatics, ICEEI 2013 (2013)
15. IFN/ENIT - Database of Arabic Handwritten words, T. U. Institute of Communications Technology, Braunschweig, Germany (2002)

Brain Tumor Treatment Advisory System

Suri Mawarne Kaidar[1], Rizuana Iqbal Hussain[2], Farah Aqilah Bohani[3],
Shahnorbanun Sahran[4], Nurnaima binti Zainuddin[5], Fuad Ismail[6], Jegan Thanabalan[7],
Ganesh Kalimuthu[8], and Siti Norul Huda Sheikh Abdullah[9]

[1,3,4,5,9] Center for Artificial Intelligence Technology,
Faculty of Information Science and Technology, Universiti Kebangsaan Malaysia (UKM),
43600, Bangi, Selangor D.E., Malaysia
[2] Department of Radiology
[6] Department of Radiotherapy & Oncology
[7] Department of Medicine
[8] Department of Pathology, Universiti Kebangsaan Malaysia Medical Centre, 56000 Cheras,
Kuala Lumpur, Malaysia
{suri.mawarne,aimnjannah}@gmail.com, farahbohani@yahoo.com,
{shah,mimi}@ftsm.ukm.my

Abstract. The process of brain tumor diagnoses involved many medical expert and large number of rules and regulation. For public, it is important to recognise the symptoms. Medical expert need the system to assist in decision making. This paper presents a brain tumor treatment advisory system (BTTAS) for public and medical experts. The aims of the system are: to educate public on brain tumor, to assist medical expert in diagnosing process, and suggested treatment. The rules for the treatment advisory system are developed with assistance of medical experts from Universiti Kebangsaan Malaysia Medical Center (UKMMC). Based on user testing evaluations, most experts have given acceptable satisfaction rate towards this application.

Keywords: Expert system, medical expert system, advisory system, brain tumor, artificial intelligent.

1 Introduction

Brain tumor is masses of abnormal cells that can grow in the brain. A brain tumor can be either malignant (cancerous) or benign (noncancerous), but in either case it can be dangerous [1] and [2]. In a report released by National Cancer Registry on all cases of cancer diagnosed in Malaysia from 1st January until 31st December 2007, brain tumor was listed as one of the ten most frequent cancers among men in Malaysia, with 3.2% from overall cases. [3]. By 2012 and 2013, estimated more than 20 000 new cases of brain tumor in US with 13 700 and 14 080 estimated death [4] and [5].

Usually the brain tumors treatment begins with surgery for diagnosis or excision. Subsequent comprehensive care may involve cooperation among neurosurgeons, neurologists, neuroradiologists, neuropathologists, radiation therapists and oncologists [6].

In this paper, we focus on Brain Tumor advisory treatment system for two type of user; medical experts and non-medical. About 10 patient data from Universiti Kebangsaan

S.A. Noah et al. (Eds.): M-CAIT 2013, CCIS 378, pp. 78–88, 2013.

Malaysia Medical Center (UKMMC) was used in developing the system. This paper is organized as the following: Section 2, we discuss the state of the art. In section 3, we describe on the methodology. Section 4 is for result and discussion. The conclusion is given in section 5.

2   State of the Art

Computer-based medical system had long been suggested to improve the quality and reduce the cost of healthcare [7]. Research on computer aided system in medical begun at the end of the 60's with several diagnosis systems which have been developed such as DENDRAL, INTERNIST a rule-based expert system to diagnose complex problem in general internal medicine and MYCIN; a rule based expert system to diagnose and recommend treatment for certain blood infections[8] and [9]. It was later followed by program such as ONCOCIN, a rule-based medical expert for oncology protocol management [10] and many other new expert system based on existing and proven intelligent applications.

Computer aided medical system is also needed for reducing the frequency amount of error especially in clinical decision making process. Although errors are common in all domains in life, error in medicine would harmed or injured a patient and hence increases the cumulative consequence [11] and [12].

Treatment advisory system generally falls in two categories: the proposed aided system can assist medical expert in diagnosing and making treatment decision or therapy plan for a particular patient case [13]. Besides, the developed system could also be a general educational tool to explain medical term for laymen[12], and as a tool to train medical practitioners and young experts [13].

3   Methodology

This project aims are to develop a knowledge-based diagnosis and advisory system for brain tumor as well as advising oncologist in handling the specific treatments required. The four core processes of this proposed advisory system are data and knowledge acquisition, knowledge representation and inference or cognition, and user interfaces. The user interfaces are designed for non-medical and medical expert.Medical expert comprises four other sub-categories of users which are neurologist, radiologist, histologist and oncologist. The treatment and therapy knowledge are acquired from UKMMC medical experts and also rule and regulation document collected by a group of UKM students. The cognition is inferred by a forward chaining inference engine.

In development phase of the BTTAS, decision table and decision tree are used as the knowledge acquisition technique to acquire expert knowledge. Using these techniques, UKM students combine the symptoms and features from four different medical fields (neurology, radiology, histopathology and oncology) and their experience to diagnose brain tumor and provide treatment suggestion.

The characteristics of the BTTAS are presented in Table 1 that states domain, knowledge resources, knowledge acquisition technique, knowledge representation technique, inference engine, explanation facility, development method, development tool, user interface and objectives of the BTTAS.

System user module in BTTAS ensures that only the authorized medical experts will conduct patient validation. Throughoncologist module, the authorized medical experts will suggest treatment after diagnosing brain tumor symptoms based on MRI images and Histopathological examinations. The user perspective module for oncologist is presented in Fig. 1. The oncologists are allowed to view the features and symptoms of brain tumor entered by neurologist, radiologist and histologist, and thus provide treatment suggestion for that particular patient.

Fig. 2 explains briefly the overview of the system/expert engine BTTAS that has been developed. BTTAS system consists of three main components: the human-machine interaction, the system programming and the database processing. Meanwhile, the system programming covers a management of user input data. WIN-PROLOG programming language is used as it can handle knowledge representation very well [14]. The last component of the engine is the database processing where there will be a server for storing all the information on brain tumor. All data are stored in the BTTAS database system to ease the maintenance processes.

Fig. 2 dictates the detailed architecture of the BTTAS for brain tumor diagnosis and treatment suggestion. Meanwhile, in the development environment, the components for storing the knowledge into the knowledge-based have also been developed. The knowledge obtained from the expert, together with other documented information is then transformed into a set of rules by the UKM knowledge engineer [15].

BTTAS Inference engine contains Diagnosis and advisory module, is capable to infer conclusions from the user input and the set of rules [15]. The Advisory Inference Engine interprets and matches data with the rules in the knowledge-based to suggest a particular treatment of brain tumor for specific patient.

Table 1. BTTAS characteristics

Items	Characteristics
Domain	Brain tumour cases in Universiti Kebangsaan Malaysia Medical Centre (UKMMC) medical expert system
Knowledge resources	Data collection from medical record office, IRIS/ OMS and MRI images from Medweb/ UKMMC Pacs sever, medical experts, medical reading in form of textbooks and research publication.
Knowledge acquisition technique	Decision table and tree
Knowledge representation technique	Rules
Inference engine	Forward chaining
Explanation facility	Relation between regulations and expertise analysis
Development method	System
Development tool	WIN-PROLOG 4320
User interface	WIN-PROLOG 4320
Objectives	To diagnose the brain tumor and treatment suggestions

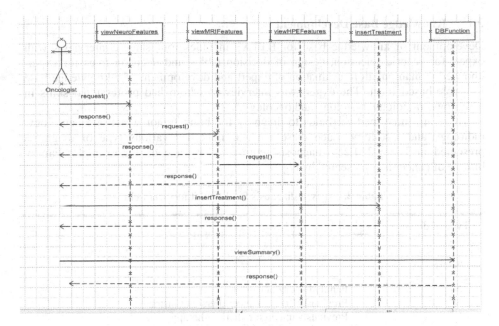

Fig. 1. Perspective User Model for Oncologist

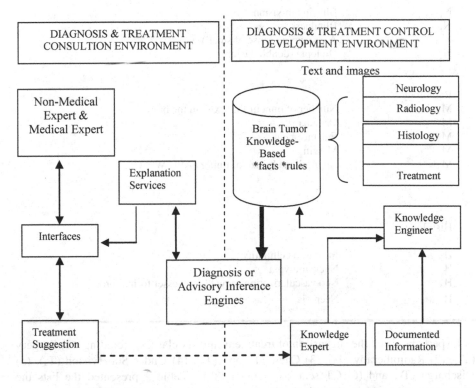

Fig. 2. The Diagnosis and advisory system for brain tumor architecture

3.1   Knowledge Base Development

Development of knowledge base for BTTAS includes four stages, interviews with expert for criteria selection, design of decision tables and decision trees for expert knowledge manipulation, evaluation of expert and development of a knowledge base in a rule-based form. The expert was interviewed whom had professional skills in neurology, radiology, histology and oncology.

There are three medical fields to diagnose the brain tumor which are neurology, radiology and histology. Each these medical field has various type of diagnosis presented in Table 2. Each diagnosis represented by a code for example; Family history of cancer is represented by N_1.

Table 2. Types of diagnosis for different medical field

Type of diagnosis	
Neurology	
N_1	Family history of cancer
N_2	Previously known cancer (personal history)
N_3	Previously undergo radiation therapy
N_4	Have chemical exposure at workplace
...	
N_{11}	Blurring in vision
N_{12}	Blindness
N_{13}	Anosmia (no sense of smell)
N_{14}	Diplopia (double vision)
...	
Radiology	
M_1	Number of tumour that exist in the brain
M_2	Oedema
M_3	Enhancement
M_4	Margin
M_5	Approximately size of tumor (AP x W x C)
...	
Histology	
H_1	Increased cellularity
H_2	Nuclear atypia
H_3	Neo vascularization (new blood vessel formation)
H_4	Necrosis
...	

Expert classified the suggestion of treatment into six classes, including: (1) Surgery (T_1), (2) Radiotherapy (T_2), (3) Chemotherapy (T_3), (4) Under Supervision (T_4), (5) Discharge (T_5) and (6) CT scan assessment (T_6). Table 3 presented the lists the treatment suggestion.

Table 3. Treatment Suggestion

Treatment Suggestion	
T_1	Surgery
T_2	Radiotherapy
T_3	Chemotherapy
T_4	Under Supervision
T_5	Discharge
T_6	CT Scan Assessment

In this stage, decision tables are used to acquire rules to build the decision tree. These tables are used to design the BTTAS knowledge base. An example decision table is generated as shown in Table 4.

Decision tables were checked by experts to confirm consistency among these rules and used the results to build the knowledge base. The examples of rules are presented as:

IF	the patient has N_2
THEN	check the patient's specific symptoms
IF	the patient's specific symptoms is N_{11}
THEN	check to radiology aspect
IF	radiology aspect is M_1
THEN	check to histology aspect
IF	histology aspect is H_1
THEN	the treatment suggestion is T_1

Decision tree is used to view the action for different rules combined of neurology, radiology and histology rules. An example is illustrated in Fig. 3.

Table 4. An example of decision tree for knowledge base

Rules	N_1	N_2	N_3	N_{11}	N_{12}	N_{13}	M_1	M_2	M_3	H_1	Treatment Suggestions
1	F	F	F	-	-	-	-	-	-	-	T_5
2	F	T	F	T	F	F	-	-	-	-	T_6
3	F	T	F	T	F	F	T	F	F	-	T_4
4	F	T	F	T	F	F	T	F	F	T	T_1

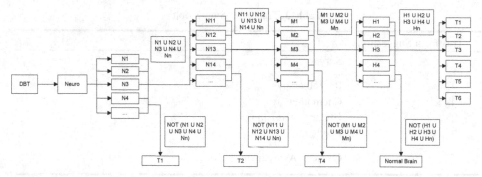

Fig. 3. An example of decision tree for knowledge base

3.2   Inference engine

Diagnosis Inference Engine will infer rules for recognising types of features of brain tumor such previously known cancer (personal history), blurring in vision, number of tumor that exist in the brain and kind and grade of tumor based on different medical field such as neurology, radiology and histology. Meanwhile, Advisory Inference Engine can infer sequence of advices based on different medical field given features of brain tumor in terms of text and graphics. The inference mechanism of the BTTAS is forward chaining. For oncologist, the inference process follows: (1) selecting the types of user which is medical expert (2) entering the patient ID, medical expert ID and password, (3) selecting the types of medical expert which is oncologist (4) viewing the features of brain tumor by neurology (5) viewing the features of brain tumor by radiology (6) viewing the features of brain tumor by histology. The inference engine checks the obtained data and information by rules stored in knowledge base and provides the six of treatment suggestions.

3.3   User Interface

User interface of the BTTAS is designed and developed by using WIN-PROLOG4320. A graphical interface based on standard Microsoft Windows image is employed in the development of BTTAS interface for user friendliness and satisfaction. These windows pop up in the middle of the screen for the attention of user.

3.4   Forward Chaining Technique in Advisory System Module

BTTAS also provides advisory system module that gives assistance on actions or treatments for the suspected brain tumor. Fig. 4 shows the interfaces on forward chaining technique for certain treatment of brain tumor (in this example, the treatment known to be surgery and provide by oncologist). This technique, the identification of a particular treatment is focus on result of brain tumor diagnosis by three medical experts which are neurologist, radiologist and histologist with the assist of the texts and images.

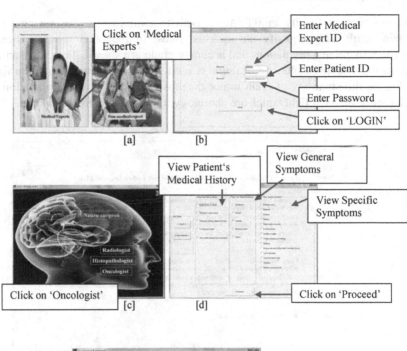

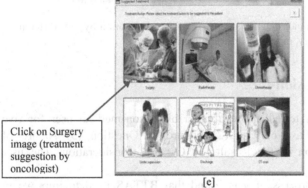

Fig. 4. An example of a series of forward chaining interfaces using WIN-PROLOG software. [a] Front page for medical and non-medical user. [b] Login page. [c] User interface for medical expert. [d] User interface for neurology. [e] User interface for treatment.

3.5   Explanation Facility

The explanation facility in BTTAS is both to explain how the system arrived at certain result and to justify the matched rules. In the other words, when rule is matched, its related regulations and accuracy grade are shown. Therefore, an user can rely to the system results. An example is used to present how explanation functions. In Fig. 5, the summary of brain tumor diagnosis consists of the explanation by four different of medical field which are neurology, radiology, histology and oncology.

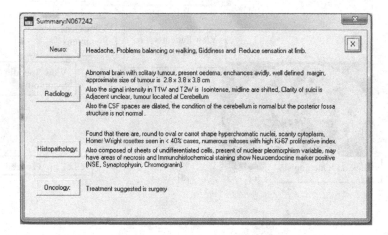

Fig. 5. The summary of brain tumor diagnosis consists of the explanation by four different of medical field which are neurology, radiology, histology and oncology

4   Result and Discussion

The developed BTTAS was tested and evaluated by 3 non-medical user for five categories which are (1) Ease of Use (2) Nature of Question (3) Nature of Explanations (4) Presentation of the Result (5) General Consideration. The user response is shown in Table 5.

Through the evaluation process, it was found that BTTAS need to improve in representing the reports generated for medical and non-medical user. The report should contain grade and type of the brain tumor, and treatment suggested based on both parameter.

Table 5. User response for typical usability design factors

Measurement	Score 1	Score 2	Score 3	Score 4	Score 5	Average score
Ease of use (not easy 1 2 3 4 5 very easy)						
Starting the system					✓ ✓ ✓	5.0
Obtaining Explanation			✓		✓ ✓	4.3
Help Facilities	✓				✓ ✓	3.6
Nature of Question (not helpful 1 2 3 4 5 extremely helpful)						
Clarity of Term				✓	✓ ✓	4.6
Clarity of Question				✓	✓ ✓	4.6
Nature of Explanations (not useful 1 2 3 4 5 extremely useful)						
WHY Explanation			✓	✓	✓	4.0
Presentation of The Result (not easy 1 2 3 4 5 very easy) (not confidence 1 2 3 4 5 extremely confidence)						
Ease to Follow					✓ ✓ ✓	5.0
Complete				✓	✓ ✓	4.6
General Consideration (low 1 2 3 4 5 high) (not useful 1 2 3 4 5 extremely useful)						
Speed of the System					✓ ✓ ✓	5.0
System is Useful				✓	✓ ✓	4.6

From the evaluation test, average score for the whole system was 4.6 / 5.0. The user tested agrees that BTTAS is useful for them.

4.1 Future Work

In the future, implementation of graphical user interface (GUI) for BTTAS is encouraged to be used in simplifying large amount of knowledge and represent in a more user-friendly interface.

To cope with rapid evolution of technologies, BTTAS development in mobile environment should be considered. The flexibility to gain access from everywhere and not restricted to a certain computer and workstation, would help BTTAS evolved more.

5 Conclusion

In this paper we have discussed the design and implementation of rule based Brain Tumor Treatment Advisory System. The system helps medical expert to diagnose Brain Tumor and assist in treatment decision making. The system also helps public to understand the basis of Brain Tumor, so they could aware with the symptoms and get earlier treatment.

Acknowledgement. This research is based on two research grants from Ministry of Science, Technology and Innovation, Malaysia entitled "Sciencefund 01-01-02-SF0694 Spiking-LVQ Network For Brain Tumor Detection" and Arus Perdana Fund AP-2012-019 entitled "Automated Medical Imaging Diagnostic Based on Four Critical Diseases: Brain, Breast, Prostate and Lung Cancer". We have obtained Ethics approval entitled "FF-342-2012 Imaging Processing of the Brain Tumor Using LVQ Classifier" from UKMMC or UKM Medical Center, Malaysia for collecting and conducting experiments on MRI Brain Tumor patient's record.

References

1. Cook, L.J., Freedman, J.: Brain Tumors, 1st edn. The Rosen Publishing Group, Inc. (2012)
2. Stark-Vance, V., Dubay, M.L.: 100 Questions & Answers about Brain Tumors, 2nd edn. Jones & Bartlett Learning (2010)
3. National Cancer Registry, National Cancer Registry Report, National Cancer Registry, Ministry of Health. Putrajaya, Malaysia (2007, 2011)
4. American Cancer Society, Cancer Facts & Figures 2012, American Cancer Society, Atlanta (2012)
5. American Cancer Society, Cancer Facts & Figures 2013, American Cancer Society, Atlanta (2013)
6. Black, P.M.L.: Brain tumors. New England Journal of Medicine 324(22), 1555–1564 (1991)
7. Clayton, P.D., Hripcsak, G.: Decision support in healthcare. International Journal of Bio-medical Computing 39(1), 59–66 (1995)
8. Fagan, L.M., Shortliffe, E.H., Buchanan, B.G.: Computer-based medical decision making: from MYCIN to VM. Automedica 3(2), 97–108 (1980)
9. de Schatz, C.V., Schneider, F.K.: Intelligent and Expert Systems in Medicine-A Review. In: XVIII Congreso Argentino de Bioingeniería SABI 2011 - VII Jornadas de Ingeniería Clínica 2011, Mar del Plata (2011)
10. Shortliffe, E.H.: Medical expert systems - knowledge tools for physicians. Western Journal of Medicine 145(6), 830 (1986)
11. Bates, D., et al.: Reducing the frequency of errors in medicine using information technology. Journal of the American Medical Informatics Association 8(4), 299–308 (2001)
12. Wellwood, J., Johannessen, S., Spiegelhalter, D.: How does computer-aided diagnosis improve the management of acute abdominal pain? Annals of the Royal College of Surgeons of England 74(1), 40 (1992)
13. Shortliffe, E.H.: Computer programs to support clinical decision making. JAMA 258(1), 61–66 (1987)

Age-Invariant Face Recognition Technique
Using Facial Geometry

Amal Seralkhatem Osman Ali[1,2], Vijanth Sagayan a/l Asirvadam[2],
Aamir Saeed Malik[1], and Azrina Aziz[1]

[1] Centre of Intelligent Signals and Imaging Research
Department of Electric and Electronic Engineering
[2] Universiti Teknologi PETRONAS
31750 Tronoh, Perak, Malaysia
Amal61@gmail.com,
{vijanth_sagayan,aamir_saeed,azrina.aziz}@petronas.com.my

Abstract. While face recognition systems have proven to be sensitive to factors
such as illumination and pose, their sensitivity to facial aging effects is yet to be
studied. The FRVT (Face Recognition Vendor Test) report estimated a decrease
in performance by approximately 5% for each year of age difference. This
research study proposed a geometrical model based on multiple triangular
features for the purpose of handling the challenge of face age variations that
affect the process of face recognition. The system is aim to serve in real time
applications were the test images are usually taken in random scales that may
not be of the same scale as the probe image, along with orientation, lighting
,illumination, and pose variations. Multiple mathematical equations where
developed and used in the process of forming distinct subjects clusters. These
clusters hold the results of applying the developed mathematical models over
the FGNET face aging database.

Keywords: FRVT, age-invariant, geometrical model, triangular features,
similarity proportion ratios, clustering; FGNET.

1 Introduction

Face recognition is a type of automated biometric identification method that
recognizes individuals based on their facial features as basic elements of distinction.
The research on face recognition has been dynamically going on in the recent years
because face recognition is involved in many fields and disciplines such as access
control, surveillance and security, criminal identification and digital library.

Automatic face detection and recognition has been a challenging problem in the
field of computer vision for many years. Though humans accomplish the task in an
easy manner, the underlying computations within the human visual system are of
remarkable complex. The apparently insignificant task of finding and recognizing
faces is the result of millions of years of regression and we are far from fully
understanding how the brain performs it. Moreover, the capability to find faces

S.A. Noah et al. (Eds.): M-CAIT 2013, CCIS 378, pp. 89–98, 2013.

visually in a scene and recognize them is critical for humans in their everyday events. Accordingly, the automation of this task would be beneficial for several applications including security, surveillance, gaze-based control, affective computing, speech recognition assistance, video compression and animation. Though, to date, no comprehensive solution has been anticipated that allows the automatic recognition of faces in real (un-affected) images [1]. In last decade, chief progresses occurred in the field of face recognition, with numerous systems capable of maintaining recognition rates superior to 90%. However real-world scenarios remain a challenge, because face acquisition procedure can experience a wide range of variations. Throughout a crime investigation, the community security agencies regularly need to match a probe image with registered database images, which may have major difference of facial features due to age deviations. Several efforts have been made to tackle this problem. Ling et al. [2] studied the aging effect on face recognition, O'Toole et al. [3] proposed a standard facial caricaturing algorithm using 3D face model, Ramanathan et al. [4] proposed a Bayesian age-difference classifier to be employed in applications such as passport renewal. These proposed techniques try to solve the problem by simulating the aging models; however, they are still far from hands-on use.

Unlike these complicated modelling methods, our system aims to perform a fast and robust aging face recognition based on a combination of geometrical and mathematical modelling. In this research, our goal is to develop a geometrical model that is age invariant. In our work we have explored the approach of using a mathematically developed geometrical model for maintaining the degree of similarity between six triangular features to address the problem of face recognition under age variations. The system to be developed is intended to operate in real time environment such as surveillance systems.

The remainder of this paper is organized as follows: Section 2 introduces the proposed face recognition geometrical model where we define the mathematical relationships between our proposed triangular features, and our tendency in constructing the systems' facial features vectors. The results and discussion of experiments are presented in Section 3. This is followed by conclusions in Section 4.

2   Proposed Geometrical Model

The proposed system is decomposed multiple stages. Face detection is the first stage at the beginning of each face recognition system. In our system, a commercial version of the conventional Viola and Jones face detector [5] is employed to detect and crop the face area that contains the main features (Eyes, Mouth, Nose, and chine). Viola and Jones detector is robust and effective in real time applications. After detecting the face area twelve facial features points are to be localized in order to extract six different triangular areas around the main facial features. Following the parameters of the triangular features i.e. (areas and perimeters) are calculated .Then those parameters are passed to a number of equations to create features vectors for the sample image. In the following those stages are illustrated in details. In Fig. 1 a block diagram of the proposed system is illustrated.

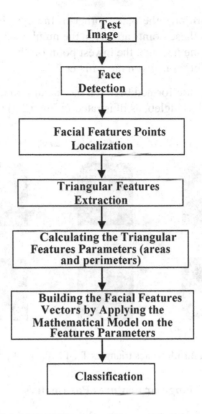

Fig. 1. Block diagram of the proposed system

2.1 Facial Features Points Localization and Triangular Features Detection

Craniofacial anthropometry which is the science that involves measurement of the skull and face, citrine landmarks and measurements are known to identify human facial characters and growth pattern. In our study we consider twelve of these land marks mostly the ones that form the circumscription of the main facial features. Those facial feature points are normally localized using Active Appearance Model (AAM) [6] which designates 64 distinctive facial points. In our model the AAM is reduced to 12 facial features points using the algorithm proposed in [6]. Craniofacial anthropometry refers to those facial features points with scientific notation to discriminate between them as follows:

- **En (endocanthion)**: the inner corner of the eye fissure where the eyelids meet. In our model these points are given the numbers 6 and 8.
- **G (glabella)**: the most prominent point in the median sagittal plane between the supraorbital. In our model these point is given the number 9.

- **ridges.ex (exocanthion)**: the outer corner of the eye fissure where the eyelids meet. In our model these points are given the numbers 5 and 7.
- **Gn (gnathion)**: in the midline, the lowest point on the lower border of the chin. . In our model this point is given the number 2.

Following six triangles are formed between the facial points and they are given the notation triangle₁ through triangle6, as illustrated in Fig. (2.a) through (2.f).

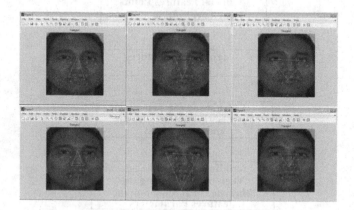

Fig. 2.a. Through **2.f.**illustrates triangles T_1, T_2, T_3, T_4, T_5, T_6 successively

2.2   Calculating the Triangular Features Parameters

After localizing the facial features points, the system will gain knowledge of the triangular vertices coordinates. After that Euclidean distances between each triangle coordinates will be calculated, which will enable the system to calculate perimeters and areas of each triangular feature. Those parameters (areas and perimeters are given the notation A and P for areas and perimeters successively followed by a subscript representing the triangle designation. For example (A_i, P_i) represent the area and perimeter of triangle number one. Finally, those parameters are used as inputs to some mathematical equations which will be discussed next, to form the features vectors for each sample image.

2.3   Deriving the Mathematical Model

In the geometry science it is known that Triangles are similar if they have the same shape, but not necessarily the same size [7]. This scientific fact inspired us to draw mathematical relationships between the six triangular features extracted during the previous stage. The Human population reached 7 Billion people around the world and thus, it is impractical to use a one-to-one comparison process for the purpose of face recognition using the measurements of our triangular features. As a different approach we were able to make use of the proportional ratio between the different triangles representing the facial features which led to fifteen different mathematical equations representing the degree of similarity between each two triangles. Based on the aforementioned geometrical theory regarding the similarity of triangles, any two triangles are considered similar even if they are of different sizes if the following mathematical relationship represented by "Eq. (1)" is satisfied:

$$A_i/ A_j = p_j^2/p_i^2. \tag{1}$$

Where A, and P represent triangles areas and perimeters successively, i and j are designations of the two triangles subject of the mathematical relationship. Eq. (1) is used to drive what is called triangles similarity proportion, which is a measurement of degree of similarity between two triangles, and it is represented by "Eq. (2)". TSP represents the triangles similarity proportion relationship.

$$TSP=A_i x p_j^2/ A_j x p_i^2. \tag{2}$$

The statistical analysis of the data collected in term of triangular features areas and perimeters had shown clearly that there is no significant difference between these measurements of different individuals. As a different approach we were able to make use of the similarity proportional ratio between the different triangles representing the facial features which led to fifteen different mathematical equations representing the degree of similarity between each two triangles. Those equation were derived using equation (2) by simply applying the formula between each two triangles, and substituting subscripts i and j by the designations of the two triangles. "Eq. (3)" through, "Eq. (17)" represents the fifteen relationships between the six triangular features shown in the following:

$(T_1,T_2)=(A_1* P_2^2/ A_2*P_1^2)$ (3)	$(T_1,T_3)=(A_1*P_3^2/ A_3*P_1^2)$ (4)	$(T_1,T_4)=(A_1*P_4^2/A_4*P_1^2)$ (5)
$(T_1,T_5)=(A_1* P_5^2/ A_5*P_1^2)$ (6)	$(T_1,T_6)=(A_1*P_6^2/ A_6*P_1^2)$ (7)	$(T_2,T_3)=(A_2*P_3^2/A_3*P_2^2)$ (8)
$(T_2,T_4)=(A_2* P_4^2/ A_4*P_2^2)$ (9)	$(T_2,T_5)=(A_2*P_5^2/ A_5*P_2^2)$ (10)	$(T_2,T_6)=(A_2*P_6^2/A_6*P_2^2)$ (11)
$(T_3,T_4)=(A_3* P_4^2/ A_4*P_3^2)$ (12)	$(T_3,T_5)=(A_3*P_5^2/ A_5*P_3^2)$ (13)	$(T_3,T_6)=(A_3*P_6^2/A_6*P_3^2)$ (14)
$(T_4,T_5)=(A_4* P_5^2/ A_5*P_4^2)$ (15)	$(T_4,T_6)=(A_4*P_6^2/ A_6*P_4^2)$ (16)	$(T_5,T_6)=(A_5*P_6^2/A_6*P_5^2)$ (17)

For each sample image enrolled in the system those fifteen relationships will be calculated and stored in a vector which will be considered as a feature vector of this specific sample image. When multiple sample images are related to the same subject, the feature vectors of these sample images will be stored in a matrix to form a class for each subject. Our approach is based on data normalization where we intend to normalize each distinct feature vectors' data using different normalization approach. The objective is to drive normal distribution over each features' data to increase the degree of uniqueness for this specific feature. After that normalization will be applied on the matrix of column vectors which represents subject's class.

Normalization refers to the formation of shifted and scaled versions of data, where the purpose is that these normalized values tolerate the comparison of consequent normalized values for different datasets in a way that eliminates the effects of certain vulgar influences. A number of normalization approaches involve only a rescaling, to arrive at values relative to some size variable. In terms of levels of measurement, such ratios only seem sensible for ratio measurements (where ratios of measurements are significant), not interval measurements (where only distances are significant, but not ratios). The aim of normalization is to make variables that come from the same population comparable to each other [8]. In the following we maintain four different normalization approaches over the datasets to study the influence of each approach over the system overall performance. In the following we illustrate the two approaches used for data normalization.

2.3.1   Column Vector Normalization
Column vector normalization in a class matrix is performed by squaring and adding the column sums in the vector and then dividing each element by the square root of the sum of squares [9].

2.3.2   Z-score Normalization
It is often useful to calculate how far, in standard deviations, a data element was from the mean. This is a very commonly used procedure and this measure has the name z-score. It is also known as a standard score. While several data sets have a fairly normal distribution, it is a very accommodating way to compare data elements from different populations— populations which may very well have contradictory means and standard deviations [9]. *Z-scores* provide a valuable measurement for comparing data elements from different data sets. The formula used for calculating z-score is as follows:

$$z = x- \mu/s. \tag{18}$$

Where:
Z: z-score; x: data element; μ: mean of the data set; s: standard deviation [9].

3   Experiments

We performed our experiments on a public aging database FG-NET [10] containing 1,002 high resolution color or gray scale face images of 82 subjects from multiple races with large variation of lighting, expression, and pose. The image size is approximately 400 x 500 in pixels. The age range is from 0 to 69 years (on average, 12 images per subject). The experiments were conducted in four rounds in each round both of the two mathematical approaches illustrated previously were applied over the features vectors separately. For each subject a distinct class was built that is represented by a matrix of features vectors. A number of well-known benchmarking classifiers were used to evaluate the performance of the system namely: K-means KNN (K-Nearest Niebuhr), and SVM (Support Vector Machine) classifiers. In the At the first round of the experiments a data set of five classes was tested using all of the three classifiers, at the second round we doubled the number of classes to ten, at the third round twenty classes were considered, at the fourth round the number of classes was doubled to forty, and finally the entire FGNET dataset which contains 88 classes was tested over the classifiers. The aim of conducting the experiments in multiple rounds using different number of classes for each round was to study the influence of increasing the number of classes over each of the three classifiers and there for decide which classifier is most convenience for our proposed model which is aimed to work with large datasets. In the next section classification results of the four rounds of experiments is illustrated to evaluate performance of each mathematical model.

4   Results Discussion

In order to show how the different features change over the years for each class i.e. subject, we plotted a surface map (mesh) for one of the FGNET subjects. Fig. 2.a show a mesh of the subject when the class' features vectors was built using the first approach. Fig. 2.b through 2.d show the mesh plots of the other three approaches.

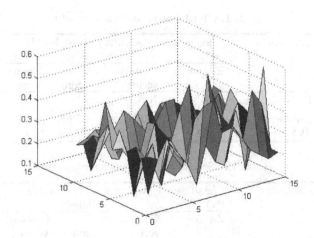

Fig. 2.A. Surface plot (mesh) of a subject class when column normalization is used

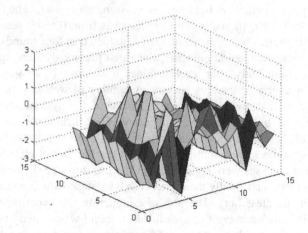

Fig. 2.B. Surface plot (mesh) of a subject class when Z-score matrix normalization ensemble is used

Fig.2 .A illustrates a mesh of a subject' class when column normalization is used to normalize the features vectors data elements. A noticeable level of transition was in the features between 5 and 10.

Finally the mesh plot of a subject class when the features data elements are normalized through Z-score matrix normalization ensemble is illustrated in Fig. 2.B. It is very obvious that all of the fifteen features change with the presence of aging effect and each of the features contribute in increasing the degree of uniqueness between the classes.

Classification results of the two mathematical approaches used to build the classes features vectors are illustrated in terms of classification accuracy, false positive rate [11] as illustrated in Tables 2 through 6.

Table 1.A. First round classification results

Classifier	Classification Accuracy (%)		False Positive Rate	
	Z-score	Column vector	Z-score	Column vector
KNN	99.8	20.4	0	79.59
K-means	72.35	30.62	27.65	69.38
SVM	72.34	10.2	27.65	8.79

Table 1.B. Second round classification results

Classifier	Classification Accuracy (%)		False Positive Rate	
	Z-score	Column vector	Z-score	Column vector
KNN	99.8	26.47	0	73.52
K-means	35.87	19.61	64.13	80.39
SVM	55.43	12.74	44.56	82.25

Table 1.C. Third round classification results

Classifier	Classification Accuracy (%)		False Positive Rate	
	Z-score	Column vector	Z-score	Column vector
KNN	99.8	16.5	0	83.49
K-means	19.79	9.71	80.21	90.29
SVM	36.81	4.85	63.18	95.14

Table 1.D. Fourth round classification results

Classifier	Classification Accuracy (%)		False Positive Rate	
	Z-score	Column vector	Z-score	Column vector
KNN	98.92	11.88	1.078	88.11
K-means	9.71	5.2	90.29	94.8
SVM	25.33	2.97	74.66	97

Table 1.E. Fifth round classification results

Classifier	Classification Accuracy (%)		False Positive Rate	
	Z-score	Column vector	Z-score	Column vector
KNN	99.17	7.61	0.82	92.38
K-means	9.71	3	90.29	96.9
SVM	5.11	2	94.88	97.93

It can be seen from the classification results that column normalization approach inadequate for building large dataset features vectors and does not asses in showing uniqueness and distinctness of data as the highest classification accuracy reported for this approach was 30.62% when the number of classes was only five, and it shows very large levels of false positives. The highest recognition accuracy 99.17% over the entire FGNET database was achieved when the Z-score matrix normalization ensemble approach was used to build the features vectors as illustrated in Table 2.E with minimum false positive rate. It can be shown from the classification results that K-means and SVM classifiers experience degradation in performance as the number of classes increases particularly SVM classifier which reported a minimum classification accuracy of 2% when the system was tested over the entire FGNET database. This makes the KNN classifier our best choice based on the classification results as it shows very small level of degradation as the number of classes increases.

5   Conclusion and Future Work

This research study proposed new geometrical features that are formed by connecting some of the facial features points defined in the anthropometric science. The main goal was to develop mathematical relationships among triangular features to accommodate for the aging variations conditions that may affect any face recognition system. For mathematical approaches were employed for the purpose of normalizing the features vectors date elements to hold normal distribution over each subject class and increase the degree of distinctness of each class over the dataset. The performance of the system was evaluated mainly in term of classification accuracy, and the maximum classification accuracy was reported when the features vectors elements were normalized using a Z-score matrix normalization ensemble approach. In our future work we are planning to test our proposed system on the MORPH aging database which is the largest publicly available face aging database with over 3000 subjects.

Acknowledgments.   The authors gratefully acknowledge the support provided by Universiti Teknologi PETRONAS, Malaysia. Also the authors wish to express their gratitude to the Dean of the Centre for Graduate Studies Associate Prof Dr. Mohd Fadzil Hassan at the Universiti Teknologi PETRONAS for participating in the excremental part of this research work and for his continuous support to the research innovation in at the Universiti Teknologi PETRONAS.

References

1. Kachare, N.B., Inamdar, V.S.: Survey of Face Recognition Techniques. International Journal of Computer Applications (0975 - 8887) 1(19) (2010)
2. Ling, H., Soatto, S., Ramanathan, N., Jacobs, D.W.: Study of Face Recognition as People Age. In: Proc. 11th Int'l Conf. Computer Vision, pp. 1–8 (2007)
3. O'Toole, A., Vetter, T., Volz, H., Salter, M.: Three-dimensional caricatures of human heads: distinctiveness and the perception of facial age. Perception 26, 719–732 (1997)
4. Ramanathan, N., Chellappa, R.: Face Verification across Age Progression. IEEE Trans. Image Processing 15(11), 3394–3361 (2006)
5. Viola, P., Jones, M.J.: Robust Real time Face Detection. International Journal of Computer Vision (2004)
6. Cootes, T.F., Edwards, G.J., Taylor, C.J.: Active appearance models. In: Burkhardt, H., Neumann, B. (eds.) ECCV 1998. LNCS, vol. 1407, pp. 484–498. Springer, Heidelberg (1998)
7. McCartin, B.J.: Mysteries of the equilateral triangle. Applied Mathematics Kettering University (2010)
8. Everett, M.G., Borgatti, S.P.: The centrality of groups and classes. Journal of Mathematical Sociology 23(3), 181–201 (1999)
9. Pandit, G.S.: Structural analysis a matrix approach. McGraw-Hill publishing company (1981)
10. FG-NET Aging Database, http://www.fgnet.rsunit.com
11. Oliver, L.H., Poulsen, R.S., Toussiant, G.T.: Estimate false positive and false negative error rates in cervical cell classification. J. Histochem. Cytochem. 25, 969 (1977)

Mamdani-Fuzzy Expert System for BIRADS Breast Cancer Determination Based on Mammogram Images

Wan Noor Aziezan Baharuddin, Rizuana Iqbal Hussain,
Siti Norul Huda Sheikh Abdullah, Neno Fitri, and Azizi Abdullah

Center for Artificial Intelligence Technology,
Faculty of Information Science and Technology, Universiti Kebangsaan Malaysia,
43600, Bangi, Selangor D.E., Malaysia
{nooraziezan,rizi7886}@gmail.com, {mimi,azizi}@ftsm.ukm.my,
neno.fitri@yahoo.com

Abstract. Breast cancer is considered as a dangerous disease attack women all over the world. A Mamdani-Fuzzy expert system is built to detect the disease in early stage by using mammogram images and data report for calcification and ultrasound data for mass size. Two input and one output which are size of mass and distribution of calcification (input) and class of BIRADS (output) have been used to develop the model. The model is able to classify 84.04 % mammogram images into the actual BIRADS. 13 images which are 13.83% wrongly classify and 2 images which are 2.13% unable to classify because of some limitation as stated in discussion.

Keywords: breast cancer, mamdani-fuzzy expert system.

1 Introduction

One of the most common cancer attack women in developing country is breast cancer [1]. This disease occur because of uncontrolled abnormal cell divided in breast [1]. This cell known as malignant tumor will form a lump, microcalcification and architectural distortion [2]. According to National Cancer Registry (NCR), there are 3,242 cases for female breast cancer has been diagnosed in 2007 [3].

Early detection of breast cancer can reduce the number of mortality rate. A Better diagnostic facility for diagnosing breast cancer have evolved based on X-rays analysis, ultrasound evaluations, mammography, ethnography and magnetic resonance imaging which give the best results in performance and input-out ratio [4].

Various methods have been applied by researchers in order to diagnose breast cancer. Chunekar, V.N et al. & Yao, X. et al.[5] and [6] use neural network approach. Liu, L. & Deng, M [7] had applied adaptive genetic algorithm (AGA) with back propagation Neural Network to get an accurate result in breast cancer classification. Pena-Reyes, C.A. & Sipper, M [8] found that fuzzy-genetic approach give high classification performance and the system involve only simple rules while focusing on Wisconsin breast cancer diagnosis (WBCD) problem.

S.A. Noah et al. (Eds.): M-CAIT 2013, CCIS 378, pp. 99–110, 2013.
© Springer-Verlag Berlin Heidelberg 2013

In this research, a Mamdani-Fuzzy Expert System model has been developed. This model used fuzzy logic approach which has been introduced by Zadeh in 1965. This method is selected because of its ability to model a complex system based on human experience and their knowledge. Two type of radiological images have been used to evaluate this model which are mammogram and ultrasound.

2   The Proposed Mamdani- Fuzzy Expert System

In development process, the expert system decision tree has been used in order to acquire knowledge from the expert as an early stage for knowledge acquisition. An interview with an expert, radiologist from radiology departments in UKM Medical Centre has been conducted in order to gain knowledge based on their experience and guideline.

The proposed mamdani-fuzzy model for maglinant breast cancer determination system architecture is illustrated as in Fig. 1.

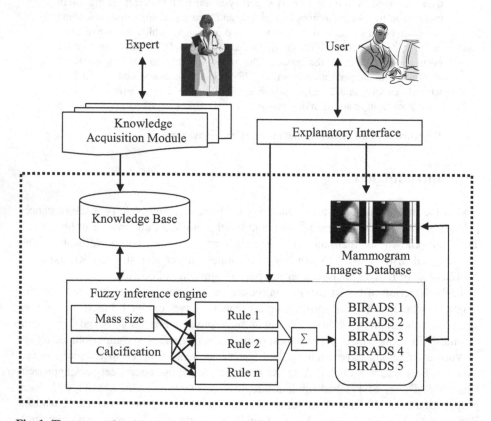

Fig. 1. The proposed architecture design based on Mamdani-Fuzzy Expert System for maglinant breast cancer determination

3 Our Proposed Mamdani-Fuzzy Model

Consider Caramihai, M., et al. [9] research in 2010, 3 steps have been used in developing fuzzy logic model, which are:

1. Input and output variables determination.
2. Define linguistic values and membership function for each fuzzy variable.
3. Define fuzzy inference rules between input and output fuzzy variables.

In our research, we use the most common fuzzy inference technique which is Mamdani method. This method involve four steps which are (1) fuzzification of the input variables, (2) rule evaluation, (3) aggregation of the rules output, (4) Final defuzzification.

The model examines two-input and three-output problems. The two crisp input variables are the size of mass (x1) and distribution of calcification (x2). The range of possible value to choose variable (universe of discourse) is determined by expert judgment.

The outputs are three BIRADS of the cancer (y) which is taken from mammogram images and data report of ultrasound and mammogram.

Linguistic variable for the model can be obtained based on tree representation below.

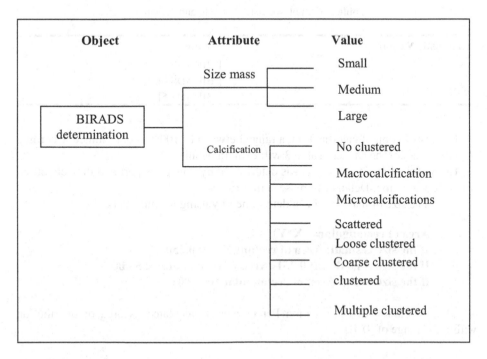

Fig. 2. Our proposed Breast Cancer decision tree representation for linguistic variable

Based on Fig. 2, object (BIRADS determination), and two attribute (mass size and calcification distribution) are used as linguistic variable. 11 of input variable and 3 output variable are identified. Table 1 will list all of input linguistic variable with their values and Table 2 will list output variable with their values.

Table 1. Input linguistic variable and the values

Linguistic Variables	Value
Size of masses ($x1$)	[small]
	[medium]
	[large]
Distribution of calcification ($y1$)	[No clustered]
	[Macrocalcification]
	[Microcalcifications]
	[Scattered]
	[Loose clustered]
	[Coarse clustered]
	[clustered]
	[Multiple Clustered]

Table 2. Output linguistic variable and the values

Linguistic Variables	Value
	[BIRADS 3]
	[BIRADS 4]
	[BIRADS 5]

For distribution of calcification, a range between 0- 100% is determined according to seriousness of the cancer. Table 3 will explain detail.

Each of the linguistic variables is determined by domain expert and they classified it into few notation. Details can be seen in Table 4.

The size of mass has been determined by using formula as below:

Area of rectangular= (X*Y)
If the size is small: Area of rectangular < 0.250
If the size is medium: 0.251<Area of rectangular < 5.000
If the size is big: Area of rectangular >= 5.001

The calculation of calcification Percentage is calculated as area of distribution within the range of [0-100]

Table 3. Percentage of seriousness based on calcification range

Image Description	Percentage of Seriousness (%)	Image sample		
Loose clustered	10	L1	L2	L3
Scattered	10	S1	S2	S3
Coarse	10	C1	C2	C3
No clustered	20	NC1	NC2	NC3
Macrocalcification	30	MA1	MA2	MA3
Microcalcifications	30	MI1	MI2	MI3

Table 3. (*continued*)

clustered	40	CL1	CL2	CL3
Multiple Clustered	50	MC1	MC2	MC3

Table 4. Details of input linguistic variable

Linguistic variable	Value of Linguistic variable	Notation	Range
Size (cm^3)	Small	S	[0, 1.25, 3.00]
	Medium	M	[2.00, 3.75, 5.50]
	Large	L	[4.50, 6.25, 7.50]
Calcification	Less	1	[0, 10, 35]
Distribution (%)	Intermediate	2	[20, 37, 60]
	More	3	[40, 60, 100]

Table 5. Details of output linguistic variable

Linguistic variable	Value of Linguistic variable	Percentage of Seriousness
BIRADS	BIRADS 3	[0, 25, 50]
	BIRADS 4	[25, 45, 55, 75]
	BIRADS 5	[50, 75, 100]

i.   Small

$$\mu SS(x1) = \begin{cases} \dfrac{3.0 - x}{3.0}x, & x \leq 3.0 \\ \\ 0 \quad , & x > 3.0 \end{cases}$$

ii.   Medium

$$\mu MS(x1) = \begin{cases} \dfrac{6.0 - x}{6.0}x, & x \leq 6.0 \\ \\ 0 \quad , & x > 6.0 \end{cases}$$

iii.   Large

$$\mu LS(x1) = \begin{cases} 0 & , & x \geq 5.0 \\ \dfrac{x - 5.0}{8.5} & ,, & x < 5.0 \end{cases}$$

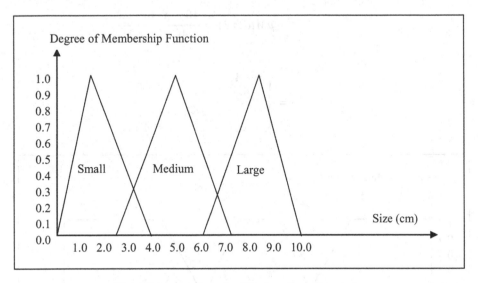

Fig. 3. Degree of membership function of input size

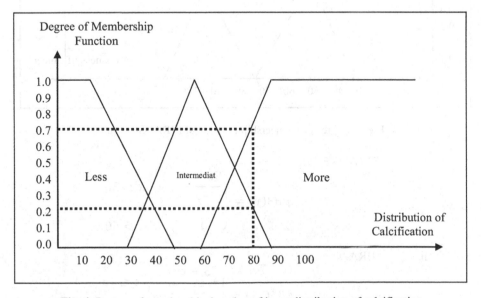

Fig. 4. Degree of membership function of input distribution of calcification

i.   Less

$$\mu LC(x2) = 1 - \frac{x}{35}$$

ii.   Intermediate

$$\mu MiC(x2) = \frac{40 - x}{40}, \frac{60 - x}{60}$$

iii.   More

$$\mu MaC(x2) = \begin{cases} \dfrac{x - 40}{100} & , \quad x \geq 40 \\ \\ 0, & x < 40 \end{cases}$$

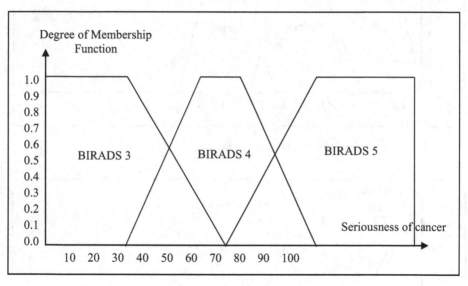

Fig. 5. Degree of membership function of BIRADS output

i.   BIRADS 3

$$\mu B3(y) = \begin{cases} \dfrac{50 - x}{50}x, & x \leq 50 \\ \\ 0 & , \quad x > 50 \end{cases}$$

ii.   BIRADS 4

$$\mu B4(x1) = \begin{cases} \dfrac{75 - x}{75}x, & x \leq 75 \\ \\ 0 & , \quad x > 75 \end{cases}$$

iii.   BIRADS 5

$$\mu LS(x1) = \begin{cases} 0 \quad , \quad x \geq 50 \\ \dfrac{x-50}{100} \text{,,} \quad x < 50 \end{cases}$$

These images are classified into a few categories. Based on analysis of sample images, we could safely assume that these images are divided into two categories, size of mass and distribution of calcification. Both of these samples are taken as input for the system and divided into three classes. Each of the class is given their own notation. Details input variables and the values as stated in table 1. Notation of calcification is classified according to the mammogram images as shown in table 2.

By looking at the report of mammogram, nine fuzzy rules are generated. Detail of the rules is shown in table 6.

Table 6. List of rules

Case	Rules	BIRADS
1	if size = S & distribution = 1	3
2	if size = S & distribution = 2	4
3	if size = S & distribution = 3	4
4	if size = M & distribution =1	3
5	if size = M & distribution = 2	4
6	if size = M & distribution = 3	4
7	if size = L & distribution = 1	4
8	if size = L & distribution = 2	4
9	if size = L & distribution = 3	5

4   Result and Discussion

After applying the fuzzy model to the mammogram image, the result as show in table below:

Based on True-Positive (TP), True-Negative (TN), False-Positive (FP), and False–Negative (FN) measures, we calculated statistical performance such as sensitivity ($\frac{TP}{TP+FN}$), specificity ($\frac{TN}{TN+FP}$) and accuracy ($\frac{TP+TN}{TP+TN+FP+FN}$). Table 7 displays the performance measures of all classifiers in terms of sensitivity, specificity and accuracy.

Table 7. Performance for Breast cancer

	Proposed (Mamdani-Fuzzy)
Sensitivity	87.01%
Specificity	81.24%
Accuracy	84.04 %

Table 8. Predicted result and actual result of mammogram images

No	Image type	Mammography	Predicted result	Actual Result
1	Microcalcification		BIRADS 5	BIRADS 5
2	Mass + Clustered		BIRADS 4	BIRADS 4
3	Mass + Scattered		BIRADS 3	BIRADS 5

Table 8. (*continued*)

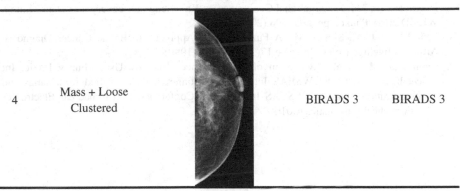

4	Mass + Loose Clustered		BIRADS 3	BIRADS 3

After applying the model to the mammogram images, it able to classify 79 images into their BIRADS, which is 84.04 % correctly. 13 image which are 13.83% wrongly classify and 2 images which are 2.13% unable to classify because of some limitation.

The limitation is because of some difficulties in understanding the term due to several ways of referencing the cancer cell.

Acknowledgment. This researcher would like to thank to Faculty of Information Science and Technology, Universiti Kebangsaan Malaysia for providing facilities and financial support under "AP-2012-019 Automated Medical Imaging Diagnostic Based on Four Critical Diseases: Brain, Breast, Prostate and Lung Cancer". We have obtained Ethics approval entitled "FF-338-2012 Breast Cancer Diagnostic Imaging System Using Mammogram and Ultrasound" from PPUKM or UKM Medical Center, Malaysia for collecting and conducting experiments on Breast cancer patient's record.

References

1. Al-Shamlan, H., El-Zaart, A.: Feature Extraction Values for Breast Cancer Mammography Images. In: International Conference on Bioinformatics and Biomedical Technology, ICBBT (2010)
2. Ganesan, K., et al.: Computer-Aided Breast Cancer Detection Using Mammograms: A Review. IEEE Reviews in Biomedical Engineering 6, 77–98 (2013)
3. National Cancer Registry, Malaysia Cancer Statistics – Data and Figure (2007)
4. Caramihai, M., Severin, I., Balan, H., Blidaru, A., Balanica, V.: Breast Cancer Treatment Evaluation Based on Mammographic and Echographic Distance Computing. World Academy of Science on Engineering and Technology 32, 815–819 (2009)
5. Chunekar, V.N., Ambulgekar, H.P.: Approach of Neural Network to Diagnose Breast Cancer on Three Different Data Set. In: International Conference on Advances in Recent Technologies in Communication and Computing (2009)
6. Yao, X., Liu, Y.: Neural Networks for Breast Cancer Diagnosis. In: Proceedings of the 1999 Congress on Evolutionary Computation, CEC 1999 (1999)

7. Liu, L., Deng, M.: An Evolutionary Artificial Neural Network Approach for Breast Cancer Diagnosis. In: Third International Conference on Knowledge Discovery and Data Mining, WKDD 2010, Phuket, pp. 593–596 (2010)
8. Peña-Reyes, C.A., Sipper, M.: A Fuzzy-Genetic Approach to Breast Cancer Diagnosis. Artificial Intelligence in Medicine 17(2), 131–155 (1999)
9. Caramihai, M., et al.: Evaluation of Breast Cancer Risk by Using Fuzzy Logic. In: Proceedings of the 10th WSEAS International Conference on Applied Informatics and Communications, and 3rd WSEAS International Conference on Biomedical Electronics and Biomedical Informatics (2010)

Comparison between Record to Record Travel and Great Deluge Attribute Reduction Algorithms for Classification Problem

Majdi Mafarja[1,2] and Salwani Abdullah[1]

[1] Data Mining and Optimization Research Group (DMO), Center for Artificial Intelligence Technology, Universiti Kebangsaan Malaysia, 43600 UKM, Bangi Selangor, Malaysia
[2] Department of Computer Science, Faculty of Information Technology
Birzeit University, P.O. Box 14, Birzeit, Palestine
mmafarja@birzeit.edu, salwani@ftsm.ukm.my

Abstract. In this paper, two single-solution-based meta-heuristic methods for attribute reduction are presented. The first one is based on a record-to-record travel algorithm, while the second is based on a Great Deluge algorithm. These two methods are coded as RRT and m-GD, respectively. Both algorithms are deterministic optimisation algorithms, where their structures are inspired by and resemble the Simulated Annealing algorithm, while they differ in the acceptance of worse solutions. Moreover, they belong to the same family of meta-heuristic algorithms that try to avoid stacking in the local optima by accepting non-improving neighbours. The obtained reducts from both algorithms were passed to ROSETTA and the classification accuracy and the number of generated rules are reported. Computational experiments confirm that RRT m-GD is able to select the most informative attributes which leads to a higher classification accuracy.

Keywords: Record to Record Travel algorithm, Great Deluge algorithm, Rough Set Theory, Classification.

1   Introduction

Attribute Reduction (AR) (or data reduction) is regarded as an important preprocesing technique in machine learning and in the data mining process [1] and [2]. It can be defined as the problem of finding a minimum reduct (subset) from the original set [3].

The attribute reduction process aims to eliminate the irrelevant and redundant attributes from high-dimensional data sets which increase the chances that a data mining algorithm will find spurious patterns that are not valid in general. Furthermore, when dealing with high-dimensional data sets, a longer time is needed to find the desired results. Liu and Motoda [3] indicate that the purposes of AR are to: (a) improve the performance (speed of learning, predictive accuracy, or simplicity of rules); (b) visualise the data for model selection; and (c) reduce the dimensionality and remove the noise.

Langley [4] divided the attribute reduction methods into two types of model (i.e., filter and wrapper) based on their dependence on the inductive algorithm that will finally use the selected subset. In a filter model, the selection process is performed independently from the induction algorithm. A wrapper model, which is

S.A. Noah et al. (Eds.): M-CAIT 2013, CCIS 378, pp. 111–120, 2013.

essentially the opposite of a filter model, uses the induction algorithm to directly evaluate the feature subsets.

The rough set theory, proposed by Pawlak [5] and [6], has been used as a simple mechanism to determine minimal subsets by locating all of the possible reducts and selecting the one with the lowest cardinality and the highest dependency. This process is a time-consuming procedure and it is only effective for small datasets. As a result, for high-dimensional datasets, many researchers now focus on employing meta-heuristics to search for better solutions instead of using the rough set theory's reduction method. Many researchers are focusing on the problem of finding a subset with minimal attributes from an original set of data in an information system [7], [8], [9], [10], [11], [12], [13], [14], [15] and [16].

In the literature, many meta-heuristic-based methods which were designed to solve the attribute reduction problem can be found, such as the Genetic Algorithm [17], [18] and [19], Particle Swarm Optimisation [20], Ant Colony algorithm [21] and [22], Tabu Search [23], Great Deluge algorithm [24], Composite Neighbourhood Structure [25], Hybrid Variable Neighbourhood Search algorithm [26] and Constructive Hyper-Heuristics [27], Bees Algorithm [28].

In this paper, we examined the effect of employing two attribute reduction methods on the classification accuracy based on 13 benchmark datasets. In this work, we investigate how the use of attribute reduction methods will influence the classification accuracy. Two attribute reduction methods were examined, Record to Record Travel algorithm (RRT) and Modified Great Deluge algorithm (m-GD).

The remainder of this paper is organized as follows: section 2 discusses the RRT algorithm, and this is followed by a detailed description of the implementation of m-GD in section 3. Section 4 presents a simulation of the proposed algorithms together with a discussion of the experimental results. Finally, concluding remarks on the effectiveness of the proposed techniques and the potential future research aspects are presented in section 5.

2   Record-to-Record Travel Algorithm

The RRT algorithm was originally proposed by Dueck [29]. It is a variant of the Simulated Annealing algorithm, with a different mechanism for accepting non-improving solutions [30]. This algorithm has a solitary parameter called the DEVIATION, which plays a pivotal role in controlling the acceptance of the worst solutions after it becomes pre-tuned. The significance of this method relates to the ease of its implementation and the required number of parameters, which influences the performance of the algorithm [29]. In this work, the RRT algorithm is applied to tackle the attribute reduction problem.

Fig. 1 shows the pseudo-code of the proposed method. The algorithm starts from a randomly generated initial solution (Sol). The best solution (Sol_{best}) is set as Sol, and the $RECORD$ is set as the fitness value of the best solution $f(Sol_{best})$. The initial solution is improved by searching its neighbourhood for a better solution (called Sol_{trial}). The neighbourhood of a solution (Sol_{trial}) is generated by a random flip-flop, where three cells are selected at random from the current solution (Sol) as in [18]. For each selected cell, if its value is '1' then it is changed to '0', which means that the feature is deleted from the current solution. Otherwise, it is changed from '0' to '1'. The cardinality of the generated trial solution should be less than the cardinality of the best solution so far, because we are trying to generate a solution with a lower cardinality and a higher quality.

Later, Sol_{trial} is evaluated (in this work, evaluation is on the dependency degree in RST). If the quality of Sol_{trial} is better or slightly worse (not more than the *DEVIATION* value) than the best value so far (the $f(Sol_{best})$), then the solution is accepted. Note that the initial value of the *RECORD* is equal to the initial fitness function. During the search process, the *RECORD* is updated with the fitness value of the best solution so far $f(Sol_{best})$. More formally, in the case of maximisation, if (Sol_{best}) is the best solution so far, and (Sol_{trial}) is the newly generated solution, then (Sol_{trial}) is accepted as the next solution if $f(Sol_{best})$ - $f(Sol_{trial})$ < *DEVIATION*, where *DEVIATION* >= *0* (is the maximum allowed *DEVIATION* that determines how much worse values than the *RECORD* will be accepted). This process is repeated until the stopping condition is met (in this work, the number of iterations is set as a stopping criterion).

Record-to-Record Travel Algorithm for Attribute Reduction (RRT)

Generate a random initial solution Sol;
Set $Sol_{best} = Sol$;
Set $RECORD = f(Sol_{best})$;
Set $DEVIATION = 0.09$;
while (stopping-criterion is not satisfied)
    Generate at random a new solution Sol_{trial} from Sol;
    Calculate $f(Sol_{trial})$;
    if $(f(Sol_{trial}) > f(Sol_{best}))$
        $Sol \leftarrow Sol_{trial}$; $Sol_{best} \leftarrow Sol_{trial}$;
        $f(Sol) = f(Sol_{trial})$; $f(Sol_{best}) = f(Sol_{trial})$;
    else if $(f(Sol_{trial}) == f(Sol_{best}))$;
        Calculate cardinality of trial solution, $| Sol_{trial} |$;
        Calculate cardinality of best solution, $| Sol_{best} |$;
        if $(|Sol_{trial}| < |Sol_{best}|)$
            $Sol \leftarrow Sol_{trial}$; $Sol_{best} \leftarrow Sol_{trial}$;
            $f(Sol) = f(Sol_{trial})$; $f(Sol_{best}) = f(Sol_{trial})$;
        end if
    else if $(f(Sol_{trial}) > RECORD - DEVIATION)$
        $Sol \leftarrow Sol_{trial}$; $f(Sol) \leftarrow f(Sol_{trial})$;
    end if
    if $(f(Sol_{trial}) > RECORD)$
        $RECORD = f(Sol_{trial})$;
    end if
end while
Return best solution;

Fig. 1. Pseudo-code of RRT for attribute reduction

3   Modified Great Deluge Algorithm

The original great deluge algorithm was applied to attribute reduction problems by Abdullah and Jaddi [24] and achieved comparable results with other methods in the literature. Mafarja and Abdullah [31] examined in their paper the ability of improving the performance of this method by changing the increasing rate (β) intelligently. They proposed a mechanism called modified great deluge for attribute reduction (m-GD).

The main idea of m-GD here is that three equaled regions are established between the quality of the initial solution ($f(Sol)$) and the maximum dependency degree which is 1 (by using RST). Based on the interval value, which is calculated as shown in Fig. 2 (a), we define three levels as follows:

$interval = estimated\_quality - f(Sol)$
$region1 = region2 = region3 = interval /3$
$level1 = level$
$level2 = level1 + interval$
$level3 = level2 + interval$

Following the example in Fig. 2, if the fitness value of the initial solution $f(Sol)$ is 0.34 and the maximum dependency degree is 1, then:

$interval = 1 - 0.34 = 0.66$ (as shown in Fig. 2 (a))
$region1 = region2 = region3 = 0.66/3 = 0.22$
$level1 = level = 0.34$
$level2 = level1 + region1 = 0.34 + 0.22 = 0.56$
$level3 = level2 + region2 = 0.56 + 0.22 = 0.78$

Each level represents the beginning of a new region in the search space, i.e., the 1st region starts from level1, the 2nd region, starts from level2 and the 3rd region starts from level3. In this method, three values for the increasing rate (β) are introduced (coded as $\beta1$, $\beta2$ and $\beta3$) to be used in updating the level in the three different regions (see Fig. 2 (b)). These values are calculated as follows:

$\beta_1 = (estimated\_quality - f(level1))/ NumOfIte\_GD$
$\beta_2 = (estimated\_quality - f(level2))/ NumOfIte\_GD$
$\beta_3 = (estimated\_quality - f(level3))/ NumOfIte\_GD$

In m-GD the level is updated depending on the region that the trial solution (Soltrial) belongs to (i.e., if the trial solution falls in region2, then the level is updated using $\beta2$ and so on). For more details refer to the pseudo code of the algorithm as represented in Fig. 3.

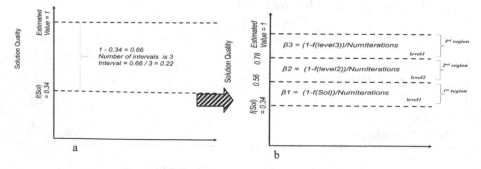

Fig. 2. Search space regions in m-GD

Modified Great Deluge for Attribute Reduction, m-GD

Generate a random initial solution Sol;
Set $Sol_{best} = Sol$;
Set $level = f(Sol_{best})$;
Set $level1 = level$;
Set number of iterations, $NumOfIte\_GD$;
Set $estimated\_quality = 1$ (as the maximum value of dependency degree in RST)
Divide the search space into 3 regions ($region = (estimated\_quality - f(level1))/3$);
Calculate the second $level$, $level2 = level1 + region1$;
Calculate the third $level$, $level3 = level2 + region2$;
Calculate $\beta_1 = (estimated\_quality - level1))/NumOfIte\_GD$;
Calculate $\beta_2 = (estimated\_quality - level2)/NumOfIte\_GD$;
Calculate $\beta_3 = (estimated\_quality - level3)/NumOfIte\_GD$;
Set iteration $\leftarrow 0$;
while (iteration $< NumOfIte\_GD$)
    Generate at random a new solution Sol_{trial} from Sol;
    Calculate $f(Sol_{trial})$;
    if ($f(Sol_{trial}) > f(Sol_{best})$) // accepting improving solutions
        $Sol \leftarrow Sol_{trial}$; $Sol_{best} \leftarrow Sol_{trial}$;
        $f(Sol) = f(Sol_{trial})$; $f(Sol_{best}) = f(Sol_{trial})$;
    else if ($f(Sol_{trial}) == f(Sol_{best})$)
        Calculate cardinality of trial solution, $| Sol_{trial} |$;
        Calculate cardinality of best solution, $| Sol_{best} |$;
        if ($|Sol_{trial}| < |Sol_{best}|$)
           $Sol \leftarrow Sol_{trial}$; $Sol_{best} \leftarrow Sol_{trial}$;
           $f(Sol) = f(Sol_{trial})$; $f(Sol_{best}) = f(Sol_{trial})$;
        end if
    else if ($f(Sol_{trial}) > level$)   // accepting non-improving solutions
        $Sol \leftarrow Sol_{trial}$; $f(Sol) \leftarrow f(Sol_{trial})$;
    end if
    if ($f(Sol_{trial}) >= level1$ & $f(Sol_{trial}) < level2$) // updating the $level$ according to the $region$
        $level = level + \beta_1$;
    else if ($f(Sol_{trial}) >= level2$ & $f(Sol_{trial}) < level3$)
        $level = level + \beta_2$;
    else if ($f(Sol_{trial}) >= level3$)
        $level = level + \beta_3$;
    end if
    iteration++;
end while
Calculate cardinality of best solution, $| Sol_{best} |$;
Return best solution;

Fig. 3. Pseudo code for m-GD adopted from[31]

4   Experimental Results

The proposed algorithms were programmed using J2EE Java, and the simulations were performed on an Intel Pentium 4 2.2 GHz computer with 2 GB of RAM and tested on 13 well-known UCI datasets, as shown in Table 1. For every dataset, the algorithm was run 20 times; the stopping conditions were the number of iterations that exceeded *NumOfIte* for both RRT and m-GD algorithms.

Table 1. UCI Datasets

Datasets	No. of Attributes	No. of Objects
M-of-N	13	1000
Exactly	13	1000
Exactly2	13	1000
Heart	13	294
Vote	16	300
Credit	20	1000
Mushroom	22	8124
LED	24	2000
Letters	25	26
Derm	34	366
Derm2	34	358
WQ	38	521
Lung	56	32

The experiments in this paper are carried out to determine the classification accuracy and the number of generated rules for all data sets based on the obtained reducts from RRT and m-GD. As in Table 2, the classification accuracy assessment is performed using the Standard Voter algorithm found in the ROSETTA library [32]. Independent tests are performed with the Voting parameter set to Simple. Table 2 show the details of the classification accuracy and the number of generated rules along with some details of the obtained minimal attributes using RRT and m-GD algorithms.

Table 2 shows the comparison of using RRT and m-GD in terms of the classification accuracy (in percentage) and the number of generated rules. The results without any attribute reduction are also presented. The results for the classification accuracy show that the m-GD method is slightly better (in the case of three data sets) than the RRT method, whereas RRT is better than m-GD in the case of two data sets. RRT and m-GD tie in the case of eight data sets. Based on the results presented for the classification accuracy, it cannot be claimed that the m-GD is consistently better than RRT. However, in relation to the minimal attributes, it can be seen that even though the classification accuracies are almost similar, the number of attributes obtained by the m-GD is fewer than that obtained by the RRT method. For example, m-GD shows an increment in the classification accuracy for the Heart, Vote, and Derm2 data sets while it simultaneously demonstrates a reduction in the dimensionality. In the case of the Derm and WQ data sets, although the classification accuracy obtained by m-GD is lower than that obtained by RRT, m-GD is better in terms of the minimal attributes.

Moreover, comparison in terms of the classification accuracy between 'without attribute reduction' and 'with attribute reduction' shows that both RRT and m-GD (that are classified under 'with attribute reduction') are able to give higher classification accuracy compared to 'without attribute reduction'. The use of all attributes does not guarantee 100% accuracy. This is most likely because the data sets with all attributes contain noise such as irrelevant and redundant attributes. Although the attributes have been reduced in the RRT and m-GD methods, good results are still able to be obtained for most of the data sets except Mushroom, Letters and Lung. These results show that attribute selection is important in producing a good quality of attribute for classification.

In terms of the number of generated rules, from Table 2, it can be seen that m-GD outperforms RRT on the case of five out of 13 data sets (i.e., Mushroom, Derm, Derm2, WQ and Lung) and ties in the case of six data sets (i.e., M-of-N, Exactly, Exactly2, Vote, Credit and LED). However, RRT manages to obtain a lower number of rules compared to m-GD in the case of two data sets (i.e., Heart and Letters). In general, m-GD produces a lower number of generated rules simultaneously with higher classification accuracy.

Table 2. Comparison between RRT and m-GD in terms of minimal attributes, classification accuracy and number of rules

Datasets	Without attribute reduction			RRT								m-GD							
				#A		Acc				#R		#A		Acc				#R	
	#A	Acc	#R	Min	Max	Min	Max	Avg	Std	Min	Max	Min	Max	Min	Max	Avg	Std	Min	Max
M-of-N	13	59.00	853	6	6	100.00	100.00	100.00	0.00	64	64	6	6	100.00	100.00	100.00	0.00	64	64
Exactly	13	36.00	839	6	6	100.00	100.00	100.00	0.00	64	64	6	6	100.00	10.000	100.00	0.00	64	64
Exactly2	13	35.00	855	10	10	71.00	71.00	71.00	0.00	606	606	10	10	71.00	71.00	71.00	0.00	606	606
Heart	13	31.00	263	6	7	31.00	41.00	33.00	4.10	234	256	6	7	69.00	69.00	69.00	0.00	244	258
Vote	16	40.00	229	8	9	53.30	70.00	66.49	6.27	135	157	8	8	70.00	70.00	70.00	0.00	135	135
Credit	20	69.00	896	8	9	70.00	73.00	72.70	0.92	725	789	8	10	66.00	73.00	71.75	2.02	725	864
Mushroom	22	100.00	7312	4	5	100.00	100.00	100.00	0.00	36	142	4	5	100.00	100.00	100.00	0.00	30	165
LED	24	7.50	1800	5	6	100.00	100.00	100.00	0.00	10	20	5	6	100.00	100.00	100.00	0.00	10	20
Letters	25	0.00	23	8	8	0.00	0.00	0.00	0.00	19	23	8	9	0.00	0.00	0.00	0.00	23	23
Derm	34	48.60	319	7	9	40.50	62.20	53.00	7.71	88	196	6	7	32.40	73.00	52.71	9.96	41	163
Derm2	34	38.90	322	9	10	41.70	52.80	47.21	3.62	260	316	8	10	41.70	61.10	52.97	5.73	253	305
WQ	38	61.50	470	13	15	61.50	63.50	62.30	1.01	453	467	12	14	61.50	65.40	61.99	1.25	449	469
Lung	56	100.00	29	6	7	100.00	100.00	100.00	0.00	21	29	4	6	100.00	100.00	100.00	0.00	15	29

Note: Minimum (*Min*); Maximum (*Max*); Average (*Avg*); Standard deviation (*Std*); Classification accuracy (*Acc*); Number of attributes (#A); Number of rules (#R)

5   Conclusion

In this paper, two single-solution-based meta-heuristic approaches for attribute reduction problems in RST, namely, RRT, and m-GD were presented. In order to address the efficiency of the proposed methods, 13 UCI benchmark data sets were used. The results showed that there was a difference between the proposed methods in terms of minimal attributes, classification accuracy and number of generated rules, where m-GD was better than RRT in some cases and m-GD produces better results in other cases. This indicates the beneficial influence of attribute reduction algorithm on the classification accuracy and the number of generated rules. As a future work, we may change the classification algorithm to study the performance of the presented attribute reduction algorithms.

References

1. Zhang, D., Chen, S., Zhou, Z.H.: Constraint Score: A new filter method for feature selection with pairwise constraints. Pattern Recognition 41, 1440–1451 (2008)
2. Hu, Q., Zhu, P., Liu, J., Yang, Y., Yu, D.: Feature Selection via Maximizing Fuzzy Dependency. Fundamenta Informaticae 98, 167–181 (2010)
3. Liu, H., Motoda, H.: Feature Selection for Knowledge Discovery and Data Mining. Kluwer Academic Publishers, Boston (1998)
4. Langley, P.: Selection of relevant features in machine learning. In: Proceedings of the AAAI Fall Symposium on Relevance, pp. 1–5 (1994)
5. Pawlak, Z.: Rough Sets. International Journal of Information and Computer Sciences 11, 341–356 (1982)
6. Pawlak, Z.: Rough sets: Theoretical aspects of reasoning about data. Springer (1991)
7. Swiniarski, R.W., Skowron, A.: Rough set methods in feature selection and recognition. Pattern Recog. Lett. 24, 833–849 (2003)
8. Liu, H., Motoda, H., Yu, L.: Feature selection with selective sampling, pp. 395–402 (2002)
9. Liang, J., Wang, F., Dang, C., Qian, Y.: An efficient rough feature selection algorithm with a multi-granulation view. International Journal of Approximate Reasoning 53, 912–926 (2012)
10. Kabir, M.M., Shahjahan, M., Murase, K.: A new hybrid ant colony optimization algorithm for feature selection. Expert Syst. Appl. 39, 3747–3763 (2012)
11. Deng, T., Yang, C., Wang, X.: A Reduct Derived from Feature Selection. Pattern Recog. Lett. 33, 1638–1646 (2012)
12. Kaneiwa, K., Kudo, Y.: A sequential pattern mining algorithm using rough set theory. International Journal of Approximate Reasoning 52, 881–893 (2011)
13. Kabir, M.M., Shahjahan, M., Murase, K.: A new local search based hybrid genetic algorithm for feature selection. Neurocomputing 74, 2914–2928 (2011)
14. Anaraki, J.R., Eftekhari, M.: Improving fuzzy-rough quick reduct for feature selection. In: 19th Iranian Conference on Electrical Engineering (ICEE), pp. 1–6 (2011)
15. Suguna, N., Thanushkodi, K.: A Novel Rough Set Reduct Algorithm for Medical Domain Based on Bee Colony Optimization. Journal of Computing 2, 49–54 (2010)
16. Liu, H., Abraham, A., Li, Y.: Nature inspired population-based heuristics for rough set reduction. In: Abraham, A., Falcón, R., Bello, R. (eds.) Rough Set Theory. SCI, vol. 174, pp. 261–278. Springer, Heidelberg (2009)

17. Wroblewski, J.: Finding minimal reducts using genetic algorithms, pp. 186–189
18. Jensen, R., Shen, Q.: Semantics-Preserving Dimensionality Reduction: Rough and Fuzzy-Rough-Based Approaches. IEEE Trans. on Knowl. and Data Eng. 16, 1457–1471 (2004)
19. Handels, H., Roß, T., Kreusch, J., Wolff, H.H., Pöppl, S.J.: Feature selection for optimized skin tumor recognition using genetic algorithms. Artif. Intell. Med. 16, 283–297 (1999)
20. Wang, X., Yang, J., Teng, X., Xia, W., Jensen, R.: Feature selection based on rough sets and particle swarm optimization. Pattern Recog. Lett. 28, 459–471 (2007)
21. Jensen, R., Shen, Q.: Finding Rough Set Reducts with Ant Colony Optimization. In: Proceedings of the 2003 UK Workshop on Computational Intelligence, pp. 15–22 (2003)
22. Ke, L., Feng, Z., Ren, Z.: An efficient ant colony optimization approach to attribute reduction in rough set theory. Pattern Recog. Lett. 29, 1351–1357 (2008)
23. Hedar, A.-R., Wang, J., Fukushima, M.: Tabu search for attribute reduction in rough set theory. Soft Computing - A Fusion of Foundations, Methodologies and Applications 12, 909–918 (2006)
24. Abdullah, S., Jaddi, N.S.: Great Deluge Algorithm for Rough Set Attribute Reduction. In: Zhang, Y., Cuzzocrea, A., Ma, J., Chung, K.-I., Arslan, T., Song, X. (eds.). CCIS, vol. 118, pp. 189–197. Springer, Heidelberg (2010)
25. Jihad, S.K., Abdullah, S.: Investigating composite neighbourhood structure for attribute reduction in rough set theory. In: 10th International Conference on Intelligent Systems Design and Applications (ISDA), pp. 1015–1020 (2010)
26. Arajy, Y.Z., Abdullah, S.: Hybrid variable neighbourhood search algorithm for attribute reduction in Rough Set Theory. In: Intelligent Systems Design and Applications (ISDA), pp. 1015–1020 (2010)
27. Abdullah, S., Sabar, N.R., Nazri, M.Z.A., Turabieh, H., McCollum, B.: A constructive hyper-heuristics for rough set attribute reduction. In: Intelligent Systems Design and Applications (ISDA), pp. 1032–1035 (2010)
28. Alomari, O., Othman, Z.A.: Bees Algorithm for feature selection in Network Anomaly detection. Journal of Applied Sciences Research 8, 1748–1756 (2012)
29. Dueck, G.: New Optimization Heuristics the Great Deluge Algorithm and the Record-to-Record Travel. Journal of Computational Physics 104, 86–92 (1993)
30. Talbi, E.G.: Metaheuristics from design to implementation. Wiley Online Library (2009)
31. Mafarja, M., Abdullah, S.: Modified great deluge for attribute reduction in rough set theory. In: Fuzzy Systems and Knowledge Discovery (FSKD), pp. 1464–1469 (2011)
32. Øhrn, A.: Discernibility and rough sets in medicine: tools and applications. Department of Computer and Information Science, PhD, p. 239. Norwegian University of Science and Technology, Trondheim, Norway (1999)

Neural Network Algorithm Variants
for Malaysian Weather Prediction

Siti Nur Kamaliah Kamarudin and Azuraliza Abu Bakar

Faculty of Technology and Information Science Universiti Kebangsaan Malaysia
46300 Bangi Selangor, Malaysia
snk.kam@gmail.com, aab@ftsm.ukm.my

Abstract. This paper studies the performance of a newly prepared time-series rainfall data by a previous researcher. The issues related are; (i) the large amount of data, and (ii) the accuracy of the prediction. In this study, the data set was obtained from Institute of Climate Change UKM, pre-processed using improved Symbolic Aggregate approximation (*iSAX*) and verified by experts. Five neural network algorithms were tested with the data set, namely standard Back-Propagation (BPNN), Back-Propagation with Momentum (BP with Mom), Quick-Propagation (QuickProp), Genetic Algorithm with neural network (GA-NN) and Particle Swarm Optimization with neural network (PSO-NN). The performances of these engines were measured according to the accuracy of prediction and the training time taken. The experimental results showed that while standard BPNN and PSO-NN achieved about the same accuracy prediction, PSO-NN is considered to be better as it showed a faster training time with acceptable prediction accuracy.

Keywords: neural network, rainfall prediction, data mining.

1 Introduction

Weather prediction is important to many sectors and area of researches. The conditions of the weather impact many sectors, for example the power usage of industrial facilities and residential facilities [1] agricultural area [1] and [2] and early warnings of natural disaster occurrences [3] and [4]. This resulted in the conduction of researches such as weather forecasting warning system [6], rainfall and flood forecasting [5] and [7].

Rainfall prediction is one of the main areas of research in the weather domain. The challenges in rainfall forecasting not only involve predicting the amount of rainfall, but it also affects the operational hydrology area for planning and management of water resources in order to handle events such as flash floods [5]. The presence of noise and uncertainties in the real world data is also a reason for researchers to experiment with different pre-processing methods [8] and different sets of time series rainfall data. The emergence of various intelligent techniques has also spurred researchers to actively conduct additional experiments for a more accurate weather prediction system.

S.A. Noah et al. (Eds.): M-CAIT 2013, CCIS 378, pp. 121–134, 2013.
© Springer-Verlag Berlin Heidelberg 2013

The problem with Malaysian weather prediction system is the use of a general system that is not customized to the Malaysian weather data, in addition to the dependency on the satellite-based system which is expensive and requires support from a complete system in order to provide useful weather data [9]. Thus, there is a need to have a system that can well handle the pattern of local data (sources: Institute of Climate Change, UKM). Issues highlighted in this study are as follows: 1. Large amount of data stream or time-series weather data cause difficulties in finding reliable patterns without loss of knowledge; 2. Accuracy of prediction. The availability of a new weather data set also provides potential and opportunity for the researcher to test the performance of the data set in rainfall prediction. In addition, the data set uses a new representation called symbolic representation by [10].

Another important factor that must be considered is the techniques to be used. The techniques used will also determine the accuracy of the forecast, subsequently highlighting the novelty of using the techniques in their research. Artificial Neural Networks (ANN) is one of the best approaches as they have demonstrated their capabilities to model linear and non-linear systems for time series data and to learn from training process with incomplete and noisy data [5] and [11]. Researchers have also been improving ANN by integrating it with other intelligent techniques. In this study, Genetic Algorithm (GA) and Particle Swarm Optimization is considered as GA helps the ANN in avoiding the local minima problem and optimization of the solutions [39] while PSO's role is to reduce the dimensionality [12] in order to overcome ANN's inconsistent performance [13]. Besides that, both PSO and GA have the ability to optimize the initial weights and hidden layers [14] and [15] in ANN architecture.

Therefore the main objective of this study is to explore the variants of Back-Propagation Neural Network (BPNN) algorithms for Malaysia's weather prediction data. The contributions of this paper are especially on the usage of rainfall data set represented in symbolic and classified by experts [16] along with the knowledge towards rainfall prediction when using the Malaysia rainfall data obtained from the Institute of Climate Change, UKM. In addition, this study also specifically aims to: 1. Investigate the performance of BPNN variants towards a new Malaysian weather data representation namely symbolic representation, and 2. To evaluate the performance of BPNN variants towards the prediction of Malaysia weather data.

The rest of the paper is organized as follows. Section 2 discusses the related works by other researchers and section 3 discusses the methodology used in this study. Next, section 4 will discuss the experimental setups followed by the reports on the results and findings of this study in section 5. Finally, section 6 will conclude this paper.

2   Related Works

Researchers begin developing weather prediction models using both statistical and intelligent techniques. In Malaysia, the Malaysian Meteorology Department (MMD) uses the Numerical Weather Prediction (NWP) tool as it is the most widely used weather forecasting tool around the world [17]. However, the quantitative precipitation forecast's accuracy is one of the aspect in the NWP model that could and should be improved [17]. Many researchers also adopted statistical methods in weather predictions to save costs as statistical methods only needs the current data to be used in the forecasting models [8], [19] and [20]. However, prediction results when applying only current data may be misleading [21] hence creating errors in predictions. Therefore, researchers who used statistical methods also recommended on the application of existing intelligent methods in addition to the statistical methods being used. For example, Artificial Neural Network (ANN) is based on the functionalities of the human brain and their learning capabilities while fuzzy logic preserves human expertise and knowledge and Genetic Algorithm (GA) is inspired by the human evolution process, also known as the Darwin theory [22].

The advantages of using ANN and the actual successful implementation of ANN in short term load prediction and agriculture motivates researchers to use this technique into other domain prediction problems. ANN with a few hidden layers with non-constant activation functions are enough to approximate weather prediction using the variables of rainfall and temperature [23] and [24]. However, the simple algorithm has one common problem in neural network algorithm i.e. the problem of over-fitting which results in the poor performance of the model [23] and [25] which occurs due to the existence of too many parameters in the forecasting process. Another characteristic of the ANN that determines the performance of the model is the number of hidden nodes in the hidden layers, therefore pre-determining the features of the forecasting model must be done so as to avoid the problem of over-fitting and under-fitting [23].

GA has also been applied in the weather prediction area of research to optimize the number of hidden nodes of their ANN architecture for rainfall prediction [26]. The use of GA to optimize the width factor or number of hidden nodes in the ANN architecture were successful as the error obtained were small, proving that the rainfall prediction model using RBF network and GA can give a satisfactory accuracy result [26]. Particle Swarm Optimization (PSO) is also another optimization technique used for determining initial weights and hidden layers [15]. The idea of the PSO is for searching the optimum solution and to share the information between each other in the colony just as ants would do in their colony. The advantage of this technique lies in their global optimization ability and its robustness [27]. Apart from that, PSO advantages also include easy implementation and quick convergence [18]. However, PSO is not without its own drawbacks. Using PSO may create local optimum problem [14] and [18] which results in low accuracy of the solution [14]. Therefore, the problem may be overcome by hybridizing it with intelligent techniques such as ANN.

Pre-processing stage in data mining process is an important part in any data mining research [7], [28] and [29] especially when it involves time series data such as the rainfall data. [16] mentioned that the key to success in pre-processing time series data are the data representation technique and the similarity measures used in one's research. In addition, the preservation of significant characteristics in the raw data are also expedient [16] for the efficiency and accuracy of the research. PAA and SAX technique is one of the most discussed time series representation technique [30], [31], [32], [33] and [34]. Although the original SAX approach based on the PAA representation has been found to allow excellent dimensionality reduction and distances measures, researchers have been known to further explore the possibilities of improving the techniques for better results [16], [32] and [34]. One research that explored the improvements of SAX and PAA is the *iSAX* which uses the RFknn algorithm, combining relative frequency and K-Nearest-Neighbour Algorithm on raw rainfall data [16]. Among the advantages of PAA is the fact that it is easy to be implemented, fast and flexible. However, it is not without disadvantages where due to the dimensionality reduction, researchers tend to miss the important features of the data especially when using time series data [16].

Among the researches which used Malaysian weather data as its experimented data, the latest work is by [10] where the data set used is obtained from the Institute of Climate Change, UKM. Their research on classifying the rainfall patterns [16] and discovering frequent rainfall patterns from symbolic rainfall data set [35] showed potential for future rainfall prediction process. By referring to their work, the rainfall data patterns is first classified into four categories namely No rain, Light, Moderate and Heavy rainfall patterns [16] followed by the detection of rainfall patterns using two processes: window sliding algorithm and case-based reasoning (CBR) method. The research aims at discovering the most frequent rainfall patterns using Allen intervals operation and measuring its support and confidence [35]. While the first research by [16] seemingly aimed at preparing the rainfall data by classifying the time series rainfall patterns into several categories, the subsequent study by [35] showed the potential experiments that can be conducted when using the prepared rainfall data set especially in rainfall prediction area. Therefore, the availability of the prepared rainfall data set is a motivation for the current researcher to apply it in the rainfall prediction experiments.

In this study, we further explore the work in [16] and [35] by using the classified rainfall data set as completed in [16]. By taking into consideration that this set of data has not been experimented upon in the rainfall prediction research, we aim to discover the performance of the data set by [16] when using the several ANN algorithm variants for rainfall prediction. The prediction in this study centers on the sequence of rainfall amount and its respective classes where the ANN algorithm will take the sequence of the rainfall amount as its input and the class as its output.

3   Research Methodology

In this study, the steps involved in completing the research are data pre-processing and models development. The following subsections explained how data are pre-processed and how models are built while the figure below shows the overall methodology in completing the research.

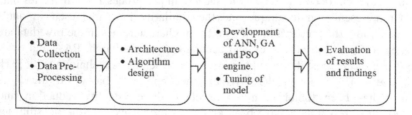

Fig. 1. Research methodology summary

3.1   Data Collection and Data Pre-processing

Data from 5 weather stations are used in this study. They are Station 16R- SK Sg. Lui, Station 17R- JPS Ampang, Station 5R- Mardi Serdang, Station 8R- Ampangan Semenyih and Station 18R- Pemasokan Ampang, Selangor area. For initial experiments, small samples from each five stations are taken to speed up the training process. [16] have classified the rainfall data into four main categories namely No rain (N), Light (L), Moderate (M) and Heavy (H). The data is first discretized into +1 and -1 values accordingly in order to speed up the training process. By observing the sequence of the prepared data from one column to the next column, a +1 value is given if the rainfall amount increased and a -1 value is given if the amount decreased. This method is a trial and error process to increase the accuracy of the experiment and to speed up the training process. The images below illustrate how data are pre-processed.

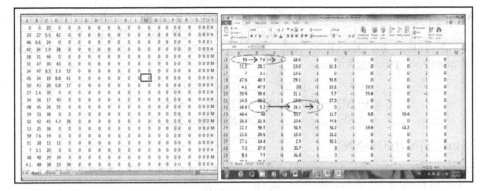

Fig. 2. Sample of prepared rainfall data (left) and discretization process (right)

3.2   Prediction Models

The models developed for this research are BPNN variants (standard BPNN, BP with Mom and QuickProp), and hybrid models (GA-NN and PSO-NN).

BPNN Variants

The standard algorithm used in each BPNN variants is as below

```
The standard pseudo code for BPNN variants are as follows.
1.  Initialize weights. (Small randomized values).
2.  While stopping condition are false, do
        o   For each training pair,
                •  Calculate output signals for hidden layers and output layers
                •  Calculate error information and weight correction term for output layer to hidden layer
                •  Calculate error information and weight correction term for hidden layer to input layer
3.  Test stopping condition.
```

Fig. 3. BPNN variants' standard pseudo code

Each model developed in this research used the number of input data as the input node and the rainfall category for each input row is used as the output node value. Since [38] declared that *one hidden layer is sufficient* for any continuous mapping from input patterns to the output patterns; therefore one-hidden layer will be used for this research. The number of hidden nodes computed is based on the equation below.

$$Number\ of\ input\ nodes\ /\ 2 = Number\ of\ hidden\ nodes \tag{1}$$

However, for BP with Mom and QuickProp, minor changes were made to the standard BPNN algorithm with the intention of improving the algorithm for better results. In BP with Mom algorithm, a momentum value (M) is inserted in the weight update (Δw_i) calculation as in the formula below where variables such as learning rate (α), error value (δ_k), hidden node value (Z_j) and previous weight correction values (Δw_{i-1}) are also taken into account.

$$\Delta w_i = (\alpha * \delta_k * Z_j) + (M * \Delta w_{i-1}) \tag{2}$$

Whereas in QuickProp algorithm, the changes made is in the calculation of hidden node (Z_j), output values (Y) and the calculation of weight update (V_{ij} and W_{jk}) made in the algorithm. In QuickProp, calculation of Z_j and Y are modified for faster computation of the algorithm.

$$
\begin{aligned}
&\text{IF } (Zin\_j < -10.0) \quad &&\text{THEN} \quad Z_j = 0.0 \\
&\text{ELSE IF } (Zin\_j > 10.0) \quad &&\text{THEN} \quad Z_j = 1.0 \\
&\text{ELSE} \quad &&Z_j = 1/(1 + exp(-Z\_in_j))
\end{aligned} \tag{3}
$$

Hybrid Model: GA-NN

In this study, GA is used to optimize the initial weights only. The algorithm for GA-NN used is as in figure below.

Generate a population of chromosomes (10 populations)
1. Extract weights for input-hidden-output (*l-m-n*) layers from each chromosome.
2. Calculate fitness of each chromosome. For each chromosome, the fitness is the error calculated using BPNN algorithm.
3. Create new population by repeating the following steps
   - Select two parent chromosome with the best fitness value (lowest error rate)
   - Cross over the parents to form two new offspring
   - Mutate new offspring by selecting random position in the offspring
   - Calculate fitness of new offspring
   - Replace chromosome with higher error rate. If offspring generate higher error, no replacements occur.
4. Repeat Step 3 to 5 until stopping condition is met.
5. Test the chromosomes and return best solution.

Fig. 4. GA-NN algorithm

Hybrid Model: PSO-NN

In this study, the PSO algorithm is also used to optimize the initial weights. One of the differences between PSO and GA is where GA has the crossover and mutation process, PSO does not. Another difference is GA uses selection process to select the best parent chromosome from the population while PSO uses update each particle in the population. The pseudo code for PSO-NN is as below.

---

1. Generate a population of particles (10 population)
2. Extract weights for input-hidden-output (*l-m-n*) layers from each chromosome.
3. Calculate fitness of each particle. For each particle, do the fitness is the error rate calculated using BPNN algorithm.
4. For each particle, do
    - Initialize velocity and position
    - Set personal best (*pbest*) and global best (*gbest*)
5. Repeat steps below until maximum iteration or stopping condition is met
    - Calculate velocity and position of each particle
    - Update velocity and position
    - Set *pbest* and *gbest*
6. Test the particles and return best solution.

Fig. 5. PSO-NN algorithm

4   Experimental Setup

4.1   Rainfall Data

The rainfall data is obtained from the Institute of Climate Change, UKM and pre-processed by a previous researcher [16]. These data have also been classified by the same researcher into four main classes namely Heavy (H), Moderate (M), Light (L) and No Rain (N). In this study, samples from five stations of daily rainfall patterns are selected to be experimented in this research. The classes of the rainfall amount are transformed into positive values of 1 (High), 0.667 (Moderate), 0.333 (Light) and 0 (No rain) and the data set are divided into according to the 90:10 training and testing respectively.

4.2   Parameter Settings

Both GA-NN and PSO-NN algorithm uses BPNN algorithm to calculate their mean squared error (MSE). The MSEs obtained are then used as the chromosomes' or particles' fitness value and they are also used as the stopping condition for both engines.

PSO-NN engine consists of several additional parameters that are fixed values according to previous researchers conducted. The minimum and maximum inertia weights, w_{min} and w_{max} are set to 0.1 and 0.9 respectively, the LR are set to 0.9 and the maximum iteration, $iter_{max}$ as 100 [36]. The maximum value of velocity is set to 2.00 [37] while the calculation for inertia, w is as according to the formula below [36].

$$w = w_{max} - [(w_{max} - w_{min}) / (iter_{max})] \times iter \tag{3}$$

4.3   Stopping Condition and Performance Measurement

The stopping conditions used in this research for all Back Propagation Neural Network (BPNN) variants are set where either MSE < 0.001 [38] or 5000 epochs are reached as according to trial and error basis. For GA-NN and PSO-NN, the stopping condition is set where if the population's fitness difference between the chromosomes are 95% similar [39] or maximum iteration of 5000 is reached, then the best solutions is said to be achieved.

All engines' performance are measured through their accuracy calculation of prediction where the calculated output are compared with the real output.

5   Experimental Results

5.1   Data Ratio Experiment Results

Using the sample data taken from the five stations (500 rows), the data were divided into the ratios between 10:90 and 90:10; training and testing respectively. By applying the best LR and Momentum parameters obtained from the previous experiments conducted in this study, the results are as follows. The LR value used is 0.5 for BPNN, BP with Mom and QuickProp while 6 is used in PSO-NN. The momentum value used is 0.1 for BPNN, BP with Mom and QuickProp.

Table 1. Experiment results on data ratio

Training (%)	Testing (%)	BPNN (%)	BPNN with Mom (%)	QuickProp (%)	GA-NN (%)	PSO-NN (%)
10	90	61.6	32.00	31.3	30.0	45.9
20	80	63.2	31.00	33.3	33.0	46.8
30	70	64.8	35.00	35.0	33.0	50.0
40	60	63.3	36.00	38.0	35.0	47.0
50	50	63.9	44.00	35.3	38.0	52.0
60	40	63.2	48.00	43.0	40.0	51.8
70	30	63.9	50.00	45.0	45.0	55.8
80	20	66.3	51.00	49.0	49.0	55.0
90	10	69.3	55.34	50.0	51.0	61.0

According to the results, percentage of each NN variants experience a small fluctuation in between the ratios, yielding accuracies with not much difference between the different ratios. However, all NN variants seem to show a certain trend where the accuracy is considered best at the lowest ratio for training (10%) and the highest ratio for training (90%), with the exception of QuickProp and GA-NN.

5.2   Stations Experiment Results

All five techniques were applied to the data from the five weather stations. The results obtained in this experiment according to the prediction accuracy are summed up in the table below.

Table 2. Prediction accuracy according to stations

Station	BPNN (%)	BPNN with Mom (%)	QuickProp (%)	GA-NN (%)	PSO-NN (%)
1	63.0	56.6	50.0	48.0	61.0
2	50.3	55.0	51.0	43.0	52.0
3	42.2	44.1	46.0	45.0	51.8
4	54.0	52.0	53.0	51.0	55.8
5	69.3	54.3	50.0	53.0	55.0
Combination of Station1 to 5	56.9	59.2	55.0	56.1	63.0
Average	55.95	53.53	50.83	49.35	56.43

According to prediction accuracy, the best technique with the highest prediction accuracy is *standard BPNN with 69.3%* by using the data from one station followed by *PSO-NN with 63%* when using the total data combined from the five stations. The usage of BPNN as a weather prediction technique has been mentioned in recent researches [7] though the technique is not new. The results depicted in this series of experiments for this study shows that BPNN is still able to give one of the best prediction accuracy despite the theory that BP with MOM and QuickProp should yield a better accuracy with the integration of momentum parameter (both BP with MOM and QuickProp) and the modification made to the algorithm (QuickProp) [40]. The results also conflicted with the theory where hybrid algorithms should yield a better accuracy prediction [22]. The results showed a lower accuracy prediction when using GA-NN and a minor higher difference when using PSO-NN. The difference between the accuracy prediction for all five engines are also depicted in the figure below (Fig. 6)

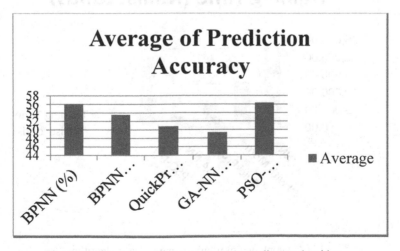

Fig. 6. Average of prediction accuracy according to algorithm

5.3   Training Time

BP with MOM and QuickProp are firstly used not only for the accuracy but also with the aim to improve the training time of the BPNN [38] and [40] as one of the major disadvantages of using BPNN is the training time. Despite the high accuracy shown by BPNN as compared to other NN variants in this research, the training time taken by BPNN is the longest as compared to other NN variants. To compare the training time, the results of the training time using each technique are in the table below.

Table 3. Prediction accuracy according to stations

Station	BPNN	BPNN with Mom	QuickProp	GA-NN	PSO-NN
Training time for combined sample data from 5 stations (seconds, *s*)	22.0410	11.7380	5.1880	1.4240	16.2070

Among the five NN variants, the fastest training time was GA-NN and QuickProp. However, since GA-NN yielded the worst prediction accuracy, it can be assumed that the GA optimization technique is not suitable with the set of rainfall data used. In comparison, PSO-NN with the second slowest training time yielded the best prediction accuracy as compared with others. By using a total of 450 rainfall patterns for training, the training time of PSO-NN as compared to BPNN is better and faster with a difference of 5.834 seconds even though their prediction accuracy is a minor 0.48% difference.

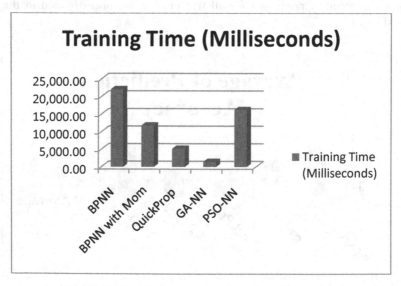

Fig. 7. Graph plot of training time for each algorithm

Theoretically, BP with Momentum and QuickProp gives faster training time, enabling the algorithm to converge faster and yield better accuracy [38] and [40] where these theories are also shown to be correct in this research. Furthermore, GA-NN has demonstrated a much faster training time as compared to the other NN variants, including PSO-NN whereas theoretically, PSO-NN should have faster training time due to the absence of crossover and mutation process that exists in GA. This may be due to the stopping condition used for both GA-NN and PSO-NN where looping will stop when the best fitness for each chromosomes or particles reach a similarity percentage of 95% [15] and [39] and the existence of the many random values used in the PSO algorithm [36].

5.4   Discussion

One of the main concerns for this research is the usage of BPNN technique as the main Neural Network algorithm in this research. Recent researches that use BPNN as their chosen NN technique has further strengthen the justification as to why BPNN is still considered as one of the best NN algorithm to be used in the weather forecasting field [7]. In addition, this research has further showed that the standard BPNN have yielded the best prediction accuracy among the NN variants as compared to the BP with Momentum and QuickProp.

ANN's advantages in self-learning and self-updating are shown in this research. Where BPNN managed to show a good performance, the hybrid between BPNN and PSO also display a satisfactory performance, more so than the standard BPNN as mentioned theoretically [13], [15] and [36]. Since this research is at the experimental stage where the data set used may be further improved in future works, only the performance of standard BPNN and PSO-NN were theoretically proven.

To summarize the findings, despite the good results achieved by previous experiments conducted by other researchers using the same technique, their results may not be comparable with the results achieved in this research. The reasons are such as (i) the set of data used for this experiment is completely different with all other experiment as the prepared data by [16] have yet to be applied by any other researches; (ii) how the data are pre-processed is also completely different with any other research. The pre-processing steps taken in this research are considered to be the most suitable with the prepared data set by [16] as determined by the series of experiments conducted by the current researcher; (iii) the nature of the data itself as observed by the current researcher. To the knowledge of the researcher, this particular set of data has been validated by the experts therefore any other major pre-processing steps taken might interfere and disturb with the nature of the data. Judging from the results achieved, it may be concluded that the prepared set of data applied in this research is not very compatible with the machine learning techniques used in this research. However, since this is the first research to use this particular set of data, there is a great potential to further conduct experiments using the same set of data with other machine learning techniques. This research may also be used as a benchmark for the rainfall data set by [16].

6   Conclusion

Among the highlighted findings of this research is the usage of the prepared rainfall data made by a previous researcher, the accuracy of the prediction made using all five engines on the rainfall data, the training time achieved by all five engines and the justification on the results. Firstly, the rainfall data set used in this research was a prepared and classified rainfall data by [16]. Since this set of prepared data has yet to be implemented in any other research related to rainfall prediction, the results obtained in this research are not comparable with any other researches' results. Following this, the second significant finding of this research is on the prediction accuracy. Even though theories states that algorithms such as BP with Mom and QuickProp should yield a better accuracy than BPNN, the research's result shows that the standard BPNN yields a better accuracy. Not only that but standard BPNN also managed to achieve one of the highest accuracy prediction besides PSO-NN, with only a minor difference of percentage. Thirdly, this research also managed to highlight and prove the theory that the hybrid algorithm will yield a faster training time as compared to the BPNN variants used. GA-NN and PSO-NN managed to train their network faster than all BPNN variants. With a difference of few seconds, PSO-NN beats standard BPNN in terms of training time.

The use of the prepared data set by [16] for prediction purpose in this research is an experiment where the current researcher aims to observe the performance of the data set when using certain NN algorithm variants as the prediction techniques. And since this research only used samples data from five weather stations, further researches may opt to use more or all of the available data. Further researches may consider expanding the pre-processing stage of the data using existing software such as Rosetta and Weka.

References

1. Zavala, V.M., Constantinescu, E.M., Anitescu, M.: Economic Impacts of Advance Weather Forecasting on Energy System Operations. In: IEEE PES, pp. 1–7 (2010)
2. Nguyen, V.G.N., Vo, T.T., Huynh, H.X., Drogoul, A.: On Weather Affecting to Brown Plant Hopper Invasion using an Agent-based Model. In: MEDES 2011, pp. 150–157 (2011)
3. Pielke Jr., R., Carbone, R.E.: Weather Impacts, Forecasts and Policy: An Integrated Perspective. Bulletin of American Meteorological Society, 393–403 (2002)
4. Li, C., Wang, Y., Liu, X.: Research on Natural Disaster Forecasting Data Processing and Visualization Technology. In: 4th International Congress on Image and Signal Processing, pp. 1775–1778 (2011)
5. Lee, R., Liu, J.: iJADE WeatherMAN: A Weather Forecasting System Using Intelligent Multiagent-Based Fuzzy Neuro Network. IEEE Transactions on Systems, Man, and Cybernatics - Part C: Applications and Reviews 34, 369–377 (2004)
6. Isa, I.S., Omar, S., Saad, Z., Noor, N.M., Osman, M.K.: Weather Forecasting using Photovoltaic System and Neural Network. In: 2nd International Conference on Computational Intelligence, Communication Systems and Networks, pp. 96–100 (2010)

7. Rahman, I.I.A., Alias, N.M.A.: Rainfall Forecasting Using an Artificial Neural Network Model to Prevent Flash Floods. In: IEEE High Capacity Optical Networks and Enabling Technologies, pp. 323–328 (2011)
8. Saima, H., Jaafar, J., Belhaouari, S., Jillani, T.A.: Intelligent Methods for Weather Forecasting: A Review. In: National Postgraduate Conference (NPC 2011), pp. 1–6 (2011)
9. Yen, W.K.: A Study of Soft Computing Approach in Weather Forecasting. Masters Thesis, Universiti Teknologi Malaysia (2010)
10. Ahmed, A.M.A.: Pattern Discovery Algorithms for Time Series Data Mining. PhD Thesis, Universiti Kebangsaan Malaysia (2012)
11. Htike, K.K., Khalifa, O.O.: Rainfall Forecasting Model Using Focused Time-Delay Neural Networks. In: IEEE International Conference on Computer and Communication Engineering, pp. 1–6 (2010)
12. Abdul-Rahman, S., Hussein, Z.A.M., Bakar, A.A.: Velocity-based Reinitialisation Approach in Particle Swarm Optimization for Feature Selection. In: 11th International Conference on Hybrid Intelligent Systems, pp. 621–624 (2011)
13. Luo, F., Wu, C., Wu, J.: A Novel Neural Network Ensemble Model Based on Sample Reconstruction and Projection Pursuit for Rainfall Forecasting. In: IEEE Sixth International Conference on Natural Computation, pp. 32–35 (2010)
14. Biao, S., Li, Y.X., Yan, W., Li, P., Meng, X., Yu, X.H.: A Modified Particle Swarm Optimization and Radial Basis Function Neural Network Hybrid Algorithm Model and its Application. In: Global Congress on Intelligent Systems, pp. 134–138 (2009)
15. Tripathy, A.K., Mohapatra, S., Beura, S., Pradhan, G.: Weather Forecasting using ANN and PSO. International J. Scientific & Eng. Res. 2, 1–5 (2011)
16. Ahmed, M.A., Bakar, A.A., Hamdan, A.R.: Improved SAX Time Series Data Representation Based on Relative Frequency and K-Nearest Neighbour Algorithm. In: IEEE 10th International Conference on Intelligent Systems Designs and Applications, pp. 1320–1325 (2010)
17. Wardah, T., Kamil, A.A., Hamid, A.B.S., Maisarah, W.W.I.: Statistical Verification of Numerical Weather Prediction Models for Quantitative Precipitation Forecast. In: IEEE Colloquium on Humanities, Science and Engineering, pp. 88–92 (2011)
18. Hong, W.C.: Rainfall Forecasting by Technological Machine Learning Models. J. Applied Maths. and Comp. 200, 41–57 (2008)
19. Wu, Y.K., Hong, J.S.: A Literature Review of Wind Forecasting Technology in the World. In: IEEE Power Tech, pp. 1–6 (2007)
20. Juneng, L., Tangang, F.T., Kang, H., Lee, W.J., Yap, K.S.: Statistical Downscaling Forecasts for Winter Monsoon Precipitation in using Multimodel Output Variables. J. Climate 23(11), 17–22 (2010)
21. Gilleland, E., Ahijevych, D., Brown, B.G., Casati, B., Ebert, E.E.: Intercomparison of Spatial Forecast Verification Methods. J. American Meteorological Society 24(5), 1416–1430 (2009)
22. Negnevitsky, M.: Artificial Intelligence: A Guide to Intelligent Systems, 2nd edn. Addison Wesley, England (2005)
23. Luk, K.C., Ball, J.E., Sharma, A.: An Application of Artificial Neural Networks for Rainfall Forecasting. Mathematical and Computer Modelling 83, 683–693 (2001)
24. Lai, L.L., Braun, H., Zhang, Q.P., Wu, Q., Ma, Y.N., Sun, W.C., Yang, L.: Intelligent Weather Forecast. In: Proceedings of the Third International Conference on Machine Learning and Cybernatics, pp. 4221–4216 (2004)
25. Philip, N.S., Joseph, K.B.: A Neural Network Tool for Analysing Trends in Rainfall. Computers and Geosciences 29, 215–223 (2003)

26. Jareanpon, C., Pensuwon, W., Frank, R.J., Davey, N.: An Adaptive RBF Network Optimised using a Genetic Algorithm Applied to Rainfall Forecasting. In: International Symposium on Communications and Information Technologies, pp. 1005–1010 (2004)

27. He, J., Valeo, C.: Comparative Study of ANN versus Parametric Methods in Rainfall Frequency Analysis. J. Hydrologic Engineering ASCE 14(2), 172–184 (2009)

28. Tan, P.N., Steinbach, M., Kumar, V.: Introduction to Data Mining. Pearson International Edition, US (2006)

29. Ju, Q., Yu, Z., Hao, Z., Ou, G., Zhao, J., Liu, D.: Division-based Rainfall-runoff Simulations with BP Neural Networks and Xinanjing Model. Neurocomputing 72, 2873–2883 (2009)

30. Gullo, F., Ponti, G., Tagarelli, A., Greco, S.: A Time Series Representation Model for Accurate and Fast Similarity Detection. Pattern Recognition 42, 2998–3014 (2009)

31. Lai, C.P., Chung, P.C., Tseng, V.S.: A Novel Two-Level Clustering Method for Time Series Data Analysis. Expert Systems with Applications, 6319–6326 (2010)

32. Chen, J., Moon, Y.S., Wog, M.F., Su, G.: Palmprint Authentication using a Symbolic Representation of Images. Image and Vision Computing 28, 343–351 (2010)

33. Lavangnananda, K., Sawasdimongkol, P.: Capability of Classification of Control Chart Patterns Classifiers using Symbolic Representation Pre-processing and Evolutionary Computation. In: 23rd IEEE International Conference on Tools with Artificial Intelligence, pp. 1047–1052 (2011)

34. Yang, H., Meng, F.: Application of Improved SAX Algorithm to QAR Flight Data. In: 2012 International Conference on Applied Physics and Industrial Engineering, pp. 1406–1413 (2012)

35. Ahmed, M.A., Bakar, A.A., Hamdan, A.R., Abdullah, S.M.S., Jaafar, O.: Discovering Frequent Serial Episodes in Symbolic Sequences for Rainfall Dataset. In: 4th Conference on Data Mining and Optimization, pp. 121–126 (2012)

36. Wu, J., Chen, E.: A Novel Nonparametric Regression Ensemble for Rainfall Forecasting using Particle Swarm Optimization Technique Coupled with Artificial Neural Network. In: Yu, W., He, H., Zhang, N. (eds.) ISNN 2009, Part III. LNCS, vol. 5553, pp. 49–58. Springer, Heidelberg (2009)

37. Kennedy, J., Spears, W.: Matching Algorithms to Problems: An Experimental Test of the Particle Swarm and Some Genetic Algorithms on the Multimodal Problem Generator. In: Evolutionary Computation Proceeding, pp. 78–83 (1998)

38. Fausett, L.: Fundamentals of Neural Network: Architecture, Algorithm and Applications. Prentice Hall Inc., New Jersey (1994)

39. Gill, J., Singh, B., Singh, S.: Training Back Propagation Neural Networks with Genetic Algorithm for Weather Forecasting. In: 8th International Symposium on Intelligent Systems and Informatics, pp. 465–469 (2010)

40. Fahlman, S.E.: An Empirical Study of Learning Speed in Back-Propagation Networks. In: CMU-CS, pp. 88–162 (1988)

Automated Evaluation for AI Controllers in Tower Defense Game Using Genetic Algorithm

Tse Guan Tan, Yung Nan Yong, Kim On Chin, Jason Teo, and Rayner Alfred

School of Engineering and Information Technology
Universiti Malaysia Sabah,
Jalan UMS, 88400, Kota Kinabalu, Sabah, Malaysia
{tseguantan,yongyn89,kimonchin}@gmail.com,
{jtwteo,ralfred}@ums.edu.my

Abstract. This paper presents the research result of implementing evolutionary algorithms towards computational intelligence in Tower Defense game (TD game). TD game is a game where player(s) need to build tower to prevent the creeps from reaching their based. Penalty will be given if player losses any creeps during gameplays. It is a suitable test bed for planning, designing, implementing and testing either new or modified AI techniques due to the complexity and dynamicity of the game. In this research, Genetic Algorithm (GA) will be implemented to the game with two different neural networks: (1) Feed- forward (FFNN) and (2) Elman Recurrent (ERNN) used as tuner of the weights. ANN will determine the placement of the towers and the fitness score will be calculated at the end of each game. As a result, it is proven that the implementation of GA towards FFNN is better compared to GA towards ERNN.

Keywords: Genetic Algorithm (GA), Artificial Neural-Network (ANN), Feed-forward Neural Network (FFNN), Elman Recurrent Neural Network (ERNN), Tower Defense game (TD game), Strategy Games, Artificial Intelligence (AI).

1   Introduction

Tower defense (TD) is a type of games that required player(s) to build tower in order to prevent certain amounts of enemies from approaching their base. It is known as real-time strategy game or turn-based game and several well-known TD games are such as Plants vs Zombies and TD in Warcraft III [1]. This genre of game provide an important testbed for Artificial Intelligence (AI) research by allowing the testing and comparison of new and experimental approaches on a challenging but well-defined problem.

In the gameplays, there are certain amounts of enemy creeps in each waves moving towards the player creeps. Player(s) should either build more defense towers or up-grade the existing tower in order to kill the enemy creeps. As a return, the player will be awarded with certain amount of resources for each kill. Thus, the player can build more as well as stronger towers to defense for the incoming waves. For a fair to the

S.A. Noah et al. (Eds.): M-CAIT 2013, CCIS 378, pp. 135–146, 2013.
© Springer-Verlag Berlin Heidelberg 2013

gameplays, the strength of the enemy creeps is improved proportionately, in order to ensure the difficulty of the game is consistent with the resources awarded. The games ended when certain amounts of waves are being eliminated or certain number of player's lives is eliminated.

The complexity of the TD games can be dissimilar based on the game design. Some of the TD games allow player(s) to build tower along the path whereas some have limited the tower to be build beside the path [1]. In other case, some TD games limited the number of towers to be built while others are having very limited type of tower to be built [1]. Most of the TD games provide different paths design. For example, Plants vs Zombies consists of five fixed lanes [2]. Field Runner and Sentinel provide several open-ended maps [3], [4]. Player(s) must carefully plan, design and allocate the towers in order to slow down as well as kill the enemy creeps that tend to reach the end point. As such, the dynamicity and complexity in TD games lead to the AI cognition for researchers to plan, design, and testing their novel idea and algorithms [5].

Eventually, there are few interesting issues involved for researching TD games. First is the design of the map. It is not an easy task in designing a map that can attract player(s) to stay longer with the game. It follows with the issue of type of towers and type of creeps design. Each tower must have its own specialty and attributes. Balancing of designing towers plays an important role. Player(s) can easily complete the game level if the towers are equipped with higher strength and ability. Otherwise, player(s) may uninstall the game if they found the game is not beatable after spending few days to the lost games. Another issue regards to the balancing of awards or resources provide to the player(s) after each kill of the enemy. Player(s) may easily lose a game if the resource given is too limited to build or upgrade towers. The last issue would be related to the testing of the gameplays. Most of the TD games have been designed without proper testing procedures on the designed maps. As the manual testing procedures lead to costly and time consuming issues. Hence, it creates an unbeatable situation for the designed map. At last, player(s) uninstall the application after few games. Basically, the inclusion of AI technology to the game can replace the manual testing procedures as well as reduce the cost and time taken for designing the game.

[6] and [7] show AI controllers could be easily generated to provide a variable level of difficulty as an opponent in Real-Time Strategy (RTS) game. [8] and [9] show the inclusion of simple Finite State Machine could reduce the cost taken to test the level of difficulty in Mario Bros and First Person Shooting games. Hence, this motives us to create AI controllers that use Genetic Algorithm (GA) and two different neural networks in this research in order to generate a possible solution to test the usability of the designed TD game's map.

In this study, a modified GA will be implemented with (1) Feed-forward Neural Network (FFNN) and (2) Elman Recurrent Neural Network (ERNN) towards a TD game in order to generate AI controllers that can be used in testing the designed map. A comparison will be carried out in order to determine which neural network will be best suit with GA in generating better controllers.

The remainder of this writing is organized as follows. In section 2, some related works will be discussed. Then, the ANN representation will be included in section 3. It follows with the modified GA discussion in Section 4. The experimental setup will be discussed in Section 5. Then, the experimental results and discussions are included in section 6. Finally, the paper will be concluded in Section 7.

2   Related Works

ANN had been applied at different areas as their ability to learn and easy to adapt to different functions and able to make prediction based on the learning data. [10] explained the usage of ANN in gaming. One of the popular ANN implementation game is Reversi [11]. The implementation of ANN enables the development of AI that discovers complex strategies that able to compute with experience human player. There are also studies on other board games such as checkers [12], tic tac toe [13] and five in a row [14]. Researchers generated superior AI controllers that are unbeatable by human expert.

In other case, [15] implements GA with ANN to increase the accuracy of multiple shooting in a ball and balancing skills between characters in duel games. [16] proposed neural-evolutionary model for case-based planning in RTS games. The study shows that ANN in GA algorithms had helps in reducing the time complexity and the characters are able to react based on current data without using heuristics-based algorithm. Besides, [17] suggest the implementation of ANN towards GA in RTS game. The authors also compared different combination of AI methods such as Back-propagation and Radial Basis Networks models.

3   Artificial Neural Networks

Artificial Neural Network (ANN) is a mathematical model that tries to stimulate the structure and functionalities of biological neural networks [18]. The starting of ANN is at 1943 when McCulloch and Pills Whor introduced the first neural model [19]. As in biological model, ANN consists of neurons that are connected among each other. The interconnection neurons are generally divided into three layers, except a multi-layer of neural structure is used. The weights used can be either in positive value or negative value. The weights also can be adjusted in order to specific output for specific input. The process of adjusting the weight can be known as learning or training [20].

There are several concerns when implementing ANN: (1) the type of problem to be solved and (2) the type of topology used in the ANN. Till now, there is no evidence that show a single type of ANN that is superior and can be suit with any type of evo- lutionary algorithms for solving all problems. Hence, most of the researches are still conducted in comparing different neural networks topology in different problem solving cases.

In this research, there are two types of neural networks involved (1) FFNN and (2) ERNN. FFNN is differs to ERNN in its learning process although both neural networks involved only three layers in their structure. FFNN involves straight forward propagation in its topology, direct from input layer to hidden and from hidden layer to the output layer.

In ERNN, the propagation process starts from input to hidden layer. Then, a learning rule is applied. After that, there are connections from the hidden layer to the context units fixed in the input layer for the first propagation process. Then, the learning will be continued straight forward from hidden layer to the output layer. The basic topology of FFNN and ERNN is shown in Fig. 1.

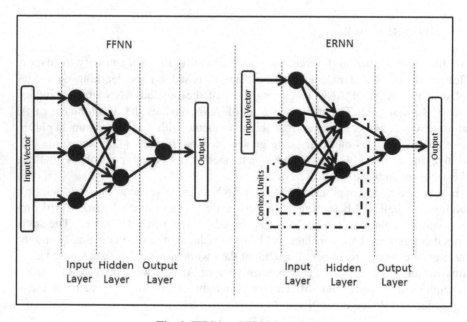

Fig. 1. FFNN and ERNN topologies

4   Genetic Algorithm

Genetic algorithm (GA) is one of the popular techniques used among all of the evolutionary algorithms. Its self-adaptation and self-organization had make GA a good techniques for solving optimization problems [21]. It had been well implemented at varies of field such as document clustering, tuning ANN weight, gaming and etc [21] and [22]. The data is presented as a fit-length chromosome. Before any of the operator is been triggered, a set of chromosome is randomly generation and it is known as individual in a population. Generally, there are three basic operators in GA that is selection, crossover and mutation.

Selection is a process of selecting fittest individuals in the populations to pass their genes on the next generations. A fitness function is been used to determine the fitness score of each individual before it can be selected. There are lots of selection methods that had been introduced and it can be classified into six categories which are proportionate selection methods, ranking selection, tournament selection, range selection, gender-specific selection and GR based selection [23]. Ranking selection is used in this research.

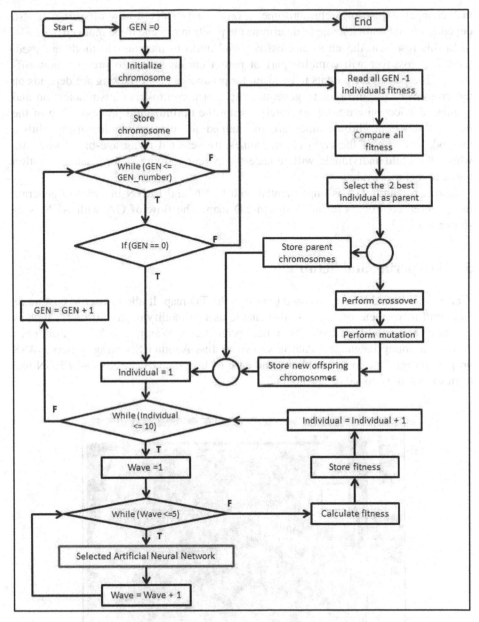

Fig. 2. Flow chart of GA toward ANNs

Reproduction started with the implementation of crossover towards two or more parent chromosomes. During this process, two new individuals are formed by swapping a sub-sequence between the parent chromosomes [24]. The length of swapping is determined by the crossover probability rate. Mutation is a process that changes a gene of a chromosome in order to produce a new offspring. It also depends on the mutation rate to determine the position of gene to be mutated.

In this research, the uniform crossover and uniform mutation methods are used. Uniform crossover will combine part of parent chromosomes to produce new offspring. The part of the parents to be taken for producing new offspring are depends on the crossover rate. In order to generate outperform controller, elitism selection and age-based selection are used separately during the optimization process. Two of the best fittest parent chromosomes are maintained for next generation using elitism method. The rest of the eight chromosomes are selected using age-based selection where the child individuals will replace the parent individual for each generation based to its fitness value.

In this research, GA is implemented with FFNN and ERNN in order to generate the required controllers to the designed TD map. The flow of GA with ANN is as shown in Fig. 2.

5   Experimental Setup

Warcraft III World Editor was used to design the TD map. It allows user to customize and build its own custom map. It also enable user to modify object attributes such as appearance and ability of the units, buildings and items. Warcraft III World Editor has its own scripting language which is known as Just Another Scripting Syntax (JASS programming). Due to its simplicity and ease of use, the GA, FFNN and ERNN had been successfully coded in few hours.

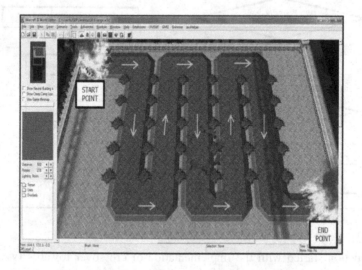

Fig. 3. Overview of Designed Tower Defense Map

The designed TD map is shown in Fig. 3. The flow of the creeps' movement is shown in arrow. The enemy path is straight forward and towers are able to be built at specific region which is limited in certain region as shown in Fig. 4. The game is designed for five waves with 20 creeps per wave. The strength and the movement speed of the creeps increase proportionate with the number of waves. Player is given with 10 lives. The game will be ended either all the creeps had been eliminated within five waves or no life of the player is remained.

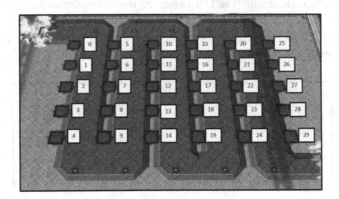

Fig. 4. Tower Building Region

Fig. 4 above shows that there are 30 regions that allow towers to be built. The list of tower regions represents the input of ANN. If a tower is built in a region then value 1 will be assigned otherwise, value 0 will be assigned.

The FFNN consists of 31 inputs (30 actual inputs and 1 bias), 16 hidden neurons (15 neurons and 1 bias neuron) and 30 output neurons. ERNN used 31 true input neurons, 31 context input neurons, 16 hidden neurons (15 neurons and 1 bias neuron) and 30 output neurons. The output of ANN shows the next tower to be built by returning 30 neurons which represents the region in the TD game. The activation function used is Binary Sigmoid function.

Eventually, there are ten chromosomes have been generated during early of the experiment. Individual chromosome will be evaluated in each generation. The evaluation function used is referred to the total number of kills minus by 90. The crossover rate used in this research is 0.7 and the mutation rate used is 0.02. There are 150 generations involved in any of the optimization processes.

6   Results and Discussions

There are a total of 20 games have been conducted, 10 games for GA and FFNN, another 10 games for GA and ERNN. The results for fitness over generation have been collected and represented using single-line graphs as shown in Fig. 5 and Fig. 6 below.

Fig. 5 and Fig. 6 show the results for one of the TD game using FFNN and ERNN, separately. The generated controllers using FFNN lost the games during early stage of the experiments. Then, the fitness score increased from -7 to 6 after generation 7th. The fitness scores maintained until generation 12th. Furthermore, the fitness score increased again from generation 13th – 16th, to the maximum fitness of the game. It shows GA has successfully maintained the optimal score until last of the generation with the integration of elitism concept in the GA.

In other case, the generated controllers using ERNN performed slightly worse compared to the FFNN case. The weakest controller reached -62 fitness score during early stage of the experiment. The GA improved the controller capability in generating -25 fitness scores during generations 9th – 13th. Then, the fitness score reached -6 during generation 16th. The fitness score reached 6 after generation 51st. Then, slowly the fitness score reached the 9 after generation 118th.

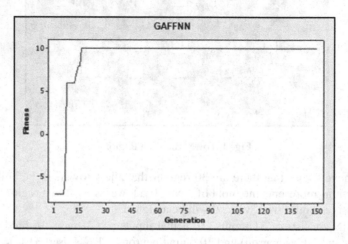

Fig. 5. One of the results with GA and FFNN used

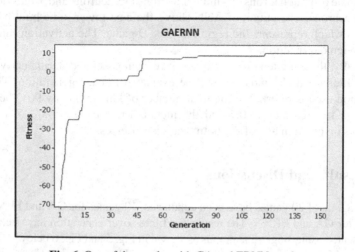

Fig. 6. One of the results with GA and ERNN used

Table 1 summarizes the average fitness scores for all the games. The average fitness score is collected by dividing the highest fitness of each generation by 150. Positive value indicates that the controllers managed to win in most of the games and negative value meant the controllers often lost the games.

Table 1. Average Fitness Over 150 Generations

RUN	GAFFNN	GAERNN
1	8.23	-1.68
2	8.57	8.55
3	7.00	7.53
4	8.26	6.65
5	8.56	4.60
6	8.40	9.00
7	7.51	-1.68
8	6.02	9.09
9	7.91	8.89
10	7.75	7.53
MINIMUM	6.02	-1.68
MAXIMUM	8.57	9.09
MEAN	7.821	5.848
STANDARD DEVIATION	0.805	4.193

Table 1 shows that the controllers generated from GA with FFNN won for all the 10 games and GA with ERNN only won 8 games. The controllers lost the games on the first run and the seventh run with the ERNN used. The controllers generated with GA and FFNN performed well in attaining of an average 7.821 fitness score which is higher compared to the controllers generated with GA and ERNN that scored an average of 5.848 fitness score. However, the maximum average fitness score obtained with GA and ERNN is higher compared to the GA and FFNN, 9.09 and 8.57 respectively. The tabulated data shows the controllers generated using GA with FFNN scored lower standard deviation compared to the GA with ERNN, 0.805 and 4.193 respectively. This shows the controllers generated using GA with FFNN were consistently performed well during the optimization stage.

Furthermore, tests have been conducted for all the optimized controllers with winning rate involved. Table 2 shows the winning rates of GA with FFNN compared to GA with ERNN. Based on the comparison, GA with FFNN controllers produced higher average winning rate compared to GA with ERNN controllers. The data also shows six out of ten of the controllers generated using GA with FFNN outperformed the controllers generated using GA with ERNN. Hence, it can be concluded that the controllers generated using GA with FFNN is slightly superior compared to the controllers generated using GA with ERNN.

Table 2. Average Winning Rate for GAFFNN and GAERNN

RUN	GAFFNN	GAERNN
1	**63.07%**	25.80%
2	59.80%	**67.27%**
3	**59.47%**	49.53%
4	**52.47%**	46.13%
5	**77.13%**	34.80%
6	60.93%	**69.40%**
7	**67.33%**	25.80%
8	49.27%	**82.00%**
9	51.87%	**77.27%**
10	**54.67%**	53.40%
Average Winning Rate	**59.60%**	53.14%

7   Conclusion and Future Works

The experimental results showed the implementation of GA had successfully tuned the weights of the FFNN and ERNN in the TD game. The FFNN and ERNN acted as machine learning to build towers without human intervention. The ANNs used also represented two different individuals in playing the TD games. The individual with FFNN used achieved better average fitness scores compared to the individual with ERNN used.

With referring to the hypothesis of this research, is that possible in reducing the cost and time used to design the TD game's map with AI involvement? Yes. The implementation of GA with FFNN and ERNN showed the designed map provided medium level of difficult to the player(s). It is impossible for a good player(s) to win the game easily. In other sense, it is also impossible for a weak player(s) to often lose the game.

In future works, game score will be considered in the evaluation instead of fitness score. The game score will be increased consistently with the increment of level of the game.

Besides, more types of tower and more types of creeps will be implemented into the map. The controller should not be limited to predict the placement of tower rather to learn how to spend the resources either to build different type of tower or upgrade the existing tower.

References

1. Avery, P., Togelius, J., Alistar, E.: Computational Intelligence and Tower Defence Games. In: IEEE Congress on Evolutionary Comp. (CEC), pp. 1084–1091 (2011)
2. Plants vs. Zombies - Walkthrough/guide, http://www.ign.com/faqs/2009/plants-vs-zombies-walkthroughguide-992681
3. Fieldrunners 2 – An Absolute Masterpiece in Every Regard, http://applenapps.com/review/fieldrunners-2-an-absolute-masterpiece-in-every-regard
4. 'Sentinel Earth 2 – Earth Defense' – Does It Live Up to the Original'?, http://toucharcade.com/2009/07/08/sentinel-2-earth-defense-does-it-live-up-to-the-original/
5. Rummell, P.A.: Adaptive AI to Play Tower Defense Game. In: The 16th International Conference on Computer Games, CGAMES, pp. 38–40. IEEE (2011)
6. Chang, K.T., Chin, K.O., Teo, J., James, M.: Game AI Generation using Evolutionary Multi-Objective Optimization. In: Evolutionary Computation (CEC), pp. 1–8. IEEE (2012)
7. Chang, K.T., Chin, K.O., Teo, J., Chua, B.L.: Automatic Generation of Real Time Strategy Tournament Units using Differential Evolution. In: Proceedings of the IEEE Conference on Sustainable Utilization and Development in Engineering and Technology, pp. 101–106. IEEE (2011)
8. Ng, C.H., Niew, S.H., Chin, K.O., Teo, J.: Infinite Mario Bross AI using genetic algorithm. In: IEEE Conference on Sustainable Utilization and Development in Engineering and Technology (2011 IEEE STUDENT). The University of Nottingham, Malaysia Campus (2011)
9. Chang, K.T., Ong, J.H., Teo, J., Chin, K.O.: The Evolution of Gamebots for 3D First Person Shooter (FPS). In: IEEE the Sixth International Conference on Bio-Inspired Computing: Theories and Applications (BIC-TA 2011). Universiti Sains Malaysia, Penang (2011)
10. Bourg, D.M., Seeman, G.: AI for Game Developers. O'Reilly (2004)
11. Moriarty, D., Miikkulainen, R.: Discovering Complex Othello Strategies Through Evolutionary Neural Networks. Connection Science 7, 195–209 (1995)
12. Chellapilla, K., Fogel, D.B.: Evolving Neural Networks to Play Checkers without Relying on Expert Knowledge. IEEE Transactions on Neural Networks 10(6), 1382–1391 (1999)
13. Fogel, D.B.: Using Evolutionary Programming to Create Neural Networks that are Capable of Playing Tic-Tac-Toe. In: IEEE International Conference on Neural Networks (ICNN), vol. 2, pp. 875–880 (1993)
14. Freisleben, B.: A Neural Network that Learns to Play Five-in-a-row. In: Second New Zealand International Two-Stream Conference on Artificial Neural Networks and Expert Systems, pp. 20–23 (1995)
15. Wong, S.K., Fang, S.W.: A Study on Genetic Algorithm and Neural Network for Mini-Games. Journal of Information Science and Engineering 28, 145–159 (2012)
16. Niu, B., Wang, H., Ng, P.H.F., Shiu, S.C.K.: A Neural-Evolutionary Model for Case-Based Planning in Real Time Strategy Games. In: Chien, B.-C., Hong, T.-P., Chen, S.-M., Ali, M. (eds.) IEA/AIE 2009. LNCS(LNAI), vol. 5579, pp. 291–300. Springer, Heidelberg (2009)
17. Wang, H., Ng, P.H.F., Niue, B., Shiu, S.C.K.: Case Learning and Indexing in Real Time Strategy Games. In: Fifth International Conference on Natural Computation, pp. 100–104. IEEE (2009)

18. Krenker, A., Bester, J., Kos, A.: Introduction to the Artificial Neural Networks. In: Artificial Neural Networks – Methodology Advances and Biomedical Applications. InTech, pp. 3–18 (2011)
19. McCulloch, W., Pitts, W.: A Logical Calculus of the Ideas Immanent in Nervous Activity. Bulletin of Mathematical Biophysics 5, 115–133 (1943)
20. Carlos, G.: Artificial Neural Networks for Beginners. In: Artificial Neural Networks – Methodology Advances and Biomedical Applications. InTech (2011)
21. Xiang, W.J., Liu, H., Sun, Y.H., Su, X.N.: Application of Genetic Algorithm in Document Clustering. In: International Conference on Information Technology and Computer Science, pp. 145–148. IEEE (2009)
22. Mitchell, M.: An Introduction to Genetic Algorithms. MIT Press (1999)
23. Sivaraj, R.: A Review of Selection Methods in Genetic Algorithm. International Journal of Engineering Science and Technology (IJEST) 3(5), 3792–3797 (2011)
24. Talib, S.H.: An Introduction to Evolutionary Computation (1998)

Artificial Bee Colony Optimization Algorithm with Crossover Operator for Protein Structure Prediction

Zakaria N. M. Alqattan and Rosni Abdullah

School of Computer Sciences, Universiti Sains Malaysia, 11800 USM, Penang, Malaysia
zqttan2@yahoo.com ,rosin@cs.usm.my

Abstract. Swarm intelligence systems are mainly introduced based on the behavior and the interactions of the insects locally with their communities and also with their environments. Artificial Bees Colony (ABC) Optimization algorithm, inspired from the honey bees' food foraging behavior, is an optimization method used for bioinformatics problems where the Protein Structure Prediction (PSP) is considered as one of these problems. For a given protein, knowing the exact action whether hormonal, enzymatic, transmembranal or nuclear receptors, etc does not depend solely on amino acid sequence but on the way the amino acid thread folds as well. This paper presents a modified ABC algorithm, where a crossover operator from the Genetic Algorithm (GA) has been added to the original ABC algorithm. To solve the PSP problem, a conformation with the lowest free energy is the target of the ABC search. The results show the ability of the modified ABC to reach the global minima for PSP problem by using around quarter of the iterations that the classical ABC needs.

Keywords: Protein structure, Prediction, Artificial Bees Colony, Genetic Algorithm.

1 Introduction

According to [1] applying swarm intelligence in bioinformatics problems is still in the beginning and there are only few numbers of researches that have been done compared to the number of researches on the evaluation methods like Genetic Algorithm application on Bioinformatics. However, there are a few researches on using Bee Colony Optimization algorithm in bioinformatics problem and especially in protein folding prediction. One of the recent studies is by [2] that suggested a search application using Honey Bee Colony Optimization for protein conformations in order to predict the 3D structure of the protein. The problem of PSP comes from the protein's structure itself. Protein structure is a complex and interesting topic at the same time. This complexity gave the protein the importance and made it as a fertile ground for the researcher to investigate. According to [3], the proteins are polymers of different amino acids combined together to build a unique sequence which presents the specialty of the proteins. There are twenty known amino acids commonly found in the proteins. These amino acids can be called as the building blocks of the proteins.

S.A. Noah et al. (Eds.): M-CAIT 2013, CCIS 378, pp. 147–157, 2013.

Depending on these blocks; the protein's biological function can be determined. According to the type and the number of the amino acids in the sequence, the protein will fold in special shape and take its three dimensional structure.

The amino acids are linked together by peptide bonds which effectively present the protein's backbone. These polypeptied chains have the freedom to rotate around themselves from the both side of the C_a atom along the protein sequence. According to [3] "Each peptide unit can rotate around two bonds: the C_a-$Ć$ and the N-C_a bonds, and by convention the angle of rotation around the N-C_a bond is called *Phi* (ϕ) and the angle around the C_a-$Ć$ bond from the same C_a atom is called *Psi*(ψ)", (Fig. 1).

Finally; since each amino acid in the protein sequence has only two main rotation angles; these angles have the main responsibility to represent the whole main chain of the polypeptide in the protein sequence, which establish the final protein 3D structure[3].

Bees Colony Optimization algorithm is one of the problem solving algorithms that have been applied on search optimization problems, it has been introduced in early 2000. The Bees Colony has many features which can be performed in order to model the intelligent search systems such as "bee dance (communication), bee foraging, queen bee, task selection, collective decision making, nest site selection, mating, floral/pheromone laying, navigation systems" [4]. In this study, a honey bee foraging behavior presented by Artificial Bee Colony (ABC) algorithm will be used in order to formulate the search process for the best protein conformation.

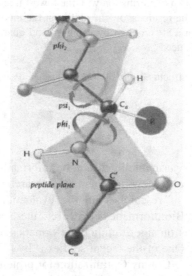

Fig. 1. The Phi (ϕ) and Psi (ψ) angles [3]

Protein representation is very important for the prediction operation. There are two main structure representations which have been used for protein folding application. The first one is called HP model which means the Hydrophobic (H) and Polar (p) amino acids residues type in the primary sequence of the protein. This model was used to present the binary sequence of the protein which contains either Hidrophobic or Polar. This sequence will finally be used to identify the structure of the protein on three dimension cubic lattice. However, the free energy computed for the conformation from this model, is the negative number of non-consecutive hidrophobic-hidrophobic connection [5].

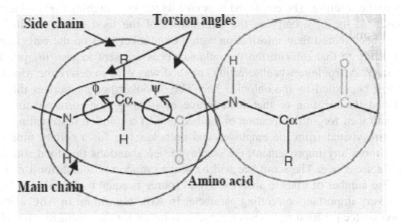

Fig. 2. Amino acid [2]

While the second representation which depends on the components of the amino acids in the protein, is the most popular representation used in the protein tertiary structure prediction. Each amino acid has two chains, namely the main and side chain. The torsion angles of the main chain are Phi (ϕ), Psi (ψ), and Omega (ω), the side chain angle is called χn angles. These angles give the amino acid the flexibility to rotate around the rotating bonds which connect the amino acid with each other in the protein sequence; these bonds are called polypeptide chain as shown in Fig 2. The protein's tertiary structure can be represented using the backbone and chain angles described earlier. The idea is to identify the protein free energy conformation using an array of angles values obtained by (as the protein is up to fold in a certain structure in which it has the lowest free energy, called the protein's native state [6]) rotating the torsion angles around the rotating bonds ([2],[7],[8],[9]).

2   Artificial Bee Colony

In the ABC algorithm, the colony of artificial bees contains three groups of bees: employed bees, onlookers and scouts. Each employed bee is sent to only one food source. This means that the number of employed bees is equal to the number of the food sources. The onlookers work after the employed bee support the onlookers with the food sources locations. The employed bee whose food source has been abandoned is called a scout. In each cycle of the search, the ABC conducts three steps: send the employed bees to the food sources and calculate their amount of nectar. Then the onlooker bees are sent to measure the nectar of the food sources after getting the information about the location of the promising food sources. The scout bee will be determined after the food source is exhausted by its employed bee and onlooker bee.

In the ABC, each food source represents a possible solution for the problem. The algorithm initialized the food sources randomly and the nectar amount of each food source was determined according to the food source quality corresponding to its vicinity to the solution. The employed bees take its job by searching the food area and then return to the hive carrying the information of the food sources visited. The employed bees shared their information with the onlooker bees on the waggling area and according to that information the onlooker bees are sent to visit the promising food sources. A "roulette wheel selection" method was used to determine which food source will be visited by the onlooker bee. The onlooker bee memorizes the nectar amount and the location of that food source or its neighbor corresponding to its quality, and then forgot the location of the bad one. If a solution representing a food source were visited from the employed and onlooker bee for a certain time called "limit" without any improvement, the employed bee abandons that food source and becomes a scout bee. The scout bee will be sent to randomly search for a new food source. The number of trials to abandon a food source is equal to the value of "limit". This is a very important controlling parameter in ABC algorithm. In ABC algorithm each food source position represents a possible solution for the optimization problem. Each solution of the optimization problem is associated with fitness value, which is the measuring parameter for its nectar amount. The ABC generates N population size of random solutions in the initial stage. N is also representing the total number of the food sources around the hive, which means that each solution represent a position of one food source and denoted as x_{ij}, where i represents the particular solution (i=1,2,...,N) and each solution is a D-dimension vector, and j represents its dimension (j=1,2,...,D).

As another foraging behavior, the bees search for a food source in a maximization search way. The bee looks for a food source with a maximum value of nectar in a short time. The goal of the foraging bee is the nectar amount, which it tries to maximize. The target of a maximization problem is to find a maximum of the objective function $F(\Theta)$, where Θ_i N. The Θ_i represent the position of the ith food source; then $F(\Theta_i)$ represents the nectar amount of that food source.

Then the onlooker bee chooses a food source depending on the probability value Pi associated with the food source. Probability value for each food source is calculated by following equation (1):

$$p_i = \frac{F(\theta_i)}{\sum_{i=1}^{N} F(\theta i)}.$$ (1)

According to the information gathered so far from the neighbor around Θi the onlooker make a comparison whether to select that food source or its neighbor. The position of the chosen neighbor is calculated by the following equation (2):

$$\theta_i(j+1) = \theta_{ij} + \beta(\theta_{ij} - \theta_{kj}),$$ (2)

where i represents the particular food source position (i= 1,2,...,N), k represent the neighbor's randomly chosen position (k= 1,2,...,N), the k value should be different from the i value . β is a random number between [-1, 1] used to estimate the neighbor food sources around x_{ij}. As the ABC search becomes close to the global optimum, the search diameter around the x_{ij} shrinks. If the fitness of the new solution (food source) is better, it takes the place of the old one in the memory. Otherwise, the old position will remain. After a number of ABC specified search cycles the food source with unimproved fitness will be abandoned, and the scout bee will be sent to find a new food source based on the equation (3). Then the food source found by the scout bee will take place of the abandoned one.

$$x_{ij} = x_{ij} + rand[0,1](x_{j\max} - x_{j\min}) . \qquad (3)$$

3   Proposed Method for PSP

In order to evaluate the performance of the modified ABC on protein structure prediction, a complete program was built. The PSP process in general contains three main phases: conformations creation, energy calculation and search algorithm implementation. Fig. 3 below illustrates the flow chart of the proposed search method for PSP problem.

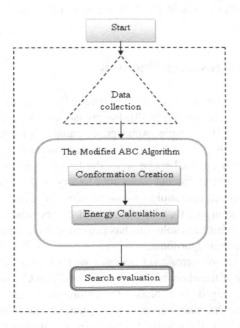

Fig. 3. The proposed PSP search method

Conformations Creation: in this phase, the food sources of the ABC algorithm is generated Then a rotation function to rotate the structure by a random value of main chain (Phi (ϕ) and Psi (ψ)) angles degree was applied. By each rotation a new conformation is added to the food sources. The number of the food sources is fixed from the beginning by the ABC algorithm parameters initialization.

Energy Calculation: for each conformation generated from the previous phase; an energy calculation function is applied. This function is used to calculate the energy of the conformation (generated protein structure), which reflects the inner interaction of the protein resident. This function is represented as the objective function for the modified ABC search algorithm. Programs like Dock [10] had used this type of scoring function. In this work, the equation used for energy calculation function contains interactions of van der Waals (ΔG_{vdw}) and electrostatic (Columbic) (ΔG_{elec}) interactions solely [11]. The energy equation $\Delta G = \Delta G_{vdw} + \Delta G_{elec}$ are presented below:

$$\Delta G = \Delta G_{vdw} \sum_{i,j} \left(\frac{A_{ij}}{r_{ij}^{12}} - \frac{B_{ij}}{r_{ij}^{6}} \right) + \Delta G_{elec} \sum_{i,j} \frac{q_i q_j}{\varepsilon(r_{ij})r_{ij}}. \tag{4}$$

Search Algorithm Implementation: The modified ABC search algorithm is used in this phase. For each cycle of the search algorithm the conformations creation and energy calculation functions are called. The crossover operator was added to the classical ABC search algorithm after the Employed bee phase. This additional step will improve the search process of the ABC corresponding to the quality of the new solutions being produced.

4   A Proposed Crossover Operator

The GA has a very important operator that the ABC doesn't have, which is the CROSSOVER operator. According to [8], the crossover operation is the heart of the method. Generally, it is the simple exchange of number of bits of a chromosome between pairs of solutions. This exchange has a large impact on the effectiveness of the search. By crossover the feathers between the two parents, a new regions will be available for search, since it was not accessible to any of the two "parent" solutions before. Through crossover operations, solutions can cooperate using the optimized features from one solution and then be mixed with others, where they can be further optimized. Cooperation between solutions has proved to have a very positive effect on the efficiency of the search algorithms.

In this proposed method, a crossover operator was added as an additional phase to the three main phases of the classical ABC algorithm. The Crossover phase has been added after the Employed bee phase to complement the differential search convergence. There are different strategies to implement the crossover operator; such as the one-point crossover, the two-point crossover, the uniform, and the orthogonal crossovers. The diversity of these strategies comes from the way that the gene has been selected for the crossover operation. In the proposed method, the uniform and the orthogonal crossovers were sequentially used [12].

Moreover, a greedy schema of the Differential Evolution (DE) method was used in the crossover operator. After carefully selecting the parents according to their fitness; the orthogonal strategy was applied for each pair. Then a pool of new offspring chromosomes will be generated. The greedy schema of the DE means that new offspring (solutions) competes with its parents for the next generation and the one with the better fitness wins, (Fig. 4). Using greedy schema for selecting one of them to the next generation in crossover operation; is the important difference between GA and differential evolution DE algorithm. That gives the DE the property to become the fast converging evolutionary algorithm [13].

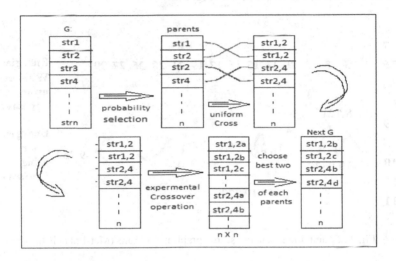

Fig. 4. The proposed Crossover procedure

However, for the PSP problem the crossover is presented by using a chromosome containing the main chain angles of the all resident amino acids in the protein; Which is the D dimensions number been presented for the food source in ABC algorithm. Assume that protein main chain angles (torsion angles) number is D, the chromosome of the uniform crossover will be an array of D dimension, (example: C[D]).

Finally, it is important to mention that cooperative ability of the GA approach which comes from its Transmigration to biological process description; gave it the advantage of modeling cooperative pathways. Thus, it is effective to use GA features in protein folding process. The protein folding is not cooperative on the dynamic level only but it is cooperative on the interaction level as well; where interactions of electrostatic, hydrophobic, van der Waals, etc., all influence the final structure [8].

5   Experimental Results

To evaluate the performance of the proposed ABC modification on PSP problem, a Met-enkaphlin experimental protein was used. This is a short sequence protein with five amino acids resident. Each amino acid has two main chain angles (Phi (φ) and Psi (ψ)). The number of the angles used that mainly influence the structure are eight angles. Those angles are the D parameters which will be added to each ABC food sources. And for the ABC parameters, the number of the colony size was set to (20), the FoodNumber which is equal to the Employbeenumberwas set to (10), and the OnlookerBeewas (10). To intensively evaluate the performance of the proposed Crossover operator, the iteration Number was set for 100 only. The number of the Runs was set to 30. The experiment was implemented on a normal PC with ADM Athlon(tm) 7750 Dual-Core processor 2.70 GHz CPU and 4.00 GB RAM. The result of this test is shown by the graph below (Fig. 5).

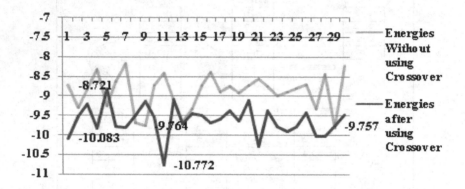

Fig. 5. Protein lowest energy search graph for 30 runs (Met-Enkephalin)

From the graph, the results show that the ABC algorithm on PSP problem has improved after adding the proposed crossover operator. The Crossover operator is dragging the search process in the earlier stages of the search to the global minima, which is the conformation with the lowest energy. Most of the conformations that have been found at the first 100 iterations during the modified ABC search algorithm are lower than **-9.0** (Kcal/mol), while most of the conformations found using classical ABC algorithm are higher than that.

Moreover, to evaluate the entire progress of the modified ABC algorithm in order to find the global minima of the search space (to find the conformation with the lowest free energy for PSP problem), the iteration number was set to 1000. The graph below expresses the ability of the modified ABC by the crossover operator to rapidly reach the global minima with an impressive iteration number (Fig. 6).

From Fig. 6, the results implicitly express the improvement of the modified ABC. The result shows the effectiveness of the modification to increase the speed of the search. From the results, the modified ABC algorithm expresses its ability to reach the global minima of the PSP problem in less iterations.

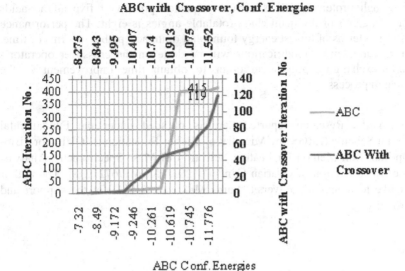

Fig. 6. Comparison between Classical ABC and Modified ABC on PSP problem

The modified ABC algorithm reached (-11.775 Kcal/mol) at iteration number (119), while the classical ABC algorithm reached (-11.779 Kcal/mol) at iteration number (415). The modified ABC outperforms the classical ABC when both were implemented on PSP search problem. It is important to mention that both algorithms reached the global minimum which has been presented by previous works. See table (1).

Table 1. The lowest energies (in Kcal/mol) obtained in previous studies

Met- enkaphlin conformation Energy	Source , Reference
-10.85	[14]
-11.71	[15]
-10,90	[16]
-11.77	This work

6   Conclusion

The CROSSOVER operator has shown its ability to enhance the search performance of the ABC algorithm. The operator was added after the Employ bee phase. There was no further change on the ABC except on the scout bee phase, which is: for each food source that exceed the "limit" send the scout bee. The experiment was applied on the Met-enkaphalinprotein which is a short sequence protein of five amino acids residents. The number of the main chain rotatable angles is eight. The performance was interesting in terms of lowest energy found and quite acceptable in terms of time. For further research, more modifications will be done on the crossover operator's selection and exchanging process in order to obtain time improvement in the algorithm search process.

Acknowledgments. Research reported here is pursued under the Fundamental Research Grant Scheme (FRGS) by Ministry of Higher Education (MOHE) for "Bio-Inspired Optimization Method for Feature Selection of Mass Spectrometry Analysis in Biomarker Identification of Ovarian Cancer" [203/PKOMP/6711268]. The author also would like to thanks to Universiti Sains Malaysia (USM) for the support and facilities provided.

References

1. Das, S., Abraham, A., Konar, A.: Swarm intelligence algorithms in bioinformatics. In: Das, S., et al. (eds.) Swarm Intelligence Algorithms in Bioinformatics. SCI, vol. 94, pp. 113–147. Springer, Heidelberg (2008)
2. Bahamish, H.A.A., Abdullah, R., Salam, R.A.: Protein conformational search using bees algorithm. In: Second Asia International Conference on Modeling & Simulation, pp. 911–916. IEEE (2008)
3. Branden, C., Tooze, J.: Introduction to protein structure. Garland, New York (1991)
4. Karaboga, D., Akay, B.: A survey: algorithms simulating bee swarm intelligence. Artificial Intelligence Review 31, 61–85 (2009)
5. Fidanova, S., Lirkov, I.: Ant colony system approach for protein folding. In: International Multiconference on Computer Science and Information Technology, pp. 887–891. IEEE (2008)
6. Chen, P., Li, J.: Prediction of protein long-range contacts using an ensemble of genetic algorithm classifiers with sequence profile centers. BMC Structural Biology 10, S2 (2010)
7. Petrescu, A.-J., Calmettes, P., Durand, D., Receveur, V., Smith, J.C.: Change in backbone torsion angle distribution on protein folding. Protein Science 9, 1129–1136 (2000)
8. Unger, R.: The genetic algorithm approach to protein structure prediction. In: Applications of Evolutionary Computation in Chemistry, pp. 153–175. Springer (2004)
9. Bahamish, H.A.A., Abdullah, R., Salam, R.A.: Protein tertiary structure prediction using artificial bee colony algorithm. In: Third Asia International Conference on Modelling & Simulation, pp. 258–263. IEEE (2009)
10. Meng, E.C., Shoichet, B.K., Kuntz, I.D.: Automated docking with grid-based energy evaluation. Journal of Computational Chemistry 13, 505–524 (1992)
11. Cui, Y., Chen, R.S., Wong, W.H.: Protein folding simulation with genetic algorithm and supersecondary structure constraints. Proteins: Structure, Function, and Bioinformatics 31, 247–257 (1998)
12. Andrade, A.V., de Errico, L., Aquino, A.L.L., de Assis, L.P., Barbosa, C.H.N.R.: Analysis of selection and crossover methods used by genetic algorithm-based heuristic to solve the LSP allocation problem in MPLS networks under capacity constraints (2008)

13. Karaboga, D., Basturk, B.: On the performance of artificial bee colony (ABC) algorithm. Applied Soft Computing 8, 687–697 (2008)
14. Eisenmenger, F., Hansmann, U.H.: Variation of the Energy Landscape of a Small Peptide under a Change from the ECEPP/2 Force Field to ECEPP/3. The Journal of Physical Chemistry B 101, 3304–3310 (1997)
15. Androulakis, I., Maranas, C., Floudas, C.: Prediction of oligopeptide conformations via deterministic global optimization. Journal of Global Optimization 11, 1–34 (1997)
16. Zhan, L., Chen, J.Z., Liu, W.-K.: Conformational study of met-enkephalin based on the ECEPP force fields. Biophysical Journal 91, 2399–2404 (2006)

Frequent Positive and Negative Itemsets Approach for Network Intrusion Detection

Anis Suhailis Abdul Kadir, Azuraliza Abu Bakar, and Abdul Razak Hamdan

Center for Artificial Intelligence Technology
Faculty of Technology and Information Science
Universiti Kebangsaan Malaysia
Bangi, Selangor, Malaysia
{anis.suhailis,aab,arh}@ftsm.ukm.my

Abstract. Recently there has been much interest in applying data mining to computer network intrusion detection. Accurate network traffic model is important for network stipulation. Significant knowledge is crucial for better accuracy in network traffic model. This paper presents the use of a Frequent Positive and Negative (FPN) itemset approach for network traffic intrusion detection. FPN approach generates strong positive and negative rules, in which produce important knowledge for building accurate network traffic model. Usually, frequent itemsets are generated based on the frequency of the presence of a particular item or itemset before generating the relevant rules. However, in FPN approach, for negative association rules, frequent absent itemsets is introduced. FPN approach has successfully enhanced the accuracy of the network traffic model by identifying volume anomaly. The experiments performed on network traffic data at the Universiti Kebangsaan Malaysia. We also report experimental results over other algorithms such as Rough Set and Naive Bayes. The results demonstrate that the performance of the FPN approach is comparable with the results of other algorithms. Indeed, the FPN approach obtains better results compared to other algorithms, indicating that the FPN approach is a promising approach to solving intrusion detection problems.

Keywords: negative association rule, associative classification, intrusion detection.

1 Introduction

Computer network intrusion detection is potentially tractable using automated classifiers. Challenges remain for developing an accurate classifier for intrusion detection model. The intrusion detection problem (IDP) is a two-class classification problem: the goal is to classify patterns of the system behaviour in two categories (normal and abnormal), using patterns of normal behaviour and pattern known attacks, which belongs to the abnormal class.

Concerning the analysis method, intrusion detection systems (IDS) are usually classified into two categories [1]; misuse IDS and anomaly IDS. A misuse or knowledge-based IDS aims at detecting the occurrence of action sequence that has been previously identified to be an intrusion. Thus, in this kind of IDS, attacks must be

S.A. Noah et al. (Eds.): M-CAIT 2013, CCIS 378, pp. 158–170, 2013.
© Springer-Verlag Berlin Heidelberg 2013

known and described proactively. Alternatively, an anomaly or behaviour-based IDS assumes that an intrusion can be detected by observing deviations from a normal or expected behaviour. The valid behaviour is extracted from previous data or information about the system. The IDS later compares the extracted behaviour model with the current activity and raises an alert each time that a certain degree of divergence from the original model is observed.

There are several classification solutions being explored for IDP such as Fuzzy [2] and [3], Bayesian [4] and [5], Support Vectors Machine (SVM) [6], Associative Classification (AC) [7] and [8] and Rough Set [8] and [9]. We present AC for IDP and motivated by negative rule. In this paper, the frequent negative together with the frequent positive itemset is introduced to produce strong negative rule and together with positive rule. The knowledge obtained from these rules enable the construction of a better classifier. The principles and characteristics of frequent positive and negative (FPN) approach will be discussed and described later in this paper. The effectiveness of this approach is evaluated and compared, based on real network data.

The remainder of this paper is organized as follows. Section 2 reviews related works on AC, and the integration between negative rules and AC. Section 3 presents the proposed approach of using a FPN for mining negative rules and later to construct the FPN classifier. In Sections 4, the experimental setup and results are discussed and in Section 5, we complete the paper with our conclusions.

2   Related Works

Exploration of a positive rules for use in classification task and the classifier is known as associative classification (AC). The use of association rules is a highly confident method for overcoming some of the constraints in the state-of-the-art classification algorithm, which is a decision tree classifier. This due to, association rule discovers associations among multiple attributes, whereas the decision tree classifier considers only one attribute at a time [10]. Subsequent research on AC is encouraging, and has led to the development of many approaches and algorithms. Classification-Based Association Rule (CBA) [10] was the first AC algorithm to employ the Apriori algorithm to find the rules for the classifier. There are several AC algorithms in the last few years beside CBA, for example Classification based on Predictive Association Rules (CPAR) [11], Classification based on Multiple Association Rules (CMAR) [12] and Multi-class Classification based on Association Rule (MCAR) [13]. Those AC are varied in several features such as rules discovery, rank rules and prune rules. In the ranking process, researchers have primarily considered support and confidence are the best factors and then followed by their preference measures. For pruning, database coverage is the most popular measure. While method to find classification association rules (CARs) is varied such as Apriori, FP-Tree, Foil Greedy and Tid-list intersection. Once the classifier is constructed, its predictive power is evaluated on the test data to forecast their class labels.

Recently, the knowledge obtained with negative rule has attracted attention from researchers working with AC. Association Rule Classification with Positive And Negative (ARC-PAN) [14], Associative Classifier with Negative rules (ACN) [15] and Multiple Target Negative Target MTNT [16] are among AC algorithm that use a negative rule when building the classifier. ARC-PAN and ACN adopted frequent itemset (FIS) approach, which is a frequency of the present item for negative rule mining. However, ACN calculated number of support for each rule without scanning the database except for the first time while ARC-PAN needed more database scan to get the value for support. Therefore, ACN claimed to have a better time in mining CARs compared to ARC-PAN. While MTNT has different features such as method used and type of NAR. MTNT used exception rule to build a classifier without additional measure for pruning neither Apriori as the basic method for mining rule. MTNT is clearly having more than one target class compared to ARC-PAN and ACN. However, the fundamental process for building classifier was similar for all method which was CBA and FIS approach.

In recent years, AC techniques have been proposed in several studies. These techniques use numerous different approaches to build a classifier. The knowledge gathered from this task successfully assists in the analysis and decision-making processes to obtain better prediction and detection in several domains [17], [18], [19] and [20]. However, the selection of quality of CARs can still be improved, either from positive or negative rules, in order to construct an accurate classifier. Normally, frequent itemsets are generated based on the frequency of the presence of a particular item or itemset [14], [16], [17], [18], [19], [19], [20], [21] and [22]. This approach is relevant to positive rule mining, but it needs to be reassessed for mining of negative rule. The FPN itemset approach is introduced to generate strong and interesting negative rule. The definition of frequent negative itemset is itemset in which had high support or frequency of being absent in a particular dataset or not being present as a classic association rule. Therefore, the strong negative items are discovered in order to generate strong negative itemset.

We will explore the potential of rules from the FPN approach to the classification model. We assumed our approach can obtain accurate information that cannot be captured by classic association rule approach. The FPN eliminates weak and inaccurate positive rules as much as possible with strong and accurate negative rules when selecting good CARs for building the accurate classifier.

3   Methodology

Based on the research work of Lee et al. [21], designing an intrusion detection system based on learning algorithm can be described in the following steps:

(1) Captured network data by using tools
(2) Process the captured network data into a suitable input format;
(3) Normalize the network flow and extract attributes or features of attack behaviour or normal usage pattern of raw data;
(4) Design and use learning algorithm to get detection rules;
(5) Integrate the detection rules into the real time IDS for detecting intrusion.

With these five steps, feature or attribute extraction and detection rules generation are two key steps.

Given a set of records, where one of the features is the class label (i.e., the concept to be learned), classification algorithms can compute a model that uses the most discriminating attribute to describe each concept. The label is the concept to be learned whether normal or abnormal connections. A classification rule learning model generates rules for classifying the connections. We found that AC has indeed selected the discriminating attribute values into the classification rules for the intrusions.

The accuracy of a classification model depends directly on the set of attributes provided in the training data. The goal of constructing a classification model is that after (selectively) applying a sequence of attribute value tests, the dataset can be partitioned into significant subsets, that is, each in a target class. Therefore, when constructing a classification model, a classification algorithm searches for attributes with a large information gain [22]. Thus, selecting the right set of system attributes is a critical step when formulating the classification tasks. Our strategy is to mine the frequent sequential patterns from the network traffic data, and then use these patterns as guidelines to select and construct IDS model.

3.1   Frequent Positive and Negative (FPN) Itemset Approach

Lets describes the concept of an association rule [23], where $I = \{i_1, i_2 ... i_m\}$ is a set of n distinct items. Let D be a set of transaction $D = \{t_1, t_2, t_3, ... t_n\}$, $t_j \subset I$, where T represents the transaction for a set of items. Let $A = \{i_1, i_2, i_3, ... i_k\}$, $A \subset I$, an itemset, be a set of items in I. The association rule takes the form $A \rightarrow B$, whereas $A \subset I$, $B \subset I$ and $A \cap B = \emptyset$. Every generated rule has its own measures, support (sup) and confidence (conf). The calculation of support is the frequency of the transaction in D containing A and B, which is also known as the probability, $P(A \cup B)$. The value of confidence is the percentage of the transaction in D containing A and also containing B; it is known as conditional probability, $P(B|A)$. The Apriori algorithm uses threshold values to filter the weak and uninteresting rules.

In the candidate generation phase, itemsets that have greater support than the minimum support are called FIS, while itemsets that have lower support than the minimum support are labelled infrequent itemsets. In Frequent Pattern Trees (FPT) method, the complete set of frequent itemsets is mined without candidate generation [24]. However, both the Apriori and FPT methods mine frequent patterns from a set of positive itemset. In the rule generation phase, frequent itemsets that have greater confidence than the minimum confidence is used to generate association rules. In each positive association rule $A \rightarrow B$ can have three forms of negative association rule: 1) $A \rightarrow \neg B$; 2) $\neg A \rightarrow B$; 3) $\neg A \rightarrow \neg B$. The negative rule indicates that there is a negative relationship between items in the association rule. A negative relationship implies the presence of items by the absence of other items in the same transactions. Brin et al. [25] discovered a correlation between the antecedent and the consequent in association rules. This correlation can be positive, negative or independent.

Negative rule also plays important roles in decision making. The opposite part of the normal pattern or knowledge is more interesting compared to strong positive rules which are predictable and common [26]. Therefore, negative rule is beneficial in detecting abnormal knowledge like fraud and intrusion. Furthermore, negative rule is able to provide more complete knowledge together with the knowledge from the positive rule in ensuring better analysis and decision. The analysis process in decision making involves two factors which are positive factor and negative factor. To make a better analysis for decision making, both factors need to be scrutinized. Therefore, the analysis will be more comprehensive and complete. Negative rule is useful in identifying items which conflict with each other or item that complement each other.

In traditional Apriori, the calculation of support for an itemset is based on the presence of the itemset in the transactions. As an alternative, this paper proposes determining the calculation of support from the absence of items or itemsets for generating negative rule, together with the frequent presence of itemset ideas. As a result, the FPN itemsets will be generated at the end of the candidate generation phase. Fig. 1 presents the FPN algorithm that was implemented in this study. Only candidates including absent itemsets in which the calculated support is above the minimum support value will be extended. For rule generation, we adopted similar steps that used in the traditional Apriori. All rules with greater confidence than minimum confidence are extracted; the rules consist of strong positive and negative rule.

For a better understanding of the FPN approach, we conducted a test with a small dataset to illustrate the mining process. Table 1 consists of a transaction dataset, whereas Table 2 shows a frequent itemset that has been discovered. L_k is denoted as all frequent k-itemsets. Table 2 lists all of the generated frequent itemsets, both positive and negative, with minimum support = 0.4. The candidate generation discontinued at L_3 because there were no more frequent itemsets with greater than the minimum support. Based on the set of frequent itemsets L_3, association rules were generated including both positive and negative rules with minimum confidence = 0.8, as shown in the right-hand column in Table 3. If we only consider frequent positive (FP) itemsets, item E will not be considered for negative rule mining as in Table 2. As a result, the generated rules will miss out the strongest items for the negative rule. Therefore, we move away from the aim of negative rule mining, in which we focused on the absence of items or itemsets.

Algorithm: Frequent Positive and Negative Itemset (FPN)
Input: D:Transactional Database, ms: minimum support, C_k: candidate itemset of size k, L_k: frequent itemset of size k
   L_1-frequent items(l)
   Initialize L_1, f, countP and countN
   scan the database and find the set of items (f)
   for each item f in F do
      for each transaction t ∈ D do
         if t contains f then
            increment the count of countP;
         else
           increment the count of countN;
      if countP≥ms or countN≥ms
         $L_1 \leftarrow L_1 \cup \{f\}$
   return L1
   For(k=1; L_k!=∅; k++) do begin
      C_{k+1}=candidates generated from L_k;
      For each transaction t in D do
         Increment the count of all candidates in C_{k+1} are contained in t
         Increment the count of all candidates in C_{k+1} are not contained in t
      L_{k+1}=candidates in C_{k+1} with ≥ ms
   End
 Return $\cup_k L_k$

Fig. 1. FPN Algorithm

	Table 1. Transaction data			Table 2. Frequent itemsets (FIS)				
TID	Items	L_1	Sup	L_2	Sup	L_3		Sup
1	A,B,D	A	5	¬A^B	4	B^D^¬E		5
2	A,B,C,D	¬A	5	¬A^¬E	4	B^D^¬F		4
3	B,D	B	7	B^C	4			
4	B,C,D,E	C	6	B^D	6			
5	A,C,E	¬C	4	B^¬E	6			
6	B,D,F	D	6	B^¬F	4			
7	A,E,F	¬D	4	C^¬E	4			
8	C,F	¬E	7	D^¬E	5			
9	B,C,F	F	5	D^¬F	4			
10	A,B,C,D,F	¬F	5	¬E^F	4			

The example shows that the number of frequent absent itemsets is greater than the number of frequent present itemsets. There are only two frequent positive itemsets in L_2, whereas there are eight frequently negative itemsets. When those frequent itemsets went through the rules generation phase, 16 rules were generated; 14 rules are negative rules and only two rules are positive rules. Furthermore, the generated negative rules are strong and considered interesting, as demonstrated in Table 3 by the values of the confidence and the support. Those rules are important and should be considered in the process of analysis to improve classification task or decision-making.

Table 3. Association Rules

Frequent +ve	Frequent +ve & -ve
D→B {100, 6}	D→B {100, 6}
B^¬F→D {100,5}	B^¬F→D {100,5}
D^¬F→B {100,4}	D^¬F→B {100,4}
B→D {85,6}	B→D {85,6}
¬A→B {80,4}	¬A→B {80,4}
¬F→B^D {80,4}	¬F→B^D {80,4}
¬F→D {80,4}	¬F→D {80,4}
	D^¬E→B {100,5}
	B→¬E {85,6}
	¬E→B {85,6}
	D→¬E {83,5}
	D→B^¬E {83,5}
	B^¬E→D {83,5}
	B^D→¬E{83,5}
	¬F→B {80,4}
	F→¬E {80,4}

3.2   FPN Classifier

Liu et al. [10] defines the AC problem as a set of training data with n attributes A_1, A_2, ... A_n and $|D|$ rows (cases). Let $C = \{c_1, c_2, ... c_k\}$ be a list of class labels. The specific values of attribute A and class C are denoted by lower case a and c, respectively. The attributes can be a categorical. A classifier is a mapping from $A \rightarrow C$, where A is the set of itemsets and C is the set of classes. The main task of AC is to construct a set of rules (model) that is able to predict the classes of previously unseen data, which are collectively known as the test data, as accurately as possible. In other words, the goal of the task is to find a classifier that maximizes the probability for each test object.

The model was implemented based on AC. Classification is to build a model (called classifier) to predict future data objects for which the class label is unknown. We proposed a new classifier and named it as FPN classifier. FPN classifier is the application of the FPN itemset approach in AC. The steps used in the FPN classifier are adapted from Classification Based Association Rules [10], can be divided into five main steps as shown in Fig 2. The FPN approach is embedded in the first step and the second step to prune weak rules together with correlation measure. As a result, interesting CARs will be generated for considering for building classifier.

- Step 1: The discovery of all frequent itemsets with FPN approach.
- Step 2: The pruning strategy for frequent itemsets to generate CARs.
- Step 3: The selection of one subset of CARs to form the classifier.
- Step 4: The ranking solution for CARs in the classifier.
- Step 5: Evaluating the accuracy of the classifier on the test data.

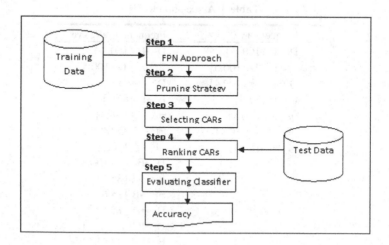

Fig. 2. FPN Classifier Process

4 Experiments and Results

In this section, we describe our experiments in building intrusion detection models on the Universiti Kebangsaan Malaysia (UKM) network traffic data. In these experiments, we applied the FPN approach to mine patterns, select attributes, and build FPN classifiers. Originally, the data consisting of UKM network traffic for the duration of 25 hours and 20 minutes with 18,572,937 transactions and 59 attributes. Bakar et al. [27] preprocessing the data with network time series procedure. The procedure involves four main methods; the discretization of raw data using entropy method, the measure of similarity method using the Principal Component Analysis (PCA), and data representation method using a combination of Piecewise Aggregate Approximation (PAA) and Symbolic Aggregate Approximation (SAA). At the end of preprocessing data, only six attributes were selected; IP address source, IP address destination, size, protocol, source port and destination port. Each attribute has five categories (i.e. 1, 2, 3, 4 and 5) as shown in Table 4. Number of transaction also been decreased by only considered five minutes duration time, which only consist 306 transactions. The dataset label or class was derived from graph plots, which have two types. The first graph type is label 1 for anomaly and the second type was label 0 for normal. The percentage of normal and anomaly transactions is 63.72% (195 transactions) and 36.27% (111 transactions).

Table 4. Attributes description for UKM network data traffic

Attributes	Category
IP address source	1,2,3,4 and 5
IP address destination	1,2,3,4 and 5
Size	1,2,3,4 and 5
Protocol	1,2,3,4 and 5
source port	1,2,3,4 and 5
destination port	1,2,3,4 and 5
Class	0 and 1

Preprocessing data are crucial for network data, which is known to be complex and huge so that the data is prepared/ready in the simple format and suitable for mining. The data dimension is smaller and it helps to make the classifier more efficient in terms of accuracy and processing time. After the data preprocessing, the dataset is needed to be ready for the classification task. A stratified 10-fold cross validation was performed on the dataset to evaluate the performance of the classifier and to evaluate the accuracy of the learning technique. The dataset will be divided into 10 folds. In each fold, the dataset will be divided into training data and test data for 10 times. The fraction of training data and test data for UKM network traffic data as in Table 5. Training data is used to develop the classification model, whereas test data is used to evaluate the classification model.

The classifier classifies new data by pattern matching with rule and gives the class for each new data. The features we constructed were determined to be general enough so that the model is able to detect new variations of the known intrusions. Building accurate and efficient classifiers for large databases depends on training data. We used association rule to generate strong CARs. Then we built our classifier by using selected CARs. The strong CARs are able to produce more accurate classifier. The best model is selected from all folds which have higher accuracy and shorter rule. In general, given a training data set, the task of classification is to build a classifier from the data set such that it can be used to predict class labels of unknown objects with high accuracy.

Table 5. The fraction of training data and test data for UKM network traffic data

Fold	Data fraction	Num. of training data	Num. of test data
1	10:90	31	275
2	20:80	61	245
3	30:70	92	214
4	40:60	122	184
5	50:50	153	153
6	60:40	184	122
7	70:30	214	92
8	80:20	245	61
9	85:15	260	46
10	90:10	275	31

In this experiment the threshold values are 1% for minimum support and 50% for minimum confidence. We also adopted correlation(corr) measures as in Formula (1) for pruning strategy and the minimum value is 1. Incorporating negation into association rule mining is very challenging. Due to the mining space and the huge number of measured rules, the number of absence items is usually enormous compared to the number of presence items. The ratio of the average number of possible items and the number of possible items with negation is huge. The total number of generated positive and negative rules is $4(3^m-2^{m+1}+1)$ of which $3^m-2^{m+1}+1$, roughly only one quarter are positive association rules [28]. Therefore, a pruning strategy is essential to the negative rule mining to reduce the mining space and time without jeopardising the quality of the negative rule. Cohen et al. [29] divided correlation strength into three categories; > 0.5 is large, 0.5-0.3 is moderate and 0.3-0.1 is small. We used 1 as the minimum value for correlation to eliminate rules that are below the minimum value.

$$Corr = \frac{P(AB)}{P(A)P(B)} \qquad (1)$$

Table 6 denotes the best results of accuracy (ACC) based on the FPN classifier for each fold. The number of positive rules (NPR), number of negative rules (NNR) and total number of rules (NR) are also recorded. The FPN classifier almost had a perfect results because the majority of the folds has 100% accuracy except for fold 8 but the accuracy is still high, which is 99.64%. Selection of the best model based on the highest accuracy and less number of rules. Therefore the best model is fold seven with accuracy 100% and the number of rules is 498. In general, FPN effectively produced classifier with 99.96% accuracy and 553 rules.

Table 6. The best accuracy for each fold

Fold	ACC	NR	NPR	NNR
1	100	737	184	553
2	100	697	177	520
3	100	725	182	543
4	100	735	188	547
5	100	727	174	553
6	100	709	183	526
7	100	679	181	498
8	99.64	668	174	494
9	100	726	183	543
10	100	732	184	548

Fig. 3 displays the comparison of number of positive and negative rules. Number of negative rules is more dominant compared to numbers of positive rules, which is four times more. The mining time is increased because of the mining space is big. However, the knowledge generated from negative rules is more important for building the accurate classifier which is nearly 100% accurate.

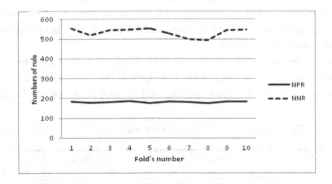

Fig. 3. Comparison of numbers of positive and negative rules

The selected rule from the all folds for UKM network traffic data *is* ¬a5^¬b1→g0 or ¬(IP address source =5) and ¬(IP address destination =1) then normal as denotes in Table 7. The rule means the traffic data with IP source other than category 5 and IP address destination other than category 1 is a normal class data. It shows that with 100% accuracy IP address source which is not in category 5 and IP address destination not in category 1 is not anomaly pattern or intrusion. This condition describes normal network traffic or log pattern.

Table 7. Selected Rules for UKM network traffic data

Rules	Rules Description
¬a5^¬b1→g0	IP address source≠5 and not IP address destination ≠1 then normal
¬a4^¬b1→g0	IP address source ≠4 and IP address destination ≠1 then normal
¬a4^¬e5→g0	IP address source ≠4 and source port≠5 then normal

There are also other rules selected from this experiment as shown in Table 7, such as ¬a4 ^ ¬b1→g0 and ¬a4^¬e5→g0. The significant attributes are important for the knowledge for identification of normal pattern. The attribute combination in the rules indicates that there is no possibility of intrusion happen in the network if IP address source is not in category 4 and IP address destination is not in category 1 or IP address source is not in category 4 and source port is not in category 5. However these rules are not dominant compared to rule ¬a5^¬b1→g0.

Table 8 provides comparisons of the accuracy results between FPN classifier and other algorithms, Rough Set and Naïve Bayes. Each row of the table shows the accuracy measure for each fold in the most left-hand column. The accuracy results for Rough Set and Naïve Bayes are obtained after from running data with data mining tool named Weka. As demonstrated in Table VIII, the results for accuracy based on the FPN approach are very encouraging, which is FPN classifier has the highest accuracy for all folds compared to other algorithms. The FPN classifier has a highest average (AVG) accuracy, 99.96%, followed by Naïve Bayes, 97.03% and Rough Set, 96.91%. The accuracy results for Rough Set and Naïve Bayes is still considered high, which is more than 90% accuracy.

Table 8 also denotes that the amount of rules for FPN classifier is more than Rough Set and Naïve Bayes, due to numbers of negative rules, which is four times more. However, those negative rules are important in building an accurate classifier.

Table 8. The comparison of FPN classifier with Rough Set and Naive Bayes

Fold	FPN Classifier				Rough Set		Naive Bayes
	ACC	NR	NPR	NNR	ACC	NR	ACC
1	100	737	184	553	96.33	133	97.03
2	100	697	177	520	97.96	147	97.03
3	100	725	182	543	95.79	138	97.03
4	100	735	188	547	97.28	157	97.03
5	100	727	174	553	96.26	141	97.03
6	100	709	183	526	96.74	161	97.03
7	100	679	181	498	96.73	99	97.03
8	99.64	668	174	494	96.20	139	97.03
9	100	726	183	543	98.60	151	97.03
10	100	732	184	548	97.20	139	97.03
AVG	99.96	714	181	532	96.91	141	97.03

5   Conclusion

The aim of this study is to improve the intrusion detection problem with negative rules. The FPN itemset approach is proposed to generate important negative rules. The novelty of this research is FPN itemset approach, which is different from traditional Apriori in the sense that FPN considers the frequent absent itemsets for strong negative rules, while Apriori only interested in frequent present items. The experiment with UKM network traffic data demonstrated that the knowledge obtained from strong negative rules which is produced from the FPN itemset approach offers an advantage to the intrusion detection by providing higher accuracy classifier. It is very valuable to get both high accuracy and explainable rules for this to improve our knowledge about the nature of the intrusion. In this paper we use associative classification for the intrusion detection problem. Intrusion detection using associative classification can yield both explainable detection rules and high accuracy.

Acknowledgment. This work was funded by the Ministry of Science and Technology Innovation, Malaysia (UKM-TT-07-FRGS0250-2010) and Public Service Department of Malaysia.

References

1. Ke, F.Y., Yan, F., Lin, Z.J.: Research of Outlier Mining Based Adaptive Intrusion Detection Techniques. In: Knowledge Discovery and Data Mining, pp. 552–555. IEEE (2010)
2. Gomez, J., Dasgupta, D.: Evolving fuzzy classifiers for intrusion detection. In: Proceedings of the 2002 IEEE Workshop on Information Assurance, vol. 6, pp. 321–323. IEEE Computer Press, New York (2002)
3. Tajbakhsh, A., Rahmati, M., Mirzaei, A.: Intrusion detection using fuzzy association rules. Applied Soft Computing, 462–469 (2009)
4. Kruegel, C., Mutz, D., Robertson, W., Valeur, F.: Bayesian event classification for intrusion detection. In: Proceedings of the 19th Annual Computer Security Applications Conference, pp. 14–23. IEEE (2003)
5. Puttini, R.S., Marrakchi, Z., Mé, L.: A Bayesian classification model for real-time intrusion detection. AIP Conference Proceedings, vol. 659. p. 150 (2003)
6. Fugate, M., James, R.G.: Computer intrusion detection with classification and anomaly detection, using SVMs. International Journal of Pattern Recognition and Artificial Intelligence 17, 441–458 (2003)
7. Li, X., Zhang, Y.: Local area network anomaly detection using association rules mining. In: 5th International Conference on Wireless Communications, Networking and Mobile Computing, WiCom 2009, pp. 1–5. IEEE (2009)
8. Xuren, W., Famei, H.: Improving Intrusion Detection Performance Using Rough Set Theory and Association Rule Mining. In: International Conference on Hybrid Information Technology, ICHIT 2006 (2006)
9. Zhang, L., Zhang, G., Yu, L., Zhang, J., Bai, Y.: Intrusion detection using rough set classification. Journal of Zhejiang University-Science A 5, 1076–1086 (2004)
10. Liu, B., Hsu, W., Ma, Y.: Integrating classification and association rule mining. In: Knowledge Discovery and Data Mining, pp. 80–86 (1998)

11. Yin, J., Han, X.: CPAR: Classification based on predictive association rules. SIAM Society for Industrial & Applied, p. 331 (2003)
12. Li, W., Han, J., Pei, J.: CMAR: Accurate and efficient classification based on multiple class-association rules. In: ICDM, pp. 369–376 (2001)
13. Thabtah, F., Cowling, P., Peng, Y.: MCAR: multi-class classification based on association rule. In: Computer Systems and Applications, p. 33. IEEE (2005)
14. Antonie, M.L., Zaïane, O.R.: An associative classifier based on positive and negative rules. In: ACM SIGMOD, pp. 64–69. ACM (2004)
15. Kundu, G., Islam, M.M., Munir, S., Bari, M.F.: ACN: An Associative Classifier with Negative Rules. Science and Engineering, 369–375 (2008)
16. Li, J., Jones, J.: Using multiple and negative target rules to make classifiers more understandable. Knowledge-Based Systems 19, 438–444 (2006)
17. Zhao, Y., Zhang, H., Wu, S., Pei, J., Cao, L., Zhang, C., Bohlscheid, H.: Debt Detection in Social Security by Sequence Classification Using Both Positive and Negative Patterns. In: Buntine, W., Grobelnik, M., Mladenić, D., Shawe-Taylor, J. (eds.) ECML PKDD 2009, Part II. LNCS(LNAI), vol. 5782, pp. 648–663. Springer, Heidelberg (2009)
18. Kamaruddin, S.S., Hamdan, A.R., Abu Bakar, A., Mat Nor, F.: Conceptual Graph Interchange Format for Mining Financial Statements. In: Wen, P., Li, Y., Polkowski, L., Yao, Y., Tsumoto, S., Wang, G. (eds.) RSKT 2009. LNCS, vol. 5589, pp. 579–586. Springer, Heidelberg (2009)
19. Zhang, Y., Jiao, J.R.: An associative classification-based recommendation system for personalization in B2C e-commerce applications. Expert Systems with Applications 33, 357–367 (2007)
20. Fugate, M., Gattiker, J.R.: Anomaly detection enhanced classification in computer intrusion detection. In: Lee, S.-W., Verri, A. (eds.) SVM 2002. LNCS, vol. 2388, pp. 186–197. Springer, Heidelberg (2002)
21. Lee, W., Stolfo, S.J., Mok, K.W.: A data mining framework for building intrusion detection models. In: Proceedings of the 1999 IEEE Symposium on Security and Privacy, pp. 120–132. IEEE (1999)
22. Chen, G., Liu, H., Yu, L., Wei, Q., Zhang, X.: A new approach to classification based on association rule mining. Decision Support Systems 42, 674–689 (2006)
23. Agrawal, R., Imielinski, T., Swami, A.: Mining association rules between sets of items in large databases. ACM SIGMOD Record 22, 207–216 (1993)
24. Agarwal, R.C., Aggarwal, C.C., Prasad, V.V.V.: A tree projection algorithm for generation of frequent item sets. Journal of Parallel and Distributed Computing 61, 350–371 (2001)
25. Brin, S., Motwani, R., Silverstein, C.: Beyond market baskets: Generalizing association rules to correlations. ACM SIGMOD Record 26, 265–276 (1997)
26. Hussain, F., Liu, H., Suzuki, E., Lu, H.: Exception rule mining with a relative interestingness measure. In: Terano, T., Liu, H., Chen, A.L.P. (eds.) PAKDD 2000. LNCS(LNAI), vol. 1805, pp. 86–97. Springer, Heidelberg (2000)
27. Bakar, A., Othman, Z.A., Muda, E.A.E., Hamdan, A.R.: The Time Series Network Traffic Anomaly Detection Using Rough Set Theory. In: Malaysian Joint Conference Artificial on Intelligent (2012)
28. Cornelis, C., Yan, P., Zhang, X., Chen, G.: Mining positive and negative association rules from large databases. In: IEEE Cybernetics and Intelligent, pp. 1–6 (2006)
29. Cohen, J., Cohen, P., West, S.G., Aiken, L.S.: Applied multiple regression/correlation analysis for the behavioral sciences. NJ Eribaum, Hillsdale (1983)

Detect Anchor Points by Using Shared Near Neighbors for Multiple Sequence Alignment

Aziz Nasser Boraik, Rosni Abdullah, and Ibrahim Venkat

School of Computer Science, Universiti Sains Malaysia, Malaysia
anba.cod09@student.usm.my

Abstract. The Multiple sequence alignment (MSA) is a fundamental step for almost all aspects of biological sequence analysis. The reliability and accuracy of sequence analyses depend on the quality of MSA. Including anchor points into multiple sequence alignment to be aligned has been proved to be a good way to increase the quality of MSA. In this paper, we have applied Shared Near Neighbors method to construct the anchor points as partial alignment columns which will be aligned for final output. These anchor points can be used as guide with DIALIGN-TX method to overcome the limitation of DIALIGN-TX to increase the accuracy of final MSA. The results showed 4-8% improvement in the six reference sets in BAliBASE 3.0 benchmark regarding to CS score compared to DIALIGN-TX. In addition, it achieved the highest overall mean Q-score and CS score comparing to other MSA methods in IRMBASE 2.0 benchmark.

Keywords: multiple sequence alignment, anchor points, protein sequences, shared near neighbors.

1 Introduction

In bioinformatics sequence alignment is a way to arrange biological sequences such as proteins, DNA or RNA. This technique is used to find out the similarity among different sequences to detect the similar regions. Sequence alignment technique can be classified into two main approaches; pairwise alignment and multiple sequence alignments (MSA). In pairwise alignment method, it is done with only two biological sequences to detect the related parts among them, while in MSA, it is usually to align more than two biological sequences. The alignment of multiple sequences actually is a fundamental step for many sequence analysis methods, e.g. function prediction, modeling binding sites, pattern identification, phylogenetic tree estimation, sequence database searching and many others [1], [2] and [3]. Hence, the quality of the result from the multiple sequence alignment is the most important factor in sequence analyses accuracy.

The way of finding the similarity among the sequences can be either global alignment or local alignment [4] and [5]. In global alignment methods, the whole input sequences will be aligned from beginning to the end to detect the best alignment, whereas local methods try to align the conserved regions within the input sequences and ignore other low similarity regions or too divergent parts.

S.A. Noah et al. (Eds.): M-CAIT 2013, CCIS 378, pp. 171–182, 2013.

The local alignment information can be included into global alignment methods. This idea was applied in DIALIGN [6], [7] and [8], which was an effective way to increase the alignment accuracy in that method, and is also have been used in some other tools like T-Coffee [9], MAFFT [10] and MUSCLE [11].

Including information to existing MSA has been shown to be a way to improve the accuracy of aligned sequences such as including anchor points from the user to specify some of the input sequences to be aligned, where these inserted sequences are supposed to homologous [12]. The information can be extracted from expert knowledge or from secondary structure predictions [13] and [14]. In addition other resources can be used to include information such as database homology searching to find local alignments [15].

In this paper, we proposed a method to detect partial columns automatically "anchor points" for existing multiple alignment software. Anchor points can be defined as pairs of aligned residues, thus many anchor points will appear like multiple alignment display. The main objective of the proposed method is to build partial alignment columns from pairwise alignments considering local multiple sequence similarities. We used Shared Near Neighbors (SNN) [16] to obtain partial alignment columns. SNN is a clustering algorithm. The similarity in SNN is based on the number of shared neighbors of two objects. Thus, we can benefit from this idea to find the strong pairwise local similarities to detect anchor points considering local multiple sequence similarities. These anchor points can be used as a guide to existing MSA tools to build the final multiple alignment. In this research, to validate the method we introduced anchor points into existing MSA tools which have an option to accept anchor points. Actually, DIALIGN-TX [8] method will be selected because it has an explicit option to include anchor points.

The basic outline of this paper is as follows. The next section provides a brief background of some MSA strategies. Section 3 describes the proposed method. Section 4 shows some experimental results of the proposed method and a brief conclusion.

2   Background and Related Work

Most of the multiple sequence alignment methods use alignment score to estimate the quality of aligned sequences. The formulations of alignment score in MSA are known as objective functions where the MSA methods try to find optimal or near-optimal alignment according to these functions such as sum-of-pairs (SP) score in dynamic programming [17] and [18]. However, under any reasonable objective function this optimization problem is NP-complete [19] to find the best solution and thus it seems to be more intractable when the number of aligned sequences increases.

The dynamic programming technique is applicable to any number of sequences in theoretically where multidimensional dynamic programming is able to find a good solution (brute-force method) but it becomes computationally expensive in both time and memory usage. Therefore, it is impractical for more than a few sequences [20].

For this reason, many heuristic approaches have been proposed to be alternative ways to find a good alignment at reasonable computational cost. One such heuristic approach is progressive strategies, where the basic idea is to build a guide tree based on the similarities among the input sequences and to grow up the MSA by repetitively align pairs of sequences using a dynamic programming algorithm or align sequence profiles along the guide tree. The advantage of this strategy is that it needs less time. However, one most significant weakness in progressive strategies is that if any errors made in early stage in any entire process will be propagated later which may decrease the quality of the final alignment.

Progressive alignment strategies have been applied by various alignment algorithms like ClustalW [5] which is one of the most popular MSA methods for the global alignment approach. In ClustalW, the final alignments are built up from a guide tree, computed by a neighbor joining algorithm [21]. The Weighted Sum of Pair score has been used to consider the weight of sequence and position dependent gap penalties. There are other popular algorithms such as T-Coffee [9], which used consistency-based objective function COFFEE [22] with progressive alignment to reduce the propagated errors caused early stage in progressive alignment. MUSCLE is based on a progressive strategy with iterative algorithm. Where the iterative approach can be started with an initial alignment solution and do refining through a series of iterations to increase the quality of aligned sequences until no more improvements can be found. MAFFT [10] is another progressive method that uses an iterative refinement of the result after applying an initial progressive alignment. in addition to rapid identification of homologous regions it uses the fast Fourier transform (FFT). Muscle [11] is another multiple sequence alignment algorithm. It first performs a progressive MSA phase followed by an iterative method. It is almost as accurate as T-Coffee. It proceeds in three parts: draft progressive using k-mer counting, revises tree from the previous iteration to improve the progressive, and do refinement part by series of deletion of each edge of the guide tree to generate two new trees, then realignment this profile.

Further improvement in MSA comes by using hidden Markov models which has been implemented in ProbCons [23] algorithm. ProbCons is an MSA based on Probabilistic and Consistency which uses combine techniques from Hidden Markov Model (HMM), progressive strategy and iterative refinement approach. In ProbCons, the expected pairwise alignments are calculated by comparing random pairwise alignments using HMM and posterior probability matrices. The guide tree is created through a greedy clustering by calculated the probabilistic value for each alignment. After that, it uses a progressive alignment approach followed by refinement iteration. ProbCons has achieved more accurate results compared to MUSCLE and MAFFT, however the run time is slower than these algorithms [23]. MUMMALS [24] followed the same approach as ProbCons, however it included secondary structure information with more complex HMMs to improve alignment accuracy.

MSAProbs [25] is another practical MSA algorithm designed by combining pair hidden Markov models and a partition function to calculate posterior probabilities. It also uses a progressive strategy for computing multiple protein sequence alignments by performing a progressive strategy along the guide tree using the transformed posterior probability. ProbAlign [26] uses the probabilistic consistency transformation method to build MSA, where residue alignment probabilities are estimated by using a partition function-based method.

DIALIGN [8] and [7] proposed another approach to overcome the weakness of the progressive alignment strategy. DIALIGN builds up the global alignment by composing the local segments that have high similarity. A general problem with the algorithmic approach in DIALIGN is that weakly conserved homologies can be easily missed, thus they may not appear statistically significant in the pairwise alignment. [27]. Therefore to overcome this kind of problem, we have proposed a method that creates a partial column to compos multiple sequence alignments from local multiple sequence similarities.

3   Proposed Method

In this paper, we have proposed a new method of constructing multiple sequence protein alignments from local pairwise alignments. We only deal with primary structure of protein sequences. Therefore, regarding to protein sequence which consists of amino acid units, we will handle with any unit of an amino acid as a single character. Thus, in this paper we will use the word character or point to indicate to amino acid.

The proposed method will start with constructing all pairwise alignments for input sequences. After that we use Shared Near Neighbors method [16] and [28] to obtain partial alignment columns from pairwise alignments. This partial columns will be used as anchor points in DIALIGN-TX method. Anchor points are specific positions in the input sequences that are forced to be aligned to each other in the final multiple alignment. Thus, DIALIGN-TX algorithm will align the reminder unaligned parts of sequences that are not included in our partial alignment. In our method, we will search for groups of positions in the sequences that are connected to each other regarding Shared Near Neighbors similarity measurement. The groups of positions which use to constructs partial alignment columns may not be consistent (see Fig.1). Therefore, we have to consider about consistency conflicts that may appear when we construct partial alignment columns which will be used as anchor points for DIALIGN-TX method.   Thus, we can get better results when using the partial alignment columns directly. The proposed method can be divided into two main steps to construct the final partial columns. Which are the all-against-all pairwise alignment and construction of partial alignment columns.

3.1   All-against-all Pairwise Alignment

The proposed method starts with constructing all pairwise alignments for input sequences by using pair-HMM approach implemented in Probcons algorithm [23] where it is a very accurate program currently available and compatible with our method [29]. Actually, we have the flexibility to use any other method to construct pairwise alignments for input sequences, but the accuracy of output may will be affected by the accuracy of the pairwise algorithm that will be used and it can benefit from more accurate pairwise methods.

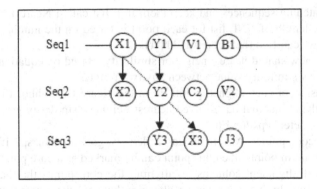

Fig. 1. Conflicts anchor points

3.2   Construction of Partial Alignment Columns

After constructing all pairwise alignments for input sequences. We used Shared Near Neighbors method [16] and [28] to obtain partial alignment columns from pairwise alignments. The Shared Near Neighbors algorithm is a density-based method. It was firstly founded by Jarvis and al. [16] which improved and used for image clustering [28] and for other research such as linguistic purposes in [30].

Shared Near Neighbor algorithm   is an alternative way to calculate direct similarity between any two points. It defines the similarity between any pair of points based on their shared neighbors (see equation 1). Regarding to protein sequence which consists of amino acid units. We will handle with any unit of an amino acid as a single character. Thus, in this paper we will use the word character or point to indicate to amino acid.

Let S={S1,S2,S3, … ,SN} be the set of N sequences over the alphabet. After performing pairwise alignment for all input sequences, each point in each sequence SN will be aligned to at least N-1 points, where any point can be aligned to a point or gap. Let S1 and S2 be two input sequences on S. Let $Xi \in S1$ and $Yj \in S2$ are two points, Let LN be the length of the N-th sequence, where $1 \leq i \leq L1$ and $1 \leq j \leq L2$.

The similarity to any two points Xi and Yj in the sequences S1 and S2 based on shared neighbors can be calculated using the following equation:

$$Similarity\ (Xi,Yj) = size(NN(Xi))\ \cap\ size(NN(Yj))(1) \tag{1}$$

Where NN(Xi) and NN(Yj) in the above equation are the list of aligned points of Xi and Yj. The size is equal to the number of points in each list. This is the basic idea behind Shared Near Neighbors algorithm for similarity measurement between any two points. Therefore, SNN can be used to detect the connected points from local alignment to construct the partial alignment columns by using this measurement. The steps of the SNN algorithms are as the following:

1. Create Nearest Neighbor List (NNL). For each point, we search for only the N-1 aligned points to the current point, where N is the number of sequences. We search

only for matching sequences and keep them in a list called Nearest Neighbor List (NNL). The length of NNL list for each point is based on the number of matching points that we can find (between 0 and N-1).

2. Construct a new shared nearest neighbor similarity. Based on equation (1), we calculate the new similarity value between any two points.
3. Set the density for each aligned point. From the Nearest Neighbor List (NNL) we can detect the points that have shared nearest neighbor similarity greater than user specific parameters epsilon Eps.
4. Search for core points. Any point has a density greater than specific parameter minPts (minimum points) then this point can be marked as a core point.
5. Create partial alignment columns by aligning the core points that have a density greater than Eps. In this step we can make sure that only the points that have strong link are aligned to each other which will be in partial columns.

Conflicts relationships between points may happen when we construct partial alignment columns (see Fig.1). To overcome this problem we can first align the high similarity points by setting the parameter $Eps = \lfloor N/2 \rfloor$. Where in this case there is no way to construct partial columns with conflicts links. After we align all high similarity points, after that we can reduce the parameter Eps to find more anchor points and set a priority to each anchor point and reduce this priority. This priority can be used by DIALIGN-TX to refine the anchor points in case there is a conflict. An example of the output of our method from reference set BBS1101 from BALIBASE 3.0 (see section 3) will be as follows :

| 1 | 5 | 2 | 2 | 1 | 0.97 |
| 4 | 2 | 28 | 27 | 1 | 0.96 |

The first two columns in the table above are the sequence number to be anchored, the third and fourth columns refer to the start positions of the anchored in the mentioned sequences, and fifth column contains a number of points that should be aligned to each other starting from the start position. The sixth column contains the specifies priority compared to other anchoring which use in case there is a conflict links between the anchor points. This output of our method will be used as input for DIALIGN-TX. These anchor points can be used as guide with DIALIGN-TX method.

4   Experimental and Results

The objective of our experiments in this section was to demonstrate the effect of including our partial columns as anchor points into a multiple sequence alignment method to the quality of the final output of MSA. In this paper, we have included the anchor points into DIALIGN-TX algorithm.

4.1   Experiment Design

To evaluate the performance of our proposed method, we have used two benchmark databases for MSA. For global protein sequence alignment, we have used the well-known standard benchmark BAliBASE 3.0 [31]. For local protein sequence alignment, we have used the benchmark IRMBASE 2.0 [7] and [8] (will be discussed in subsection 4.2). We have evaluated the alignment results in terms of Q-score. Therefore, the performance of MSA method can be measured based on two different ways: Q-score (sum-of-pairs score) and true column score (CS). Q-score software can be downloaded from [32]. Q-score can be defined as the number of pairs of amino acid that are aligned correctly in our alignment divided by the number of aligned pairs of amino acid in the reference alignment. The Q-score value is between 0 and 1. a maximum of Q-score value is 1, and when the algorithm fails to align any correct pair of amino acid then the score will be 0. The CS (true column score), also known as the column score, is the number of columns that are aligned correctly in our method divided by the number of aligned columns in the reference.

In general, the Q-score measurement is almost more appropriate to evaluate MSA than the CS score because if there is a single residue is missing to be aligned ,then the score will ignore all correctly aligned residues in that alignment column. However, there are some situations where the column score can be more meaningful than the Q-score. An example of this case in BAliBASE, the reference sets containing orphan sequences.

4.2   Data Sets

As we mentioned before, we have used two benchmark databases to evaluate MSA as the following:

BAliBASE 3.0
The BAliBASE is the database that contains high quality protein multiple sequence alignments. This database has been designed to use as evaluation reference and also a comparison for programs of protein multiple sequence alignment. This database contains 6255 protein sequences with full length and homologous sections. It can be used to compare the quality of global alignments. This database contains 386 reference alignments. BAliBASE reference alignments are grouped into five reference sets as the following:

First reference (RV11 & RV12): consists of equidistant protein sequences with 2 different levels of conservation. RV11 database has the very different protein sequences (less than 20% identity) and there are 76 reference alignments. The data set RV12 has sequences of medium divergent (between 20% to 40% identity) and there are 88 reference alignments.

Second reference (RV20): consists of protein sequences that are highly divergent called "orphan" sequence and there are 82 reference alignments.

Third reference (RV30): consists of subgroups with less than 25% identity between groups. It has 60 reference alignments.

Fourth reference (RV40): Consists of protein sequences with N/C-terminal extensions. It has 49 reference alignments.

Fifth reference (RV50): Consists of sequences of protein with internal insertions. It has 31 reference alignments.

The evaluation on BAliBASE 3.0 will be only scored with core blocks which indicate to biologically correct alignments of the input sequences.

IRMbase 2.0

IRMBASE 2.0 [7] contains 192 alignment reference classified in four groups (ref1, ref2, ref3 and ref4). Where each group contains 48 alignment references. This benchmark consists of simulated conserved motifs that are implanted from non-related and random sequences. IRMBASE was created to test the local multiple protein alignment algorithms.

4.3   Results

We have computed the Q-score and true column score (CS) for all alignments in BAliBASE 3.0 and IRMBASE 2.0. We have compared our strategy to the eight leading multiple alignment programs. Clustalw, MSAprobs, ProbCons, PicXaa, Muscle, Mafft, Probalign and T-coffee. The default parameters were used for all methods. First, we evaluated the accuracy of our proposed method using the BAliBASE data set. Tables 1 contain the Q-scores of multiple sequence alignment obtained by DIALIGN-TX with our anchor points, and DIALIGN-TX without our anchor points as well as some modern aligners. The proposed approach consistently achieved a considerable improvement over the last version of DIALIGN-TX. This is true for both, Q-score and the total column (CS) score (table 2).

For Q-score, our proposed method achieved enhancement 2-5% more. For references RV40 and RV50, alignment methods that can more effectively to detect true aligned points, where our method achieve a considerable improvement over DIALIGN-TX. However, we can see that in RV20, the result from DIALIGN-TX with anchor points got less effectively detect for similarities compared with other references, where RV20 consists of protein sequences that are highly divergent.

Table 1. Q-score of DIALIGN-TX with anchor points on BaliBase 3.0 compared to other usual aligners

Method	RV11	RV12	RV20	RV30	RV40	RV50	Overall
DalignTX with anchor points	60.43	91.60	92.40	81.85	89.42	87.10	83.80
DialignTX	55.49	89.43	89.09	76.90	83.80	82.33	79.51
Clustalw	58.15	88.41	88.77	77.14	78.93	76.91	78.05
Mafft	69.18	93.62	93.50	87.97	92.14	90.14	87.76
MSAprobs	**74.63**	**94.86**	**94.36**	**88.19**	92.54	**90.91**	**89.25**
Muscle	65.75	92.32	91.51	84.24	86.48	85.29	84.26
PicXAA	70.60	94.65	93.30	86.81	**93.23**	89.38	88.00
Probalign	71.05	94.64	93.54	86.44	92.20	89.09	87.83
Probcons	73.65	94.64	93.73	87.62	90.32	90.06	88.34
T-COFFEE	73.16	93.44	93.20	85.75	89.20	90.21	87.49

For the CS score, the improvement is significant where it has been increased around 4-8%. The proposed method could pick up the local similarities among the aligned sequences that are often missed by DIALIGN-TX methods.

Further analysis of the alignment results indicated that the proposed alignment method achieved close to the best result for RV12, RV20, RV40 and RV50 reference sets, in terms of Q-scores. However, our proposed method with DIALIGN-TX is still outperformed on many of the BAliBASE test sequences by T-COFFEE, Muscle, Probcons, Probalign, PicXAA,, Probcons, Mafft and MSAprobs; the latter is the most method is currently the best performing regarding the mean Q-score and CS score overall all multiple aligner on BAliBASE data set.

Table 2. True column score (CS) of Dialign-TX with anchor points on BaliBase 3 compared to other usual aligners

Method	RV11	RV12	RV20	RV30	RV40	RV50	Overall
DalignTX with anchor points	37.41	80.15	43.43	45.57	53.83	53.87	52.38
DialignTX	31.72	76.00	34.69	39.91	45.17	45.66	45.53
Clustalw	32.53	75.59	33.86	38.18	39.82	36.49	42.75
Mafft	48.35	84.35	48.87	60.60	60.65	58.28	60.18
MSAprobs	**53.70**	**87.43**	**54.07**	**63.43**	62.54	**61.38**	**63.76**
Muscle	43.32	82.00	42.22	47.68	45.32	47.50	51.34
PicXAA	47.86	86.81	43.64	58.79	**63.58**	53.93	59.10
Probalign	48.15	86.77	46.69	59.72	60.73	54.35	59.40
Probcons	51.45	86.91	51.00	60.11	54.03	58.92	60.40
T-COFFEE	51.92	85.55	49.19	55.95	54.82	60.36	59.63

In tables 3 and 4, the accuracy of the proposed method was evaluated based on IRMBASE 2.0 alignment benchmark. The average Q-score and CS scores of the proposed methods are given in the first column of Table 3 and Table 4.

Table 3. Q-score of DIALIGN-TX with anchor points on IRMbase 2.0 compared to other usual aligners

Method	ref1	ref2	ref3	ref4	Overall
DalignTX with anchor points	**90.03**	94.33	**94.51**	**94.72**	**93.40**
DialignTX	89.41	**94.90**	93.75	93.73	92.94
Clustalw	7.14	10.61	19.86	26.35	15.99
Mafft	87.63	91.96	89.86	88.50	89.49
MSAprobs	78.20	83.02	87.59	85.10	83.48
Muscle	30.36	34.54	54.00	57.39	44.07
PicXAA	89.89	89.34	89.21	86.97	88.85
Probalign	82.41	82.73	83.57	22.18	67.72
Probcons	79.04	87.23	86.84	88.12	85.31

The result shows that the proposed method with DIALIGN-TX achieves the highest overall mean Q-score and CS scores. It achieved the best result in ref1, ref3 and ref4 for Q-score measurement. Moreover, for CS score, the proposed method achieved the best result in ref4. We can see that DIALIGN-TX shows the best overall performance on IRMBASE 2.0 in ref2, ref3 and ref4 compared with other methods. The proposed method also could get some improvement in some references. The results from IRMBASE 2.0 data set can give us an idea about different algorithms how they deal with locally conserved motifs. The result proves that our method could detect local multiple sequence similarities among the aligned sequences.

Table 4. True column score (CS) of Dialign-TX with anchor points on IRMbase 2.0 compared to other usual aligners

Method	ref1	ref2	ref3	ref4	Overall
DalignTX with anchor points	76.75	84.14	86.28	**87.46**	**83.66**
DialignTX	76.36	**86.10**	**86.45**	85.00	83.48
Clustalw	0.07	2.51	5.69	10.10	4.59
Mafft	**80.38**	82.22	80.46	77.44	80.13
MSAprobs	55.68	55.54	69.76	68.81	62.45
Muscle	7.09	12.86	28.40	38.14	21.63
PicXAA	79.49	75.59	76.55	74.01	76.41
Probalign	55.24	58.48	65.80	18.53	49.52
Probcons	51.11	63.51	66.68	72.23	63.38

4.4   Discussion

In this paper, we have proposed a new method to compose multiple sequence protein alignments from local pairwise alignments. We first use a Probcons algorithm to produce all pairwise alignments for all input sequences. After that we used Shared Near Neighbors to obtain partial alignment columns from pairwise alignments considering local multiple sequence similarities. These partial columns will be used as anchor points in DIALIGN-TX method. Consistency conflicts may accrue when we construct partial alignment columns. Therefore we first search for high similarity points which can be used to build the partial column without any conflict, after that we search for lower similarity and set priority based on the similarity measurement in SNN. This priority value can be used by DIALIGN-TX to refine the anchor points in case there is a conflict.

The major drawback of DIALIGN-TX is using pairwise similarities as a basis for MSA without considering local multiple sequence similarities. We overcome this limitation by using SNN algorithm and focus on measurement similarity for more than two points. This appears to be a promising approach to get over the limitations of DIALIGN-TX. The result showed that we obtained improvement in the six reference sets in BAliBASE 3.0 over the last version of DIALIGN-TX by using our method and It seems that using multiple local alignments instead of pairwise fragments which have been used in DIALIGN can be a good way to increase the quality of MSA.

In future work, we plan to extend our method to find more related anchor points by reducing amino acid alphabet.

Acknowledgment. This research is supported by UNIVERSITI SAINS MALAYSIA and has been funded by the Research University Cluster (RUC) grant titled by "Reconstruction of the Neural Microcircuitry or Reward-Controlled Learning in the Rat Hippocampus" (1001/PSKBP/8630022).

References

1. Edgar, R.C., Batzoglou, S.: Multiple sequence alignment. Current Opinion in Structural Biology 16, 368–373 (2006)
2. Notredame, C.: Recent progress in multiple sequence alignment: a survey. Pharmacogenomics 3, 131–144 (2002)
3. Kemena, C., Notredame, C.: Upcoming challenges for multiple sequence alignment methods in the high-throughput era. Bioinformatics 25, 2455–2465 (2009)
4. Thompson, J.D., Linard, B., Lecompte, O., Poch, O.: A comprehensive benchmark study of multiple sequence alignment methods: current challenges and future perspectives. PloS One 6, e18093 (2011)
5. Thompson, J.D., Higgins, D.G., Gibson, T.J.: CLUSTAL W: improving the sensitivity of progressive multiple sequence alignment through sequence weighting, position-specific gap penalties and weight matrix choice. Nucleic Acids Research 22, 467–480 (1994)
6. Morgenstern, B., Dress, A., Werner, T.: Multiple DNA and protein sequence alignment based on segment-to-segment comparison. Proceedings of the National Academy of Sciences of the United States of America 93, 12098–12103 (1996)
7. Subramanian, A.R., Weyer-Menkhoff, J., Kaufmann, M., Morgenstern, B.: DIALIGN-T: an improved algorithm for segment-based multiple sequence alignment. BMC Bioinformatics 6, 66 (2005)
8. Subramanian, A.R., Kaufmann, M., Morgenstern, B.: DIALIGN-TX: greedy and progressive approaches for segment-based multiple sequence alignment. Algorithms for Molecular Biology: AMB 3, 6 (2008)
9. Notredame, C., Higgins, D.G., Heringa, J.: T-Coffee: A novel method for fast and accurate multiple sequence alignment. Journal of Molecular Biology 302, 205–217 (2000)
10. Katoh, K., Misawa, K., Kuma, K., Miyata, T.: MAFFT: a novel method for rapid multiple sequence alignment based on fast Fourier transform. Nucleic Acids Research 30, 3059–3066 (2002)
11. Edgar, R.C.: MUSCLE: multiple sequence alignment with high accuracy and high throughput. Nucleic Acids Research 32, 1792–1797 (2004)
12. Morgenstern, B., Prohaska, S.J., Pöhler, D., Stadler, P.F.: Multiple sequence alignment with user-defined anchor points. Algorithms for Molecular Biology: AMB 1, 6 (2006)
13. Deng, X., Cheng, J.: MSACompro: protein multiple sequence alignment using predicted secondary structure, solvent accessibility, and residue-residue contacts. BMC Bioinformatics 12, 472 (2011)
14. Subramanian, A.R., Hiran, S., Steinkamp, R., Meinicke, P., Corel, E., Morgenstern, B.: DIALIGN-TX and multiple protein alignment using secondary structure information at GOBICS. Nucleic Acids Research 38, W19–W22 (2010)

15. Thompson, J.D., Plewniak, F., Thierry, J., Poch, O.: DbClustal: rapid and reliable global multiple alignments of protein sequences detected by database searches. Nucleic Acids Research 28, 2919–2926 (2000)
16. Jarvis, R.A., Patrick, E.A.: Clustering Using a Similarity Measure Based on Shared Near Neighbors. IEEE Transactions on Computers C-22, 1025–1034 (1973)
17. Needleman, S.B., Wunsch, C.D.: A general method applicable to the search for similarities in the amino acid sequence of two proteins. Journal of Molecular Biology 48, 443–453 (1970)
18. Waterman, M.S.: Identification of Common Molecular Subsequences. Journal of Molecular Biology, 195–197 (1981)
19. Wang, L., Jiang, T.: On the complexity of multiple sequence alignment. Journal of Computational Biology: A Journal of Computational Molecular Cell Biology 1, 337–348 (1994)
20. Lipman, D.J., Altschul, S.F., Kececioglu, J.D.: A tool for multiple sequence alignment. Proceedings of the National Academy of Sciences of the United States of America 86, 4412–4415 (1989)
21. Saitou, N., Nei, M.: The neighbor-joining method: a new method for reconstructing phylogenetic trees. Molecular Biology and Evolution 4, 406–425 (1987)
22. Notredame, C., Holm, L., Higgins, D.G.: COFFEE: an objective function for multiple sequence alignments. Bioinformatics 14, 407–422 (1998)
23. Do, C.B., Mahabhashyam, M.S.P., Brudno, M., Batzoglou, S.: ProbCons: Probabilistic consistency-based multiple sequence alignment. Genome Research, 330–340 (2005)
24. Pei, J., Grishin, N.V.: MUMMALS: multiple sequence alignment improved by using hidden Markov models with local structural information 34, 4364–4374 (2006)
25. Liu, Y., Schmidt, B., Maskell, D.L.: MSAProbs: multiple sequence alignment based on pair hidden Markov models and partition function posterior probabilities. Bioinformatics 26, 1958–1964 (2010)
26. Roshan, U., Livesay, D.R.: Probalign: multiple sequence alignment using partition function posterior probabilities. Bioinformatics 22, 2715–2721 (2006)
27. Corel, E., Pitschi, F., Morgenstern, B.: A min-cut algorithm for the consistency problem in multiple sequence alignment. Bioinformatics 26, 1015–1021 (2010)
28. Ert, L., Steinbach, M.: Finding Clusters of Different Sizes, Shapes, and Densities in Noisy, High Dimensional Data, pp. 47–58 (2003)
29. Blackshields, G., Wallace, I.M., Larkin, M., Higgins, D.G.: Analysis and comparison of benchmarks for multiple sequence alignment. In Silico Biology 6, 321–339 (2006)
30. Ert, L., Steinbach, M.: Finding Topics in Collections of Documents: A Shared Nearest Neighbor Approach. Performance Computing, 1–20 (2002)
31. Thompson, J.D., Koehl, P., Ripp, R., Poch, O.: BAliBASE 3.0: latest developments of the multiple sequence alignment benchmark. Proteins 61, 127–136 (2005)
32. QSCORE multiple alignment scoring Software, http://www.drive5.com/qscore

A PSO-Based Feature Subset Selection
for Application of Spam /Non-spam Detection

Amir Rajabi Behjat[1], Aida Mustapha[1], Hossein Nezamabadi-pour[2],
Md. Nasir Sulaiman[1], and Norwati Mustapha[1]

[1] Faculty of Computer Science and Information Technology, Universiti Putra Malaysia,
43400 UPM Serdang, Selangor, Malaysia
rajabi.amir6@gmail.com, {aida,nasir,norwati}@fsktm.upm.edu.my
[2] Department of Electrical Engineering, Shahid Bahonar University of Kerman
P.O. Box 76169-133, Kerman, Iran
nezam@mail.uk.ac.ir

Abstract . The difficulties of email spam detection system associated with high
dimensionality in feature selection process and low accuracy of spam email
classification. However, in machine learning, Feature selection (FS) as a global
optimization problem decreases irrelevant and redundant data and creates a set
of acceptable results with high accuracy. This paper presents a feature selection algorithm based on particle swarm optimization (PSO), which decreases
dimensionality and improves the accuracy of spam email classification. PSO as
a computational model fallows the social behavior of bird flocking or fish
schooling. The proposed PSO-based feature selection algorithm searches the
feature space for the best feature subsets. The evolution of feature selected is
determined by a fitness function. The classifier performance and the length of
selected feature vector as a classifier input are considered for performance evaluation using Ling-Spam and SpamAssassin databases. Experimental results
show that the PSO-based feature selection algorithm was presented to generate
excellent feature selection results with the minimal set of selected features to be
caused by a high accuracy of spam email classification based on Multi-Layer
Perceptron (MLP) classifier.

Keywords: feature selection, PSO algorithm, Spam email classification, MLP
classifier.

1 Introduction

Electronic mails (e-mails) are the best way for communication in the world. Recently,
this technology has been a serious problem over the internet. This important problem is
called spam email or junk email that is delivered by different protocol such as simple
mail transfer protocol (SMTP) [1]. The high number of these emails consumes band-
width resource when using network resource. Moreover, they are able to quickly block
or make full server storage space for large sites that have thousands of users. On the
other hand, the high number of spam emails wastes valuable time for important communication needs. Consequently, the spam email as an unwanted email threats govern-
ments [2] and [3]. In general, spam detection is related to a classification problem with

S.A. Noah et al. (Eds.): M-CAIT 2013, CCIS 378, pp. 183–193, 2013.
© Springer-Verlag Berlin Heidelberg 2013

two classes, spam and non-spam. Recently, most of the spam detection models follow machine learning techniques that classify spam emails [4] and [5]. One of the problems that threats spam email classification is the choice of optimal input feature subsets for classifier to be done by feature selection process. However, High data dimensionality problems depended on feature selection decreases the efficiency and accuracy of all classifiers such as neural networks [6]. They believe that limiting the feature space and decreasing the huge number of features in the message can prevent a high dimensionality. In fact it is better to identify features with concepts the document deals with, or with the problems the document challenges [7]. The irrelevant features can influence the accuracy of the learned MLP classification, the time needed to learn a classifier, the number of examples needed for learning, and the cost associated with the features [10] and [8]. Today, The Particular Swarm algorithm (PSO) like as evolutionary algorithm has been applied to select relevant features in spam email detection and optimizes classifier parameters such as artificial neural network. Moreover, PSO is able to implement and solve an extensive range of optimization problems [9]. Therefore, due to the efficiency of PSO in steering large search spaces for optimal solutions (feature space), PSO algorithm is used in this study to optimize feature selection problem in spam email detection system.

In literature, feature selection has been described in different studies, [12], [13], [14], [9] and [11], but the lack of feature and dimensionality reduction increased operation time and classification errors.

In this study, a PSO algorithm is applied to select discriminatory features of body, header and attached files of spam/non-spam emails using MLP classifier in order to decrease dimensionality and increase classification accuracy. In general, the contribution of this study is:

— Evaluation PSO algorithm using SpamAssassin and LingSpam datasets and comparing its performance with other heuristic algorithm such as immune and GA algorithm based on feature selection method in spam email detection system.
— The classification of spam emails with the minimum number of features for reduction of computational complexity and time that increase the classification accuracy.

The remainder of this paper is organized as follows: the principle of particle swarm optimization is given in Section 2. Section 3 presents basic Multi-layer perceptron neural network. Section 4 and 5 describe feature extraction and feature selection processes. In section 6 the proposed PSO based on feature selection algorithm is presented. Section 7 describes experimental results from using the proposed feature selection method and section 8 summarizes this paper.

2   The Principles of Particle Swarm Optimization (PSO)

Dr. Eberhart and Dr. Kennedy have proposed Particle Swarm Optimization in 1995 [14]. This algorithm is based on the behavior and swarming of bird flocking or fish schooling that is able to optimize different fields such as data clustering and optimization of artificial neural network [12],[16],[17]. In this algorithm, each particle is such as a point in D-dimensional space, so the ith particle is represented as $X_i = (x_i1, x_i2, ... x_{is})$. Due to PSO calculates the best fitness rate (pbest) according to previous position of each particle, thus this rate is $P_i = (p_i1, p_i2, ... p_is)$ for any particle.

The global best and velocity of particle i are 'gbest' and $V_i = (v_i1, v_i2, ... v_{is})$ respectively Fig.1. The manipulation of each particle is continued as below:

$$vid = w*v_{id}+c1*rand \, ()*(p_{ad}-x_{id}) +c2*Rand \, ()*(p_{ad}-x_{id}) \tag{1}$$

$$x_{id} = x_{id}+ v_{id} \tag{2}$$

Where w is the inertia weight, c1 and c2 are the stochastic acceleration weighting that leads particles toward pbest and gbest positions. rand () and Rand () are the random functions between [0,1]. V_{max} shows the velocity of each particle [9],[18].

2.1   Binary Particle Swarm Optimization (BPSO)

The BPSO follows the action of chromosomes in genetic algorithm, so it is coded such as a binary string. In the specific dimension, the particle velocity is used like as probability distribution that has the main role to produce the particle position randomly. Updating the particle position follows below equation.

$$\text{If } rand_3 < \frac{1}{1+e^{-v_i^{i+1}}} \text{ then } X_i^{i+1}=1; \text{ else } X_i^{i+1} = 0 \tag{3}$$

In each dimension, a bit value {1} shows the selected feature can participate for the next generation. On the other hand, a bit value of {0} is not required as a relevant for next generation [18].

3   Multi-Layer Perceptron Neural Network (MLPNN)

Today, most researchers have been using MLPNN for its ability to learn complex data structures and work fast with large amount of data. A set of small processing units build neurons of multi-layer back propagation that are arranged in different layers, namely input, hidden and output layers. In fact, these layers are organized to minimize appropriate error functions by a set of parameters such as mode of learning, information content, activation function, target values, input normalization, initialization, and learning rate. The error propagation is content of forward pass and backward pass, so forward pass fixes network weights and backward pass adjusts weights according to error-correction tools. Lastly, the actual results are compared by adjusting the weights during the learning process to accomplish the classification [20], [21], [22], [23] and [24].

4   Feature Extraction

Feature extraction can have an effective role in development of classification process. In this study 156 extracted features by [26] are considered a set of words features, symbols and converted images to the text, which give a clear difference between two spam and non-spam classes. Due to the aim of feature extraction to identify the kind of email or message in the data set (spam or non-spam), in [27] the stop words and HTML tags are discarded that decrease high dimensionality such as 'he', 'she', 'it', 'that' and 'a'. On the other hand, Feature selection of this study is based on genetic algorithm (BPSO), thus the numeric vector of features is a necessary need that is done after feature extraction process.

5   Feature Selection

Feature selection includes a set of K relevant extracted features from D dataset. Considering the dataset as the entire set of emails means that the feature can be included words, image, HTML tags pre-processed [25] and [31]. In other words, feature selection chooses a set of necessary subset features to classify features of a large dataset [26]. One the important purpose of feature selection is to reduce the high dimensionality of the feature space and select some features with high weights. Each email or message consist a set of numeric or categorical attributes (h(1),h(2),h(3),...,h(m),h(m+1)) where m shows predictive attributes and h(m+1) is a class of emails, namely spam and non-spam.

6   Proposed PSO Based Feature Selection Algorithm in Spam Detection System

The binary particle swarm Optimization (BPSO) searches the best relevant features of a dataset in the feature space. The particle position can be a candidate solution, which is divided into two classes (spam and non-spam). The evaluation of the algorithm and feature selection process depends on fitness function that indicates the expected fitness on the selected features.

6.1   Initialization

The initial population is produced by distribution of particles in the search space and the coding of each particle is done randomly, so they produce a binary vector such as n= 1, 2,..., m; where m is length of the feature vector extracted. The gens of this m-dimensional vector signify the feature selection. As a matter of fact, '1' means the selected feature can participate in the classification and '0' rejects the attendance of feature in this process. The velocity of each position is calculated by equations (1-2) and is condensed in a range [0, 1] by Eq.3.

6.2   Fitness Function

In this stage, the selected particles are evaluated and returned. In fact, in each iteration, fitness function evaluates the ability of particles that id related to increase the separation of classes among the different classes. The following fitness function shows the performance and ability of algorithm for selection of the best features.

$$\text{Fitness} = \alpha * \gamma \ (F_i(t)) + \beta * \frac{|N|-|F|}{|N|} \tag{4}$$

Where $F_i(t)$ show the selected feature by particle I in iteration t, $\gamma \ F_i(t)$ is the quality of classifier for selected features |F| indicates the length of selected features and |N| is the total number of features. β and α are the most important parameters that show the importance of feature length and classification quality, so $\alpha \in [0, 1]$ and $\beta = 1 - \alpha$. the selected features are classified during classification process that separate features into the classes, namely spam and non-spam classes. This process is done by MLPNN classifier as an evaluator of PSO fitness function. Actually in n-dimensional dataset, PSO is able to choose any number of features and identifies the value of each feature.

After that, MLP classifier evaluates this produced data dimension. MLP fitness value is measured by 10-fold cross validation method. For doing this method, this study separates data into the 10 parts and did training set with 9 parts; however the testing part is followed by 1 part of data. In each iteration, 90% of the data is used for training while the 10% is used for its test. Totally, the number of classifier runs is 10 times for training and testing that the average of these runs indicates the desired accuracy for PSO-based feature selection in spam detection system. In this study, the accuracy of classifier plays an important role in feature selection, because the accuracy of spam detection system is based on classification accuracy that decreases the rate of errors, so the parameters of PSO algorithm are set as $\alpha = 0.80$, $\beta = 0.13$. The population size = 20, normally, C1 and C2 are set to 2 and the weight values are 1. In addition, the training stage of MLP classifier is tested with different neurons from 3 to 15. The transfer functions of hidden layer and output layer are labelled 'tansig' and 'purelin', respectively. The training function is 'trainlm'. The network is trained for a maximum of 60 epochs to 0.01 of error goal. Furthermore, updating particles is done using two ways, namely 'pbest' and 'gbest'. pbest is the best fitness or solution of each particle in each iteration that is stored. On the other hand, gbest is the best fitness of each particle, when the whole population or swarm becomes a neighbor of that particle. The algorithm of PSO-based feature selection in spam detection system is prepared in Fig.1.

7   Experimental Result

7.1   Data Sets

The platform used in this study is based on Intel Pentium IV 2.7 GHz CPU, 4 GB RAM, windows 7 operating system. The development environment is Matlab (R2010b). To make the comparison between feature selection methods applied in [19] [26], the same classification datasets are used to be classified by MLP based on BPSO feature selector. During the classification process, 90% of data are applied in training set and 10% are used for testing set. In fact, features used in the training and testing sets are extracted from SpamAssassin and LingSpam datasets that during the pre-processing step is discarded their stop words and HTML tags of emails such as 'he', 'she', 'it', 'that' and 'a'. Note that the features selected in these datasets are extracted from subject, body, and picture and files attached of spam and non-spam emails and are normalized before running on the algorithm. The characteristics of these datasets are summarized in Fig. 2.

LingSpam Data Set: The Ling-Spam dataset is content of 481 spam emails and 2171 non-spam emails with the standard bag-of-words representation. In this dataset, each e-mail is divided into two sections: the words found in the subject header and the words found in the main body of the message. In this study, we mixed both parts, because some features of subject and body can have a direct effect on classification process.

SpamAssassin Data Set: The SpamAssassin (http://spamassassin.apache.org) applied in this study is content of 6000 spam and non-spam emails. Basically, this dataset is divided into three parts of email, namely body of email, subject of email and attached pictures of the email that in this study BPSO selects 70 robust features from these parts that is able to increase the accuracy of MLP classifier. Not that the pictures attached in this data set are converted to the text file, after that the features election is done on them.

The performance of feature selection algorithms is dependent on classifier accuracy and fitness function. In order to increase the accuracy of classifier, we set MLPNN parameters as the nodes of hidden layer that are tested from 3 to 15, output layer followed two nodes, first node indicates spam e-mail and second node is non-spam e-mail, the transfer functions of hidden layer and output layer are 'tansig' and 'purelin', respectively, the training function is 'trainlm'. Performance function is MSE and the network is trained for a maximum of 60 epochs to 0.01 of error goal. The parameters of PSO are set such that $\omega_{max} = 0.8$ and $\omega_{min} = 0.4$, C1=2 and C2=1.5 as learning are 2, maximum and minimum velocities are 7 and -7 (V_{max} and V_{min}) respectively and t_{max} as the number of iteration is 200. Although tuning the classifier and feature selection algorithm parameters are challenging steps that are achieved after testing and selecting different parameters, there are different studies that help to find the best parameters for PSO based feature selection [17] and [18].

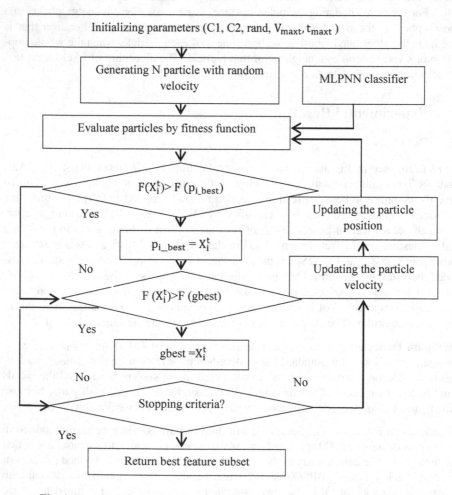

Fig. 1. The algorithm of PSO-based feature selection in spam detection system

7.2   Performance Measurement

Three classic measures such as accuracy, recall, and precision and miss rate are applied to evaluate performance of detection system. Accuracy measures the corrected massages. Recall is a percentage of spam massages that are identified correctly to the total categorized spam massages. Precision is defined as a proportion of properly-classified spam massages to the classified spam massages. Miss rate is percentage of classified non-spam massages incorrectly to the original number of non-spam massages. In this paper, we form a relationship with [25] and show the results of Accuracy, Recall, Precision and miss rate in Fig. 3.

7.3   Results and Discussion

In this study, the PSO algorithm searched d feature from N features in the feature space. The performance of fitness function is identified by classic measurements such as accuracy, recall, precision and miss rate. In the next step, PSO algorithm searches best features to be able to increase fitness function (classification performance). Based on Fig. 1 and Fig. 3 the PSO based feature selection in comparison with other feature selection algorithms applied by different classifier such as SVM, BP Neural Network, Naive Bayesian and Linear Discriminant increases performance with less iteration. In fact this algorithm proved that has an ability to better feature selection with 26 for SpamAssassin and 34 for LingSpam datasets. Furthermore, it turns out from Fig. 4 that MLP classifier decrease the miss rate comparing the classifiers applied in previous studies based on PSO as a feature selection method.   Fig. 3 presents the results of different classifier and feature selection method applied in [27], [28], [29], [30] and [18]. The training and testing stages are done by different size of features, so the result with less number of features shows a better accuracy in comparison with increasing the number of features. The results indicate that PSO is so competitive with other algorithms such as GA and IG [27] and [13]. However, accuracy of the other algorithms can be close to that PSO algorithm. Based on Fig. 3, the result shows that PSO has an ability to decrease the number of features and high dimensionality comparing another algorithm in other studies. Fig.1 plots the average performance of classification process based three measurement methods (Accuracy, Recall and Precision). This figure illustrates a reasonable performance and the upper peak in accuracy comparing applied classifiers in previous studies. Then, each subset is chosen for testing and the remaining subsets are applied as training sets. In fact this stage is continued where each subset is selected as a test set one time.   PSO based feature selection is run 10 times and the average of 10 runs is calculated as an accuracy rate. Our method in this study obtains better performance. To illustrate, Naive Bayesian

Number	Dataset	Dataset Elements	Training Data	Testing Data	Components	Number of Feature	Classes
1	SpamAssassin	6000	5400	600	(H, S)	26	2
2	LingSpam	2589	2330	259	(H, S)	34	2

Fig. 2. The Characteristics of LingSpam and SpamAssassin Datasets

Method		Accuracy (%)	Recall (%)	Precision (%)	Miss Rate (%)
Linear Discriminant		98.55	91.49	99.77	0.04
SVM (Linear Kernel)		99	96.26	97.72	0.45
SVM (RBF Kernel)		99.2	81.90	99.00	0.17
BP Neural Network		99.13	95.84	98.93	0.21
MLPNN	SpamAssassin	99.98	100.00	98.87	0.02
	LingSpam	99.79	97.68	98.45	0.06

Fig. 3. Performances of Linear Discriminant, SVM, and BP Neural Network and MLP NN on LingSpam and SpamAssassin datasets

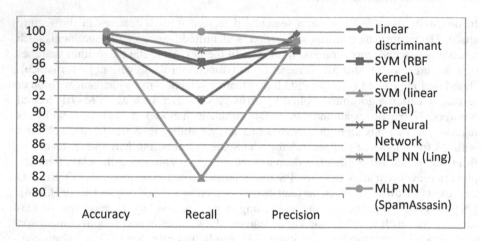

(a)   Accuracy, Precision, Recall

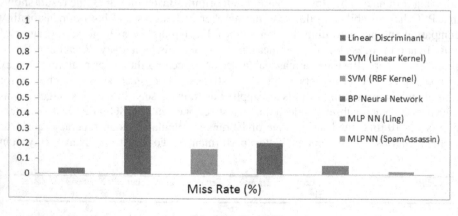

(b)   Miss Rate (%)

Fig. 4. Comparison of performances on corpus LingSpam among Linear Discriminant, SVM (Linear Kernel), SVM (RBF Kernel), BP Neural Network, MLP NN

classifier in [27] classified 100 robust features with 99% accuracy, while the proposed PSO in this study selects 56 features with 99.79% accuracy rate based MLP. The accuracy result in Fig. 3 shows that PSO algorithm is competitive with the heuristic algorithms such as GA and IG method.

In the feature selection problem, this study selected 56 relevant features within 156 features extracted in [27] in order to decrease high dimensionality and increase the accuracy of applied MLPNN classifier. This study compared its results with other methods and show that these results can be competitive, especially for GA algorithm. Our result recorded the high performance of MLPNN classifier with the reasonable number of features comparing other methods. However, there is a challenge associated with feature extraction of the proposed datasets. Actually, this study used extracted features in [26] that some of them could not be effective features during the feature selection process to reduce accuracy, on the other hand most of the authors did not report feature applied in their studies to limit the comparison of applied result in this study with them.

8   Conclusion

In this work, the PSO based feature selection algorithm selects relevant features in the feature space to solve feature subset selection problems. Our algorithm is a binary version of the PSO algorithm to code feature subsets in binary strings. Furthermore, this study applied MLPNN classifier to identify the best fitness function and high accuracy for increasing detection rate in the spam detection system. The PSO result showed that the proposed algorithm is enabling to decrease high dimensionality in the large space with the high number of features. On the other hand, PSO algorithm can be a competitive method in accuracy with other heuristic algorithms for feature selection problems. A future expansion of this study is to extract the number of relevant features that increase accuracy more than the obtained accuracy in this study. In addition, an application of other heuristic algorithms can improve the classification accuracy for spam detection systems and can be a competitive way with the proposed algorithms.

References

1. Wu, Q., Wu, S., Liu, J.: Hybrid model based on SVM with Gaussian loss function and adaptive Gaussian PSO. Engineering Applications of Artificial Intelligence, 487–494 (2010)
2. Sang, M.L., Dong, S.K., Ji, H.K., Jong, S.P.: Spam Detection Using Feature Selection and Parameters Optimization. In: International Conference on Complex, Intelligent and Software Intensive Systems, Krakow, Poland, pp. 883–888 (2010)
3. Nitin, J., Bing, L.: Analyzing and Detecting Review Spam. In: Seventh IEEE International Conference on Data Mining, Omaha, NE, pp. 547–552 (2007)
4. Michalak, K., Kwasnicka, H.: Correlation-based Feature Selection Strategy in Neural Classification. In: Sixth International Conference on Intelligent Systems Design and Applications, Washington, DC, USA, pp. 741–746 (2006)

5. Matthew, Chung, K.P.: Using phrases as features in email classification. The Journal of Systems and Software, 1036–1045 (2009)
6. Huang, C.L., Wang, C.J.: A GA-based feature selection and parameters optimization for support vector machines. Expert Systems with Applications, 231–240 (2006)
7. Zahran, B.M., Kanaan, G.H.: Text Feature Selection using Particle Swarm Optimization Algorithm. World Applied Sciences Journal 7, 69–74 (2009)
8. Alper, U., Alper, M., Ratna, B.C.: mr 2PSO: A maximum relevance minimum redundancy feature selection method based on swarm intelligence for support vector machine classification. Information Sciences, 4625–4641 (2011)
9. Tu, C.-J., Li, Y.C., Jun, Y.C., Cheng, H.Y.: Feature Selection using PSO-SVM. International Journal of Computer Science (2007)
10. Wang, I.R., Youssef, A.M., Elhakeem, A.K.: On Some Feature Selection Strategies for Spam Filter Design. In: IEEE Electrical and Computer Engineering, Canadian, pp. 2186–2189 (2006)
11. Islam, R.M., Chowdhury, M.U., Zhou, W.: An Innovative Spam Filtering Model Based on Support Vector Machine. In: International Conference on Computational Intelligence for Modeling, Control and Automation, Vienna, Austria, pp. 348–353 (2005)
12. Sirisanyalak, B., Sornil, O.: Artificial Immunity-Based Feature Extraction for Spam Detection. In: International Conference on Software Engineering, Artificial Intelligence, Networking, and Parallel/Distributed Computing, pp. 359–364 (2007)
13. El-Alfy,EL.M.: Discovering Classification Rules for Email Spam Filtering with an Ant Colony Optimization Algorithm. In: IEEE Evolutionary Computation, Trondheim, pp. 1778–1783 (2009)
14. Kennedy, J., Eberhart, R.: Particle swarm optimization. In: IEEE International Conference on Neural Networks, Perth, WA, pp. 1942–1948 (1995)
15. Valle, Y.D., Venayagamoorthy, G.K., Mohagheghi, S., Hernandez, J.C., Harley, R.G.: Particle Swarm Optimization: Basic Concepts, Variants and Applications in Power Systems. In: IEEE Evolutionary Computation, pp. 171–195 (2008)
16. Lai, C.H.: Particle Swarm Optimization-Aided Feature Selection for Spam Email Classification. IEEE, Kumamoto (2007)
17. Ramadan, R.M., Abdel-Kader, R.F.: Face Recognition Using Particle Swarm Optimization-Based Selected Features. International Journal of Signal Processing, Image Processing and Pattern Recognition 2(1), 51–66 (2009)
18. Soranamageswari, M., Meena, C.: Statistical Feature Extraction for Classification of Image Spam Using Artificial Neural Networks. In: International Conference on Machine Learning and Computing, Bangalore, pp. 101–105 (2010)
19. Soranamageswari, M., Meena, C.: An Efficient Feature Extraction Method for Classification of Image Spam Using Artificial Neural Networks. In: International Conference on Data Storage and Data Engineering, Bangalore, India, pp. 169–172 (2010)
20. Vafaie, H., Jong, K.D.: Genetic Algorithms as a Tool for Feature Selection in Machine Learning. In: Fourth International Conference on Tools with Artificial Intelligence, Arlington, VA, pp. 20–23. IEEE (1992)
21. Perez, F.M., Gimeno, F.J.M., Jorquera, D.M.M., Abarca, J.A.G.M., Morillo, H.R., Fonseca, I.L.: Network Intrusion Detection System Embedded on a Smart Sensor. IEEE Industrial Informatics 58 (2011)
22. Tretyakov, K.: Machine Learning Techniques in Spam Filtering. Data Mining Problem-oriented Seminar, MTAT.03.177, pp. 60–79 (2004)

23. Carpinteiro, O.A.S., Lima, I., Assis, J.M.C., de Souza, A.C.Z., Moreira, E.M., Pinheiro, C.A.M.: A Neural Model in Anti-spam Systems. In: Kollias, S.D., Stafylopatis, A., Duch, W., Oja, E. (eds.) ICANN 2006. LNCS, vol. 4132, pp. 847–855. Springer, Heidelberg (2006)
24. Pazoki, A., Pazoki, Z.: Classification system for rain fed wheat grain cultivars using artificial neural network. African Journal of Biotechnology 10(41), 8031–8038 (2011)
25. Ruan, G., Ying, T.: A three-layer back-propagation neural network for spam detection using artificial immune concentration. Soft Computing 4(2), 139–150 (2010)
26. Androutsopoulos, I., Koutsias, J., Chandrinos, K.V., Spyropoulos, C.D.: An experimental comparison of Naive Bayesian and keyword-based anti-spam filtering with personal e-mail messages. In: Proceedings of the 23rd ACM SIGIR Conference on Research and Development in Information Retrieval, New York, USA, pp. 160–167 (2000)
27. Androutsopoulos, I., Koutsias, J., Chandrinos, K.V., Paliouras, G., Spyropoulos, C.D.: An evaluation of Naive Bayesian anti-spam filtering. In: 11th European Conference on Machine Learning, Barcelona, Spain, pp. 9–17 (2000)
28. Clark, J., Koprinska, I., Poon, J.: A neural network based approach to automated e-mail classification. In: Proceedings of International Conference on Web Intelligence, pp. 702–705. IEEE (2003)
29. Koprinska, I., Poon, J., Clark, J., Chan, J.: Learning to classify e-mail. Information Sciences 177, 2167–2187 (2007)
30. Alper, U., Alper, M.: A discrete particle swarm optimization method for feature selection in binary classification problems. European Journal of Operational Research 206(3), 528–539 (2010)
31. Martin, S., Nelson, B., Sewani, A., Chen, K., Joseph, A.: Analyzing Behavioral Features for Email Classification

The Effect of Normalization
for Real Value Negative Selection Algorithm

Mohamad Farhan Mohamad Mohsin, Abdul Razak Hamdan,
and Azuraliza Abu Bakar

Faculty of Information Science & Technology,
Universiti Kebangsaan Malaysia, Selangor, Malaysia
farhan@uum.edy.my, {arh,aab}@ftsm.ukm.my

Abstract. The preliminary information of data being normalized into 0 and 1 is essential for an accurate data mining result including real value negative selection algorithm. As one class classification, only the self sample is available during normalization; therefore there is less confidence it fully represents the whole problem when the non-self sample is unknown. The problem 'out of range' arises when the values of data being monitored exceed the boundary as the setting in the normalizing phase. This study aimed to investigate the effect of normalization technique and identify the most reliable normalization algorithm for real value negative selection algorithm mainly when the non-self is not available. Three normalization algorithms – the min max, soft-max scaling, and z-scores were selected for the experiment. Four universal datasets were normalized and the performance of each normalization algorithm towards real value negative selection algorithm were measured based on five key performance metrics-detection rate, specificity, false alarm rate, accuracy, and number of detector. The result indicates that the real value negative selection is highly relied on type of normalization algorithm where the selection of appropriate normalization approach can improve detection performance. The min max is the most reliable algorithm for real value negative selection when it consistently produces a good detection performance. Similar to Z-score it also has similar capability however min max seems to a better approach in term of higher specificity, lower false alarm rate, and fewer numbers of detectors. Meanwhile, the soft max scaling is found not suitable for real value negative selection algorithm.

Keywords: Negative Selection Algorithm, Min Max, Soft Scaling, Z-Score.

1 Introduction

Normalization is a part of data transformation where data are normalized into appropriate form for mining. Also known as standardization of variables, the purpose is to rescale the input value into specific output range such as 0.0 to 1.0. For certain data mining algorithms, this process is compulsory to perform. Such algorithms like neural network, clustering, and distance based algorithms, the normalized input variables could impact their learning speed and produce better mining result. Another algorithm

S.A. Noah et al. (Eds.): M-CAIT 2013, CCIS 378, pp. 194–205, 2013.
© Springer-Verlag Berlin Heidelberg 2013

that relies on the normalized input variables is the real value negative selection algorithm of artificial immune system.

Real value negative selection algorithm (r-NSA) uses real value vector of features U with n attributes to represent the problem space donated as $U = [0,1]^n$. The aim of r-NSA is to generate a set of mature T-cell detector in the complementary space for continuous data where all data is required to be in the normalized range 0.0 to 1.0. The approach differs from its preliminary version of where detector is represented as a binary detector. Tailored to biological negative selection process, r-NSA offers different learning mechanism where it requires only self sample to produce a detection model. Throughout the learning, the self sample will discriminate any randomly generated immature T-cell that has similar shape to the self sample. As the result, a model representing set of mature T-Cell is generated. For future implementation, r-NSA relies on the mature detectors to detect abnormal features. This type of learning is beneficial for the problem where only normal sample is available in most of the time (Liang et al., 2012).

To generate detector, the self sample representing the population of normal group is required to be normalized between 0 and 1. There are various normalization algorithms available however the most commonly chosen approach among researchers is min-mix normalization as it is a direct approach to transform value into 0 to 1. In data mining, this approach is appropriate to apply when the minimum and maximum boundary is available and no extreme value dominated the input data [6]. Therefore, to have a better representation in r-NSA, both boundary limits must be precisely determined in order to represent the whole population.

Since r-NSA is heavily relied only on self sample during training, there is less confidence the self sample fully represents the whole problem mainly when the unseen non-self sample is unknown. The problem 'out of range' arises if the values of data being monitored exceed the boundary as used in the training phase. It is an advantage if the non-self sample is available during training however in most situations; the non-self sample is hard to be seen and the boundary of non-self sample can be taken into consideration. As practice in many r-NSA researches, the suggestion from expert is used to determine the boundary limit likes allocating +20% extra reservation space for min and max values [3]. Any value that lies outside the expert's given range (if value exceeds the boundary range), a default normalized value 1.0 (when it exceeds the maximum) or 0.0 (if less than minimum boundary) is given.

Although the min-max is the most selected approach, there are very less researchers tested other normalization approaches which are claimed to be more robust towards data distribution effect. In this paper, we experiment several normalization approaches including min max, soft-max scaling, and z-scores for r-NSA. We investigated the effect of normalization towards detection performance and seek the most reliable normalization algorithm for r-NSA. We define the reliable if the normalization algorithm able to produce a consistent detection result and do not harm the information content even the non-self sample is not considered during the training. To meet the requirement of r-NSA, we also extended the standard z-score function with logistic function such that the function able to generate normalized value from 0.0 to 1.0. For experiment, several UCI datasets with different characteristic were chosen and the experiment was divided into two groups; (1) by considering the self and non-self; with the assumption the non-self sample is available during normalization (2) by considering only self sample without non-self information. Then the normalized dataset is mined with V-Detector algorithm [7].

This paper is organized as follows. Section 2 outlines the real negative selection algorithm. Then, the normalization method is discussed in section 3. It will be followed by a discussion on the experiment setup of the study in Section 4. In Section 5, the experiment and result of the study will be presented and final sections conclude this work.

2   Real Value Negative Selection Algorithm (r-NSA)

Negative selection of immune system involves the T-cells maturation and discrimination of self/non-self in thymus. Inspiring from both principles, NSA was introduced into the world initially in computer virus detection [4] and later in various areas related to anomaly detection problem such as fraud [5], fault [9], and intrusion [8, 11]. The basic concept of NSA is depicted in Fig. 1.

r-NSA is a branch of NSA. It uses real value vector of features U with n attributes to represent the problem space; it is donated by $U = [0,1]^n$. The self sample is donated as $Self \subseteq U$ and the $Nonself \subseteq U$ be the complementary of $Self$, such that $Self \cap Nonself = \emptyset$, $Self \cup Nonself = U$. The aim of r-NSA is to generate a set of mature T-cell detector in the complementary space for continuous data. Each detector $d = (c_d, r_d)$ has a center $c \in [0,1]^n$ and assigned with a non-self radius $r_d \in \mathbb{R}$. An element z is said to be lie under the coverage of non-self if the $distance(c_d, z) < r_d$. Similar to d, the self element $s = (c_s, r_s)$ also has a center c_s and a self radius r_s where an element z lies close to the self center will be considered as self elements [2].

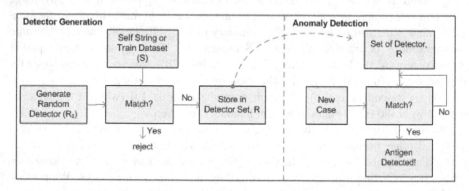

Fig. 1. The basic concept of NSA [4]

3   Normalization Approaches

Methods of normalization are many. In this study, we tested 3 normalization algorithms – min max, soft max scaling, and z-score. The min max and soft max scaling method produce the normalized value between 0 and 1 while the standard z-score does not have specific output range. To make it simpler, the basic representation is given as follows. A set of dataset, D with n-number of records and m-attributes is represented in $n \times m$-matrix. The x_{nm} of the metric represents the n-th record with m-th attribute value for the n-record. D donates a raw dataset before normalization while $\bar{\bar{D}}$ is after normalization.

$$D = \begin{bmatrix} x_{11} & x_{12} & x_{1m} \\ x_{21} & x_{22} & x_{2m} \\ ... & ... & ... \\ x_{n1} & x_{n2} & x_{nm} \end{bmatrix}$$

Based on the matrix, we define the type of normalization used in this study as follows:

a.   Min Max Normalization (MM)

MM is a linear normalization on the original self data as shown in equation (1). Based on the min and max value of attribute m, each value x of attribute m is map to a new range where the new_Max is set to 1.0 and new_Min as 0.0. The MM has a drawback called 'out of range' if the unseen non-self value lies outside the original self data range. As solution in this study, a reservation space value +20% will be reserved for min and max values. Any value that exceed the max or min boundary, the default normalized value will be given; 1.0 when it exceeds the max boundary and 0.0 if below the min boundary.

$$\bar{\bar{x}}_{nm} = \frac{x_{ij} - \min(m)}{\max(m) - \min(m)} (new\_Max - new\_Min) + new\_Min \tag{1}$$

b.   Soft Max Scaling (SMS)

SMS is mostly applied in many neural network projects where the final normalization output is between 0 and 1. There are two steps involved. First, the value to be normalized x_{nm} is first transform into new range depend on the distance between mean $\bar{x}_m$ and standard deviation σ_m and the values located far from the mean are map to exponentially greater degree. As shown in equation (2), this function assumes all data is 99.7% linearly distributed such that $\lambda = 3$ represents data has 3σ.

$$\check{x}_{nm} = \frac{(x_{nm} - \bar{x}_m)}{\lambda(\frac{\sigma_m}{2\pi})}, \lambda = 3 \tag{2}$$

In the second step, the transformed value $\check{x}_{nm}$ from equation 2 will be presented into the logistic function since the $\check{x}_{nm}$ is not within 0 and 1. Acknowledged as sigmoid function, the function generates s-curve shape where all the values are plotted within the 0.0 and 1.0. The logistic function is depicted in equation (3).

$$\bar{\bar{x}}_{nm} = \frac{1}{1 + e^{-\check{x}_{nm}}}$$ (3)

C.   Z-Score

Similar to SMS, the value for x_{nm} is normalized based on mean $\bar{x}_m$ and standard deviation σ_m. This approach is useful when the actual min and max value of attribute m is unknown and when outliers dominate the data. The equation 4 shows the ZS formula. The original output of ZS is not within the 0.0 to 1.0; therefore it is not suitable for r-NSA.

$$\check{x}_{nm} = \frac{(x_{nm} - \bar{x}_m)}{\lambda(\sigma_m)}, \lambda = 1$$ (4)

To make ZS available for r-NSA, the value $\check{x}_{nm}$ will be transformed into 0 and 1. Inspired from SM, the original ZS output is rescaled into 0.0 to 1.0 using logistic function in equation (3) and then we labeled the modified Z-Score as MZS. For this study, we tested the MZS performance with different data distribution over mean; where $\lambda = 3$ shows data has 3σ or 99.7% linearly distributed from mean and $\lambda = 1$ representing the original ZS.

4   Experimental Setup

This section explains how this study was conducted. The steps are pictured in Fig. 2.

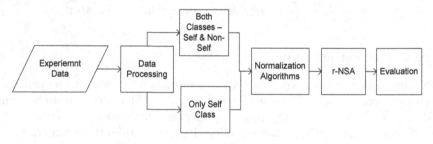

Fig. 2. The general research methodology

Four datasets with different characteristic were selected as experiment data; three were taken from UCI datasets [10] – the Iris (IRS), Pima Indians Diabetes Database (DBC), and Glass (GLS) while the Biomedical (BIO) dataset taken from [7]. The IRS and BIO are the benchmark data in r-NSA research. To meet the requirement as one class classifier, the IRS dataset that contains three class targets (Setosa, Verginica, Versicolor) were separated into 3 sub datasets – the IRS-SNS (Setosa), IRS-VRS (Versicolor), IRS-VGN (Verginica) where in each sub dataset, consist of only two classes- the self and non-self. For example, when verginica is considered as self sample, the other two classes will be considered as non-self. All dataset were pre-processed where all unknown numeric attributes were replaced with mean value. Then they were presented into the normalization algorithms. Table 1 summarizes the experiment data used in this study.

Table 1. The experiemnt data

Dataset	Total				
	Record	Attribute	Self	Non-self	Target
BIO	209	5	134	75	2
DBS	768	6	500	268	2
GLS	215	10	163	51	2
IRS (IRS-SNS, IRS-VRS, IRS-VGN)	150	5	50	100	3

For the purpose of normalization, the experiment was divided into two groups. Firstly, both classes – self and non-self group. In this group, the information of both classes such the min-max value was considered available during training. Meanwhile, for the second group, only self sample information was taken into consideration for normalization. After that, all the normalized dataset were mined with V-Detector algorithm [7]. The detector radius was set to 0.001 and 50% of self data selected for detector maturation. The experiment was run for 50 times and their performance averages were recorded for further analysis.

In the evaluation stage, the effect of normalization was evaluated based on five performance metrics- the highest detection rate (**DR**), specificity (**SPS**), accuracy (%) and lower false alarm rate (**FAR**), and number of detector (**ND**). In practice, each normalization algorithms generates different result when applied to a specific data. The identification of the most reliable normalization algorithm is difficult to determine when the algorithm generates different result in each performance metrics such as has good score in DR and % however less performed in SPS, FAR, ND. In anomaly detection study, a good model must have a balance result between DR, FAR, and SPS.

In this study, the preference matrix approach [1] was implemented. This approach suggests the best model based on the accumulative score of each performance metrics. The matrix consists of columns representing the performance metrics (DR, SPS, FAR, %, ND) and each normalization algorithms (1MZS, 3MZS, SMS, MM) are stored in rows. The score is given based on the priority of each metric; highest (DR, SPS, %) or lowest priority (FAR, DM). The score 1 is given for the best mining result according to priority metric. The last column of the preference matrix represents the total scores of all priorities. The lowest score indicates the best normalization approach.

5   Experiment and Result

In this section, the effect of normalization towards r-NSA is presented. The results are divided into two sections based on how data was normalized;

- both classes – the self and non-self group
- single class- only the normal class group

a.   Both Classes – The Self and Non-self Group

In this group, the information of self and non-self group is considered available during normalization.   The results shown in Table 1 indicate the effect of normalization approach towards r-NSA after 50 runs and they are separated according to the performance metrics. The last column of each sub Table 2 ** indicate the accumulative score of how many times the normalization algorithm rank at the highest result.

Table 2. The r-NSA mining result based on the normalization of self and non-self group

	DR								SPS						
	BIO	DBS	GLS	IRS-SNS	IRS-VGN	IRS-VRS	**		BIO	DBS	GLS	IRS-SNS	IRS-VGN	IRS-VRS	**
1MZS	88.1	49.4	88.0	100.0	98.0	93.4	3		80.0	77.5	84.6	91.2	85.6	85.2	-
3MZS	76.7	33.4	85.8	100.0	89.8	89.3	1		94.6	88.3	93.6	91.2	94.8	98.3	2
SMS	95.5	39.1	75.3	99.7	96.6	78.6	1		60.9	77.4	83.2	97.6	84.0	60.3	1
MM	75.7	16.9	71.1	100.0	97.9	93.6	2		98.0	98.0	94.7	96.3	78.0	84.8	3

	FAR								%						
	BIO	DBS	GLS	IRS-SNS	IRS-VGN	IRS-VRS	**		BIO	DBS	GLS	IRS-SNS	IRS-VGN	IRS-VRS	**
1MZS	20.0	22.5	15.4	8.8	14.4	14.8	-		82.9	67.7	85.4	97.0	93.9	90.7	1
3MZS	5.4	11.7	6.4	8.8	5.2	1.7	2		88.2	69.1	91.8	97.0	91.4	92.3	1
SMS	39.1	22.6	16.8	2.4	16.0	39.7	1		73.3	64.0	81.3	99.0	92.4	72.5	1
MM	2.0	2.0	5.3	3.7	22.0	15.2	3		90.0	69.7	89.1	98.8	91.3	90.7	2

	ND						
	BIO	DBS	GLS	IRS-SNS	IRS-VGN	IRS-VRS	**
1MZS	64	129	57	16	26	27	-
3MZS	30	80	38	16	19	17	1
SMS	115	128	48	5	17	35	1
MM	10	44	23	8	25	26	3

Based on the table, 1MZS is performed well in term of DR where it achieves the highest score out of 6 experiments and closely followed by MM. The other approaches also have similar capability to generate good detection rate when their results are comparable except MM in DBS dataset. In term SPS and FAR, the MM able to correctly classify self sample in most datasets with a significant different compared to other approach, excluding 3MZS. The 3MZS leads the highest SPS and FAR on two IRS datasets. Meanwhile, the accuracy % of each approach is recorded accurate when their accuracies generate a comparable performance. Besides that, the ND metrics indicates MM is the most efficient algorithm when producing lowest number of detector and able to detect anomaly with good detection result.

The result from Table 2 is further investigated to select the most reliable normalization algorithm for r-NSA. The selection is made by considering all performance metrics.   In Table 3, the results of Table 2 are represented in the preference matrix table and the score is given based on the priority of each performance metric. From the table, the best normalization algorithm for r-NSA can be determined based in the lowest accumulative score as showed in the last Table 3 column **.

Table 3. The preference matrix based on normalization of self and non-self class

	BIO							DBS					
	DR	SPS	FAR	%	ND	**		DR	SPS	FAR	%	ND	**
1MZS	2	3	3	3	3	14	1ZS	1	3	3	3	4	14
3MZS	3	2	2	2	2	11	3ZS	3	2	2	2	2	11
SMS	1	4	4	4	4	17	SM	2	4	4	4	3	17
MM	4	1	1	1	1	8	MM	4	1	1	1	1	8

	GLS							IRS-SNS					
	DR	SPS	FAR	%	ND	**		DR	SPS	FAR	%	ND	**
1MZS	1	3	4	3	4	15	1MZS	1	3	3	3	3	13
3MZS	2	2	2	1	2	9	3MZS	1	3	3	3	3	13
SMS	4	4	3	4	3	18	SMS	2	1	1	1	1	6
MM	3	1	1	2	1	8	MM	1	2	2	2	2	9

	IRS-VER							IRS-VRS					
	DR	SPS	FAR	%	ND	**		DR	SPS	FAR	%	ND	**
1MZS	1	2	2	1	2	8	1MZS	1	2	2	2	2	9
3MZS	4	1	1	3	3	12	3MZS	3	1	1	1	1	7
SMS	3	3	3	2	1	12	SMS	4	4	4	3	3	18
MM	2	4	4	4	4	18	MM	2	3	3	2	2	12

From the Table 3, MM is the most reliable normalization algorithm for r-NSA where the algorithm scores well in three datasets – BIO, DBS, and GLS. Meanwhile, the original Z-Score (1MZS) records the best score for IRS-VER and IRS-VRS dataset. The experiment also reveled that 3MZS is able to generate a consistent result for r-NSA when placed at the second rank for 5 datasets out of 6. For SMS, it overtakes other approaches in IRS-SNS dataset however it has the lowest performance in other dataset. This indicates that SMS is not suitable for r-NSA even though the information of both classes is considered for normalization. The information in Table 4 summarizes the most reliable normalization algorithm for r-NSA. The table lists the first and second rank of the best score.

Table 4. The most reliable normalization algorithm for r-NSA when considering self and non-self sample during normalization

	BIO	DBS	GLS	IRS-SNS	IRS-VER	IRS-VRS
1MZS					①	①
3MZS	②	②	②		②	②
SMS				①		
MM	①	①	①	②		
				① - rank 1, ② - rank 2		

b.   Single Class- Only the Normal Class Group

The aim of this experiment is to investigate the performance of r-NSA when data is normalized based on the information only from the self sample. In this case, we assume that the non-self sample is not available during training. For MM, the min max value is set to +20% from the actual self sample value while for MZS and SMS, the normalization is based on the mean and standard deviation of the self sample. The full results are presented in Table 5.

Table 5. The r-NSA mining result based on the normalization of self group

	DR							SPS						
	BIO	DBS	GLS	IRS-SNS	IRS-VGN	IRS-VRS	**	BIO	DBS	GLS	IRS-SNS	IRS-VGN	IRS-VRS	**
1MZS	89.4	34.4	67.5	93.8	69.8	74.3	3	53.0	76.2	68.6	79.8	62.6	58.0	-
3MZS	81.4	27.6	87.8	89.6	95.6	97.0	2	86.3	90.2	87.6	88.0	79.5	81.8	1
SMS	69.3	27.5	48.5	67.4	18.5	26.3	-	51.5	73.1	65.2	51.8	59.9	60.3	-
MM	71.7	21.6	44.0	48.6	98.3	78.9	1	96.4	94.0	96.5	78.3	86.2	91.4	5

	FAR							%						
	BIO	DBS	GLS	IRS-SNS	IRS-VGN	IRS-VRS	**	BIO	DBS	GLS	IRS-SNS	IRS-VGN	IRS-VRS	**
1MZS	47.0	23.8	31.4	20.2	37.4	42.0	-	66.1	61.6	68.3	89.2	67.4	68.9	-
3MZS	13.7	9.8	12.4	12.0	20.5	18.2	1	84.6	68.3	87.6	89.1	90.3	92.0	4
SMS	48.5	26.9	34.8	48.2	40.1	39.7	-	57.9	57.2	61.3	62.2	32.3	37.7	-
MM	3.6	6.0	3.5	21.7	13.8	8.6	5	87.5	68.7	84.0	58.5	94.3	83.1	2

	ND						
	BIO	DBS	GLS	IRS-SNS	IRS-VGN	IRS-VRS	**
1MZS	152	129	64	19	51	45	-
3MZS	42	84	45	25	22	22	-
SMS	160	127	61	44	55	52	-
MM	9	62	24	24	18	20	5

From Table 5, the best DR is scored by the Z-Score (ZS) which indicates its ability to detect anomaly even the non-self information is unknown. Out of 6 dataset, the highest score of 5 dataset are lead by 1MZS and 3MZS. The MM also achieved best DR but only in IRS-VGN data. Although the DR of MM is not good as ZS, the ability of MM in classifying the self sample as self and not as anomaly is proven when generating the best SPS and FAR result. Meanwhile, in accuracy %, 3MZS is the highest approach and it is followed by MM. Similar in both class experiments, the MM has generated the lowest number of detector ND for all dataset with a significant result. The experiment also revealed that the SMS is not suitable for r-NSA when it badly performed in all dataset.

The next step is to identify the most reliable normalization algorithm for r-NSA when only the self sample class is available during training. The information in Table 5 is represented in the preference matrix (Table 6) and the best normalization approach is summarized in the last column.

Table 6. The preference matrik based on normalization of self class

BIO	DR	SPS	FAR	%	ND	**
1ZS	1	3	3	3	4	14
3ZS	2	2	2	2	2	10
SM	4	4	4	4	3	19
MM	3	1	1	1	1	7

DBS	DR	SPS	FAR	%	ND	**
1ZS	1	3	2	3	3	12
3ZS	2	2	3	2	2	11
SM	3	4	1	4	4	16
MM	4	1	4	1	1	11

GLS	DR	SPS	FAR	%	ND	**
1ZS	2	3	3	3	4	15
3ZS	1	2	2	1	2	8
SM	3	4	4	4	3	18
MM	4	1	1	2	1	9

IRS-SNS	DR	SPS	FAR	%	ND	**
1ZS	1	2	2	2	1	8
3ZS	2	1	1	1	3	8
SM	4	4	4	3	4	19
MM	3	3	3	4	2	15

IRS-VER	DR	SPS	FAR	%	ND	**
1ZS	3	3	3	4	3	16
3ZS	2	2	2	2	2	10
SM	3	4	4	3	4	18
MM	1	1	1	1	1	5

IRS-VRS	DR	SPS	FAR	%	ND	**
1ZS	3	3	3	3	3	15
3ZS	1	2	2	1	2	8
SM	4	4	4	4	4	20
MM	2	1	1	2	1	7

Based on the preference matrix in Table 6, the ZS and MM have similar capability to generate good detection performance when the non-self sample is not considered during normalization. The MM becomes the best model for all dataset excluding IRS-SNS. Meanwhile ZS is performed well in two data set; the IRS-SNS and DBS. Although the ZS and MM indicates similar capability, the performance of MM seems to be better than ZS since it able to generate consistent SPS, lower FAR, and fewer detectors. The information in Table 7 summarizes the most reliable normalization algorithm for r-NSA. The table lists the first and second rank of the best result.

Table 7. The most reliable normalization algorithm for r-NSA when considering only self sample during normalization

	BIO	DBS	GLS	IRS-SNS	IRS-VER	IRS-VRS
1MZS		②		①		
3MZS	②	①	②	①	②	②
SMS						
MM	①	①	①	②	①	①

① - rank 1, ② - rank 2

6   Conclusion

Three types of the normalization algorithm namely min max, soft max scaling, and Z-score were experiment and their effects towards r-NSA performance were studied. From the analysis, it is proven that r-NSA is highly relied on type of normalization algorithm where the selection of appropriate normalization approach can improve detection performance. Although the MM, SMS, and MZS similarly change data representation between 0.0 and 1.0, the results were varies among them. The information depicted in Fig. 3 indicates how data had changed after they were presented into three different normalization algorithms. It represents the attribute 1 of BIO dataset and the first picture in the figure shows the original attribute 1 before normalization. This study investigated the performance of normalization algorithm in two scenarios; (1) both self and non-self samples are available during training (2) only self sample available.

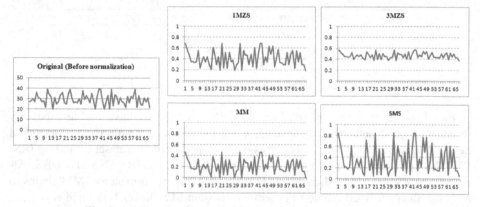

Fig. 3. Attribute 1 of BIO dataset before and after normalization

Based on the result, MM is the most reliable algorithm for r-NSA when it consistently produces good detection performance for r-NSA for both scenarios. Similar to MM, MZS also has capability to increase r-NSA detection performance. Nevertheless MM seems to be a better approach in term of higher specificity, lower false alarm rate, and fewer numbers of detectors. In term of detection rate, MM is well performed mainly when the information of self and non-self is available. Meanwhile, MZS generates good detection rate in both scenarios however less effective in maintaining the specificity and false alarm rate. It is important to note that the appropriate allocation to reserve extra value for min and max in MM is required to make sure MM can generates good normalization set particularly when non-self sample is not available. The study also reveals the SMS algorithm is not suitable for r-NSA mostly when information of the non-self class is unknown.

As conclusion, we conclude that MM and MZS are the most reliable algorithms for r-NSA. Since the conclusion is derived from the best mining result ranking, we will further strengthen the finding with statistical analysis in the future work.

References

1. Al Shalabi, L., Shaaban, Z.: Normalization as a Preprocessing Engine for Data Mining and the Approach of Preference Matrix. In: International Conference on Dependability of Computer Systems, DepCos-RELCOMEX 2006, pp. 207–214 (2006)
2. de Castro, L.N., Timmis, J.: Artificial immune systems: a new computational intelligence approach, vol. 1. Springer, Britain (2002)
3. Dasgupta, D., Krishnakumar, K., Wong, D., Berry, M.: Negative Selection Algorithm for Aircraft Fault Detection, pp. 1–13 (2004)
4. Forrest, S., Perelson, A.S., Allen, L., Cherukuri, R.: Self-Nonself Discrimination in a Computer. In: Proceedings of the 1994 IEEE Symposium on Security and Privacy, p. 202. IEEE Computer Society (1994)
5. Gadi, M.F.A., Wang, X., do Lago, A.P.: Credit Card Fraud Detection with Artificial Immune System. In: Bentley, P.J., Lee, D., Jung, S. (eds.) ICARIS 2008. LNCS, vol. 5132, pp. 119–131. Springer, Heidelberg (2008)
6. Han, J., Kamber, M.: Data Mining: Concepts and Techniques. Morgan Kaufmann Publishers, San Francisco (2006)
7. Ji, Z., Dasgupta, D.: Real-Valued Negative Selection Algorithm with Variable-Sized Detectors. In: Deb, K., Tari, Z. (eds.) GECCO 2004. LNCS, vol. 3102, pp. 287–298. Springer, Heidelberg (2004)
8. Kotov, V.D., Vasilyev, V.I.: Artificial immune system based intrusion detection system. In: Proceedings of the 2nd International Conference on Security of Information and Networks, pp. 207–212. ACM, Famagusta (2009)
9. Laurentys, C.A., Ronacher, G., Palhares, R.M., Caminhas, W.M.: Design of an Artificial Immune System for fault detection: A Negative Selection Approach. Expert Systems with Applications 37, 5507–5513 (2010)
10. Murphy, P.M.: UCI repositories of machine learning and domain theories (1997)
11. Wen-zhong, G., Guo-long, C., Qing-liang, C.: Improved negative selection algorithm for network anomaly detection on high-dimensional data. Journal of Computer Applications 29, 805–807 (2009)

Base Durian Ontology Development Using Modified Methodology

Zainab Abu Bakar and Khairul Nurmazianna Ismail

Department of Computer Science,
Faculty of Computer and Mathematical Sciences
Shah Alam, Malaysia
zainab@tmsk.uitm.edu.my ,nurmazianna@gmail.com

Abstract. Most ontology development methodologies practice on ontology reuse. However, those methodologies fail to define precisely the ontology reuse activities. Since the process of ontology engineering should be iterative, a base ontology is a fundamental of a new ontology. This paper highlights building a new base ontology from analysis of 40 webpages of a durian domain using Modified Methodology and is known as Durian Base Ontology (DuriO). The extracted terms are reorganized and converted into Web Ontology Language (OWL). SPARQL is employed in the experiment to obtain the desired answer by evaluating the functionality of DuriO using competency questions. Result shown that, DuriO can be employed successfully in semantic search engine. However, DuriO is base type ontology which many aspects such (class, property and instance) can be further incorporated.

Keywords: ontology methodology, base ontology, DuriO, semantic web.

1   Introduction

Ontology has been introduced in the field of philosophy. However, computer science field also uses the ontology in other perspective. Although there is no universal consensus on the definition of ontology but it is generally accepted that ontology is a specification of conceptualization [1]. The success of Semantic Web relies heavily on formal ontologies to structure data for comprehensive and transportable machine understanding [2]. Ontology aim at getting knowledge in a generic way and provide a commonly agreed of a domain [3]. Ontology can take the simple form of taxonomy or a vocabulary with standardized machine interpretable terminology supplemented with natural language definition [4]. Normally, ontology often written using semantic markup languages such as Resource Description Framework (RDF) and Web Ontology Language (OWL). There are three ways to start construction of ontology namely reuse other ontology that suit your needs, leverage information asset that already have value for your organization and to engineer ontology from scratch [5]. Each of these ways has its own drawbacks. Existing work is not always a perfect fit to your needs and ease of customizing it for your needs often depends a lot on how the original work designed [6]. Meanwhile by using your organization information

S.A. Noah et al. (Eds.): M-CAIT 2013, CCIS 378, pp. 206–218, 2013.
© Springer-Verlag Berlin Heidelberg 2013

artifacts to build ontology, it will be a confidential issue if the ontology needs to be published and reuse by other knowledge engineer. Moreover developing ontology from scratch is a resource intensive process and costly [7]. Among the construction ways, reuse of existing ontology is always been suggested in ontology development methodology.

2   Motivation

In Malaysia, agriculture is one of interest in Vision 2020 under the National Key Economic Areas (NKEA) [8]. The government gives many initiatives to assist farmers to increase agricultural yield and thus enhancing the national economy. Planting durian is one of the strategies that can increase the income of agriculture economy. Durian fetches good price both locally and abroad. Frozen durian flesh is exported to China by the Malaysian Government. The frozen durian is processed by a local government agency, Federal Agriculture Marketing Authority (FAMA) Batu Kurau, Perak, [9]. There are many varieties of durian grown by farmers in Malaysia and many parts of world. Besides Malaysia, Thailand is a great competitor in the durian market. In addition, there is also some durian tree in Brazil but it is cultivated not for the market. Since the price of quality durian is good and with the availability of techniques of planting and marketing from the web and documentation produced by local government agency, this information must be represented, stored, linked, and retrieved by farmers and other interested users.

There is information on durian plantation available the web, but the information is hard to find. This is a weakness of keyword search engine. A search engine is a tool in World Wide Web (WWW) which its design to search for information on the web [10]. The information may formatted in web pages, images and other types of files. Common search engine such as Altavista, Yahoo and Google use keyword-based search to retrieved information. User types the text query into the search engine application. Search engine searches into its enormous database for the keyword [11]. Every engine has its own technique of collection system to fill its database. Indexing system is used to organize the database for fast searching. It returns a list of hit pages that includes relevant as well as irrelevant pages [10]. Survey shows that practically 25% of web users are incapable to discover useful information in the first set of URLs result that are produce by search engine [12]. Motivate from those problem, Semantic Web proposes to overcome the difficulties listed above by making web content understand by machine [13].

3   Problem Statement

Ontology which is applied in Semantic Web applications allows relevant information accessible from the WWW. Most ontology developments apply reuse techniques [14]. Ontology reuse can be defined as the process in which available (ontological) knowledge is used as input to produce new ontologies. Researcher point out that building ontology by reusing existing ontologies is more cost effective if the technique is applied appropriately [15]. However, the reuse technique is not described in details in most ontology development. Therefore, Modified Methodology apply reuse ontology in this paper is describe in details.

4   Semantic Web

Basically, WWW contains data in five types of file such as compression, document, audio, images and movies. The same data may be kept in different types of file and those data do not have relations. Because of these various representations on the web, it may cause data interoperability [16]. From the survey done by Woodruff [17], 76.3 % data from the WWW is in HTML markup language. HTML tags in HTML markup language only afford the ability to view the information but not to understand it.

The next generation web call the Semantic Web, aims at making its content not only machine readable but also machine understandable. For example, a statement such as 'durian is marketable plantation in Malaysia' can be understood by a normal person but computers has difficulty to understand the statement. Statements are built with from syntax rules. The syntax of a language explains the rules for structure the language statements. As a result, the semantic web deals with how syntax becomes semantic. In addition, the semantic web is not about links between web pages but it describes the relationships between things and the properties of the things [18].

There are four approaches to semantic search and different semantic search engines use one or more of these approaches[19]. The objective of semantic search is to use meaning to improve the accuracy of retrieved answer in search engine. The first approach uses contextual analysis to help to disambiguate queries. for example, the word "plant", refers to agriculture or factory or something else. Meanwhile the second approach focuses on reasoning such as given set of facts that are represented in the system, additional facts can be inferred from them. Third approach emphasize on understanding of natural language. These engines process the content and create index of the content.The queries submit by user is identify the intent of the information. The fourth approach uses ontology to represent knowledge about a domain and expand the queries. In this approach, when a user enters a query for a word "durian," the system adds terms from fruits ontology because a durian is a kind of fruit to make the search more focused as well as more broad This approach is used by a large number of semantic search systems [19].

Thus, domain ontology development for durian plantation is essential for the success of semantic search engine and Malaysia Vision 2020.

5   Ontology Development Methodology Background

There are various techniques in ontology development methodologies. It is processes that composed of a series of activities that are perform in order to build ontology. Ontology development is a state of the art rather than a science, thus there is not a single correct ontology methodologies to develop ontology [20]. Therefore, analysis of ontology development methodologies should be taken in account to study the phases of available ontology development methodologies to fit base ontology development.

5.1   Analysis of Ontology Development Methodology

Domain ontology or domain-specific ontology is a model for a specific domain, which represent part of the world. Particular meanings of terms applied to the domain are provided by domain ontology. Ontology development methodologies were studies and each of them was established under different scenarios with various focuses and limitations [21]. In this paper, the reuse technique in ontology development methodologies is studied.

Table 1 is the summary of four ontology development methodologies namely Enterprise Ontology, TOVE, Methontology and Ontology Development 101. It shows that the activities in most methodologies suggest of reuse existing ontology but they fail to define in details the process of ontology reuse. From the table also, it shows various methodologies have been described as making competency questions; replicate situations that require ontology and feasibility study as the methods to identify purpose of ontology development. Meanwhile using phase such as ontology capture, specify terminology, development specification and enumerate key-words to list the possible

Table 1. Summary of Ontology Development Methodologies

Name	Details of the methodology available	Domain related	Ontology Reuse (related phase)	Phases
Enterprise Ontology (Uschold and King, 1995)	Activities and technique proposed are not exactly specified and its information is very little available.	Supply chain domain	Yes (Integration existing ontology phase)	• Identify the purpose • Ontology capture • Coding • Integrating existing ontology • Evaluation and documentation
Toronto Virtual Enterprise (TOVE) (Gruninger and Fox, 1995)	Activities and technique proposed are not exactly specified and its information is hard to find.	Business enterprise domain	No	• Capture motivating scenarios and formulize informal • Specify terminology of the ontology within a formal language • Formulize the competency questions • Specify axioms and definitions for the terms in the ontology • Establish conditions for characterizing the completeness of the ontology

Table 2. (*continued*)

Methontolo gy (Fernandez-Lopez et al. ,1997)	Activities and technique proposed are not specified and easy to find.	Several domains such as chemical ,environment pollutants and Reference-Ontology	Yes (Integration phase)	• Pre-development: Environment and feasibility study • Development: Specification • Development: • Conceptualization • Development: Formalization • Development: Integration • Development: Implementation • Post-development: Maintenance Plus additional 5 supportive activities and 3 management activities
Ontology Developme nt 101 (Noy and McGuinnes s, 2001)	Activities and technique proposed are exactly specified and possible to find.	Several domains such as government budgetary Ontology and wine ontology	Yes (Using existing ontologies phase)	• Determine the domain and scope • Consider using existing ontologies • Enumerate key-terms • Define classes and class hierarchy • Define properties of classes • Define facets of slots • Create instances

term of the domain ontology. Therefore, most ontology development methodologies have the same goal but contain different methods and practice reuse ontology technique.

6   Modified Methodology for Base Ontology

According to Lopez, Methondology ontology development methodology can be categorized as a lot of details information that can be as reference. The methodology is also said to be the most mature but still lack of details in pre-development phase [22]. Therefore, Methondology ontology development methodology was chosen and some modification on the activities for DuriO development compatibility of DuriO development as in Table 2.

Table 2.   States Comparison between Methondology and DuriO Development

Methondology	Modified Methodology
Plan	Determine scope
Specification	Analysis of webpages
Conceptualization	Reuse graph knowledge model
Formalization	
Integration	Discussion with expert
Implementation	Implementing the ontology
Maintenance	Validate competency questions

Table 2 shows the different phases between Methondology and Modified Methodology. Both methodologies use different method and phases name but to achieve the same output. For example in second phase, Methondology use specification as the name that produce possible terms as the output. While Modified Methodology use 'analysis of webpages' as the name to achieve the same output in specification. Noticeably, Methondology has general phases but Modified Methodology has focus phases.

6.1   Determine Scope of DuriO

Roughly, planning on building something must have a specific purpose. Same goes to ontology methodology. As explain before, various ontology methodology have been described as making competency question, replicate situations that require ontology or feasibility study as the methods to identify the purpose of ontology development. In Ontology Development 101, the ways to determine the scope of the ontology is to sketch a list of questions that a knowledge base based on the ontology should be able to answer [23]. In the durian plantation domain, the following are the competency questions:

1.   What are the durian plantation stages?
2.   What are the durian Malaysia varieties?

6.2   Enumerate Important Terms

Enumerate important terms is an important activity for all involved term should be spell out. Often, this activity use brainstorm technique and written each of the terms that appear. There is also semi-automatic term extraction from the domain handbook, where technically it produces controlled terms.

However, in this paper the terms are extracted from 40 selected HTML documents based on Malaysia Agriculture Research and Development Institute (MARDI) guideline (item 1-10). MARDI is a statutory body which has been mandated to conduct research in agriculture, food and agro-based industries in Malaysia. MARDI research endeavors for almost 40 years had fruitfully generated many new crop varieties and clones, animal breeds and its management practices [24].

A study done by Zainab and Khairul on the analysis of 40 HTML documents is examined [25]. There are ten general crops characteristic obtained from MARDI as a guideline. These characteristics are the most available features in crops. Analysis of 40 HTML documents regarding durian resulted another eleven other characteristics of durian to be added to the general characteristic of crop as shown in Table 3 (item 11-21). These characteristics can be a terms in DuriO.

Table 3. Durian Characteristic. [25]

No.	Crops characteristic	No.	Durian Characteristic
1.	Description	11.	Belief
2.	Varieties	12.	Nutrition
3.	Uses	13.	Health
4.	Propagation	14.	Odor and Flavor
5.	Culture and management practices	15.	History
6.	Pests	16.	Market
7.	Disease	17.	Session
8.	Fertilizer	18.	Price
9.	Harvest practices	19.	Serving fresh fruits
10.	Post-harvest practices	20.	Recipe
		21.	Growing stage

As for the base ontology, DuriO had twenty-one terms as a fundamental for ontology development and it will be expanded later as future work.

6.3   Define Classes, Hierarchy and Properties Using Knowledge Model Approach

Most ontology methodologies define classes, hierarchy and properties phases are execute differently. The prior resulting term are formed into classes and hierarchy. Properties created the link of the terms. All these activities generate some tables such as classes table, hierarchy table and properties table. While developing DuriO, those activity performed simultaneously by using a knowledge model approach. Knowledge modeling is a process of creating a computer interpretable model of knowledge or standard specifications about a kind of process and/or about kind of facility or product. The advantage of using knowledge model approaches it able to displays classes, hierarchy and properties in one view. Using a single view, redundancy name on various class and hierarchy can be minimizing.

To build the knowledge model, there is a set of guideline to produce a good knowledge model. It explains the steps to build a knowledge model, the criteria to determine the concepts, instances and properties [26]. There are several knowledge models that have been produced as banana, grapes, mango, potatoes and tomatoes. Banana and mango knowledge model are selected to be study to produce durian knowledge model since they in same category namely fruit [27]. Knowledge model of banana and mango were studied on their classes, hierarchy and properties similarities. Those similarities are the basic needs for fruits domain ontology.

Table 4. Details on Domain Related Knowledge Model

Domain knowledge model	Class similarity

Banana plantation

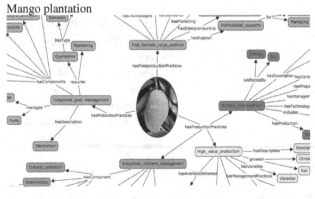

Consist of four classes. The classes have subclasses and summarize as related subclasses.

Class name	Related Subclasses
high_value _productio n	Orchard_ma nagement
Integrated _nutrient_ manageme nt	Fertilizer
Integrated _pest_man agement	Pest and disease
Post_harve st_value_a ddition	Marketing and processing

Mango plantation

Mango plantation knowledge model consist the same four classes as in banana plantation knowledge model with additional one new class as in below table. The classes have subclasses and summarize as related subclasses.

Class name	Related Subclasses
Nursery_ manageme nt	Vegetative process

Table 4 shows the banana knowledge model has four major classes. These classes have a growing subclass and properties among each other. If assessed, it produces a complex and complete knowledge model. It may be easy to understand in this phase but will give a big challenge in the implementation phase. Meanwhile in mango knowledge model, it has five main classes. It consist the same four classes as in banana knowledge model with additional one new class namely Nursery_management. Meanwhile in durian knowledge model (Table 5), it has four main classes as a base knowledge model. This is to reduce the difficulties in the implementation phase later. The prior fruits and durian characteristics are also added into the durian knowledge model.

Durian knowledge model is the result of combination of important term from 40 HTML documents and also the analysis of classes in banana and mango knowledge model. With this approach, the properties can be creating simultaneously to link relationship between classes and it does reduce the redundancy problem rather than table approach.

Table 5. Details on DuriO Knowledge Model

Durian knowledge model

Consist of four classes. The classes have subclasses and summarize as related subclasses.

Classes	Related subclasses
DurianDescription	Basic information
DurianProductionPractices	Production process
PestManagement	Pest and disease
PostHarvestPractices	Marketing and durian product.

6.4   Create Instance

Once the terminology box (T-Box) that contains classes, hierarchy and properties completed, instances should be included into the ontology. In this phase, the domain expert involvement is very helpful for the purpose of producing ontology with correct and accurate information. Meetings and discussions with the durian plantation expert from MARDI is crucial phase. After few revision and discussion with the experts, the DuriO overview is shown below.

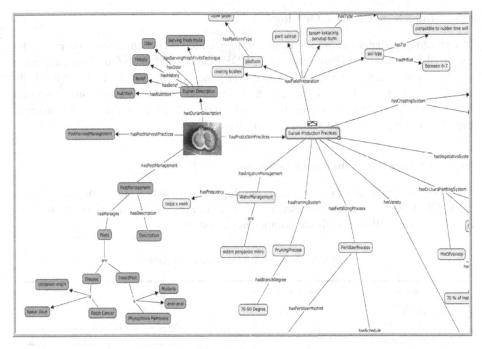

Fig. 1. DuriO knowledge model snapshot

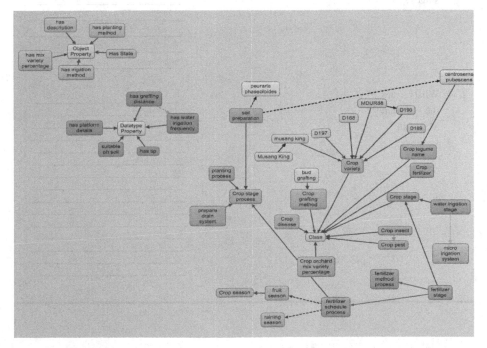

Fig. 2. Visualization of DuriO

6.5   Implement the Base Ontology

In order to implement the ontology, an ontology editor is chosen. The ontology can be exported into different formats including RDF Schema and Web Ontology Language (OWL). Particularly, the DuriO have implemented in OWL and verified its consistency by using TopSPIN inference engines. The triples of DuriO are store in a triple-store. Durio can be viewed using visualization tools after the storing process. Fig. 2 is the Durio overview using visualization tool.

6.6   Validate the Competency Questions

In order to verify and validate the ontology as regards competency questions, SPARQL Query Language is used on DuriO. Following are the two example of SPARQL queries that models the competency questions.

Table 6. Competency Questions Validation

No	Competency questions	SPARQL query	Answer	
			s	o
1.	What are the durian plantation stages?	1 select ?s ?o 2 {?s owl:hasCropStage ?o}	durianMalaysia	pestManagementStage
			durianMalaysia	landPreparationStage
			durianMalaysia	waterIrrigationStage
			durianMalaysia	vegetativeStage
			durianMalaysia	pruningStage
			durianMalaysia	fertilizerStage
			durianMalaysia	orchardStage
2.	What are the durian Malaysia varieties?	1 select ?s ?o 2 {?s owl:hasVariety ?o. 3 ?s rdfs:label "durian malaysia" 4 }	durianMalaysia	teka
			durianMalaysia	MDUR79
			durianMalaysia	D190
			durianMalaysia	D99
			durianMalaysia	D197
			durianMalaysia	D24
			durianMalaysia	IOI
			durianMalaysia	D168
			durianMalaysia	MDUR88
			durianMalaysia	bukitMerah
			durianMalaysia	MDUR78
			durianMalaysia	D189
			durianMalaysia	musangKing
			durianMalaysia	D145

7 Discussion

In order to develop DuriO presented in this paper, the methodology outlined in Table 2 has been followed. This methodology purposely builds ontology from reuse ontology technique and using Modified Methodology ontology development as guidelines. The most important task in the methodology is the building the knowledge model. Then, it is important to assign all the necessary time to carry out a good knowledge model. This knowledge model has to be agreed on by the domain experts. Base ontology is an evolving ontology to be complete. Therefore, validation process on every evolve stage must be consider to maintain the consistency.The reuse ontology technique is detailed out in Modified Methodology. As agreed by researcher, ontology reuse technique will reduce the cost since it avoids rebuilding existing ontologies.

However, this technique is only suitable for base ontology as a starting point of ontology development. Thus, evolving ontology development should be practiced to make DuriO complete as possible.

8 Conclusion

In this paper, we have shown how ontologists could develop base ontology from reuse ontology technique using Modified Methodology. Since the process of ontology building should be iterative, a base ontology which is defined here as a good starting draft of ontology. As the result, DuriO might be incomplete in some aspect but it could facilitate the ontology development process with evolving improvements. This approach has been used to define a DuriO for semantic search engine.

Acknowledgement. The research is sponsored by Malaysian Government and Universiti Teknologi MARA (UiTM) Malaysia under the Large Research Grant Scheme (LRGS) LRGS/TD/2011/UITM/ICT/01.

References

1. Ding, Y., Foo, S.: Ontology research and development. Part 1 - a review of ontology generation. Journal of Information Science 28, 123–136 (2002)
2. Maedche, A., Staab, S.: Ontology learning for the Semantic Web. IEEE Intelligent Systems 16, 72–79 (2001)
3. Kapoor, B., Sharma, S.: A comparative study ontology building tools for semantic web applications. International Journal of Web & Semantic Technology (IJWesT) 1, 1–13 (2010)
4. Lau, R.Y.K., Yuefeng, L., Yue, X.: Mining Fuzzy Domain Ontology from Textual Databases. In: International Conference on Web Intelligence, pp. 156–162. IEEE/WIC/ACM (2007)
5. Allemang, D., Hendler, J.: Semantic web for the working ontologist: effective modeling in RDFS and OWL. Morgan Kaufmann (2011)

6. How to: extend an ontology, http://topquadrantblog.blogspot.com/2011/03/how-to-extend-ontology.html
7. Bontas, E.P., Mochol, M., Tolksdorf, R.: Case studies on ontology reuse. In: Proceedings of the IKNOW 2005 International Conference on Knowledge Management (2005)
8. Performance Management & Delivery Unit, http://etp.pemandu.gov.my/Overview_of_NKEAs_-@-Overview_of_NKEAs.aspx
9. Eksport Durian Sejuk Beku ke Republik Rakyat China, http://www.fama.gov.my/c/document_library/get_file?uuid=2e2cb4c8-cf8d-4863-b523-229338a74b1c&groupId=7382656.html
10. Mukhopadhyay, D., Banik, A., Mukherjee, S., Bhattacharya, J., Kim, Y.-C.: A Domain Specific Ontology Based Semantic Web Search Engine (2004)
11. Antoniou, G., Christophides, V., Plexousakis, D., Doerr, M.: Semantic Web Fundamentals (2003)
12. Search beyond Google, https://www.technology.com/articles/print_version/roush0304.asp
13. Berners-Lee, T., Hendler, J., Lassila, O.: The Semantic Web. Scientific American 284, 5 (2001)
14. Leung, N.K.Y., Lau, S.K., Fan, J., Tsang, N.: An integration-oriented ontology development methodology to reuse existing ontologies in an ontology development process. In: Proceedings of the 13th International Conference on Information Integration and Web-based Applications and Services. ACM, Ho Chi Minh City (2011)
15. Pinto, H.S., Martins, J.P.: A methodology for ontology integration. In: Proceedings of the 1st International Conference on Knowledge Capture. ACM, Victoria (2001)
16. Cruz, I.R., Huiyong, X., Feihong, H.: An ontology-based framework for XML semantic integration. In: Proceedings of the International Database Engineering and Applications Symposium, IDEAS 2004, pp. 217–226 (2004)
17. Woodruff, A., Aoki, P.M., Brewer, E., Gauthier, P., Rowe, L.A.: An investigation of documents from the World Wide Web. Computer Networks and ISDN Systems 28, 963–980 (1996)
18. W3C Semantic Web, http://www.w3.org/RDF/FAQ
19. Sudeepathi, G., Anuradha, G., Babu, P.M.S.P.: A survey on Semantic Web Search Engine. International Journal of Computer Science Issues 9, 241–245 (2012)
20. Jones, D., Bench-Capon, T., Visser, P.: Methodologies for ontology development. In: Proc. IT&KNOWS Conference of the 15th IFIP World Computer Congress, pp. 20–35. Citeseer (1998)
21. Wache, H., Voegele, T., Visser, U., Stuckenschmidt, H., Schuster, G., Neumann, H., Hübner, S.: Ontology-based integration of information-a survey of existing approaches. In: IJCAI 2001 Workshop: Ontologies and Information Sharing, pp. 108–117. Citeseer (2001)
22. Fernández López, M.: Overview of Methodologies for building ontologies. In: Proceedings of the IJCAI 1999 Workshop on Ontologies and Problem-Solving Methods, KRR5 (1999)
23. Grüninger, M., Fox, M.S.: Methodology for the Design and Evaluation of Ontologies (1995)
24. Malaysia Agriculture Research and Development Institute, http://www.mardi.my/en/web/guest/kata-aluan-ketua-pengarah.html
25. Bakar, Z.A., Ismail, K.N.: Classification of durian characteristics for semantic representation from web documents. In: 2012 IEEE Symposium on E-Learning, E-Management and E-Services (IS3e), pp. 1–5. IEEE (2012)
26. Sini, M., Yadav, V., Singh, J., Awasthi, V.: Knowledge models in agropedia indica (2009)
27. Knowledge Models, http://agropedia.iitk.ac.in/

Enhanced Arabic Information Retrieval:
Light Stemming and Stop Words

Jaffar Atwan[1], Masnizah Mohd[1], and Ghassan Kanaan[2]

[1] Faculity of Information Science and Technology Universiti Kebangsaan Malaysia,
Malaysia, Bangi
jaffaratwan@gmail.com, mas@ftsm.ukm.my
[2] Amman Arab University, Faculty of Computer Science and Informatics, Amman, Jordan
ghkanaan@aabfs.org

Abstract. Stemming is a process of reducing inflected words to their stem, base or root from a generally written word form. For languages that is high inflected like Arabic. Stemming improve the retrieval performance by reducing words variants. The effectiveness of stop words lists with light stemming for Arabic information retrieval (General stopwords list, Khoja stopwords list, Combined stopwords list), were investigated in this paper. Using vector space model as the popular weighting scheme was examined. The idea is to combine (General and Khoja) stopwords lists with light stemming to enhance the performance, and compare their effects on retrieval. The Linguistic Data Consortium (LDC) Arabic Newswire data set was used. The best performance was achieved with the Combined stopwords list, with light stemming.

Keywords: Arabic, Information Retrieval, Stemming, Stopword.

1   Introduction

English language get the focus in the field of the information retrieval (IR) research, recently research work for Arabic language in IR had been developed and good efforts was spent. But, this work and efforts still not sufficient with what has been done in English language, which has been the main field of IR for a long while. This is despite the fact that the Arabic language is one of the major six languages of the United Nations, the mother tongue of around 220 million people. In addition, because it is the language of the Qur'an, it is also the second language for many Muslims and Muslim countries around the world [1].

This paper attempts to compare the use and effect of light stemming with different stopwords lists for Arabic information retrieval (AIR). Using light stemming and vector space model with weighting scheme, and three stopwords lists are implemented in order to determine the effect of stop words elimination with light stemming on an AIR system.

The light stemmer Light10 [2], remove the prefixes and suffixes of the word based on a list of suffixes and prefixes of different length. Light10 is the most widely and

S.A. Noah et al. (Eds.): M-CAIT 2013, CCIS 378, pp. 219–228, 2013.

popular Arabic stemmer and which is included in the Lemur toolkit one of the most popular information retrieval tools. [2] has applied this stemmer run with different systems over TREC 2002 data and findings indicate that light stemmer is outperform the others using Khoja stopwords. The weighting scheme to be used are the TF*IDF weight. Three stop words lists will be used, a general stopwords list[3], Khoja stopwords list [4], and a combined list.(Khoja and General) Although stemming is an important factor when dealing with AIR, light stemming was implemented in this study in order to test the effect of stop words with on stemming and the overall performance.

The motivation of this work is to investigate the most suitable stopwords list that is appropriate for AIR. Furthermore, to establish for future work in order to test new stopwords list(s) with new Arabic stemmer that is consist of Larkey and Khoja combined together[5].

2   Related Work

2.1   Arabic Language

Arabic language is one of the international languages; It is a language for more than 20 countries, and one of the major languages spoken in United Nation. Arabic is the language of Holy Quran, which is the holy book of the Islamic world (hundreds millions of Muslims). A tremendous number of Arabic internet applications and users have grown during the past ten years along the Arabic world.

Arabic language is one of the most complex morphological languages in the world. Arabic has 28 alphabet letters which is written from right to left. From one root a big number of words can be generated. These characters can have diacritical marks on them (called Damma, Fathah, Kasra, Shaddah) which decide how a word should be pronounced. If these diacritical marks are improperly used, this may cause errors in the pronunciation and hence change meanings of words. Arabic has two genders (feminine and masculine), three cardinalities (singular, dual and plural), three grammatical cases (nominative, genitive and accusative) and two tenses (perfect and imperfect) [5].

However, a modern standard Arabic language now in use; in internet, books, news, TV, and other types of media. Most of Arabic information researchers working with modern Arabic   a few work have been   done in natural language processing side for diacritic Arabic, and also a few work done for Arabic dialects which is very different from country to another even in the same country[5] and [6]. Here in this paper we introduced the effect of stopwords for modern standard Arabic language IR system.

2.2   StopWord

Stopwords are very common words that appear in the text that carry little meaning; these words are part of how we described nouns in text, and express the concepts like location or quantity. These stopwords have two different impacts on the information IR process. They can affect the retrieval effectiveness because they have a very high frequency and tend to diminish the impact of frequency differences among less common words, which affecting the weighting process. The removal of the stopwords also changes the document length and subsequently affects the weighting process. They can also affect efficiency due to their nature and the fact that they carry no meaning, which may result in a large amount of unproductive processing [9]. The removal of the stopwords can generally improve retrieval effectiveness [6]

Identifying a stopwords list or a stopwords list that contain such words in order to eliminate them from text processing is essential to an IR system. Stopwords lists can be created in different ways. Words that are most frequently used [7], or based on syntax which need an expert based on a personal judgment. Although, we can translate stop words lists from English language to Arabic language to be used in AIR system. They can also be created using a combination of the syntactic and most frequently used in corpus and translation from English language to achieve the benefits of both approaches [5].

There is no general standard stopwords list to use in an IR experiment for the Arabic language. The stopwords list used in the Lemur Toolkit is the one created by [4] when she was creating her Arabic stemmer and is relatively short (168 words). This list was used by [8] and [2]. [9] used a list they created by translating an English list and augmenting it with high frequency words from the corpus creating a rather large list contained, 1,131 words. However they do not discuss the effect of the list. [5]create a list of stop list using three methods which are translating from the English language to Arabic language, identification of common words in arbitrary Arabic documents, and manual search of synonym of the previously identified stop words. They divided the stop words into two categories which are useless (i.e. meaningless words) and useful (which can be used to detect the syntactic meaning of the subsequent words). But, these lists are not presented very well in their work.

[10] has created domain dependent list which has three problems. The waw letter "و" which means "and" precede some words and there is duplication of words. This letter is removed from the words with other letters in the normalization process with good stemming algorithm. Another problem is some of their words are not considered a stopword due to its frequently appeared like Cairo which is a city name. Although, the stopwords list depends on the domain it may not be suitable for any corpus.

[3] investigated three stopwords lists which are General Stopwords list, Corpus-Based Stopwords list, Combined Stopwords list, .and the effects of term weighting schemes, and stopwords on Arabic retrieval, using the Lemur Toolkit over LDC Arabic newswire. However [3] did not implement the stemming in his study in order to isolate the effect of stopwords from any other factors. [3] also, claims the performance of a general stopwords list or a combined list was relatively close and recommended. He claim the use any of these stopwords list is recommended but the general stopwords list is certainly preferred if we are dealing with different corpus.

2.3   Stemming

Stemming or conflation is a technique that finds the relationship between multiple forms of a word. This technique reduces the word to its stem or root, to its abstract form without any addition in affixes (prefix, suffix or infix) Table 1 shown that many words are derived from the same root or stem. Four different approaches to Arabic stemming can be identified manually using constructed dictionaries, algorithmic light stemmers which remove prefixes and suffixes, morphological analyses which attempt to find roots, and statistical stemmers, which group word variants using clustering techniques[8].

Table 1. Many words are derived from the same Root Ktb كتب

Arabic Word	English Meaning	Arabic Root(Stem)	English Root(Stem)
مكتب	Office	كتب	Ktb
كاتب	Writer	كتب	Ktb
مكتبة	Library	كتب	Ktb
مكتوب	Written	كتب	Ktb

Here in our study, we proposed the light stemmer [2]. Light10, with new list of stop words. Light stemmer; light10 was designed to strip off strings that were frequently found as prefixes or suffixes, but infrequently found at the beginning or ending of stems. Also, it doesn't remove any infix and did not deal with the word morphology or pattern as in[4]. [11] introduced the Al-Stem light stemmer at TREC 2002,and demonstrated that it was less effective than Light10.

In [4] Khoja removes suffixes, infixes and prefixes and uses pattern matching to extract the roots. In the stemmer an aggressive stemming done for words reducing it to their roots. [4] removed diacritics, stop-words, punctuation marks, numbers, the conjunction prefix "و" (and), and the definite article "ال" (the) are all removed. Input words are also checked among a large list of prefixes and suffixes, and the longest of these is stripped off, if found. The resulting word is then compared to a list of patterns and if a match is found, the root is produced The prominent Arabic light stemmer is Aljlayl [12] Light stemming and it does not deal with patterns or infixes; it is simply the process of stripping off prefixes and/or suffixes.

In [9] introduced a light stemmer similar to Light10, but that removed more prefixes and suffixes. It was shown to be more effective than Al-Stem, but was not directly compared to Light10.

Light stemmer [2] Light10, has benefited from the discussion with Aljlayl [12] the prominent Arabic light stemmer. Also, Light10 as we mentioned before in light stemming does not deal with patterns or infixes; it is simply the process of stripping off prefixes and/or suffixes. They follow same process in Khoja and Aljlayl in general for normalization and affixes removing [2] try different versions of Light stemmer but they claim the best one is the Light10. In Light10 [2] remove the list of suffixes and prefixes from the Arabic words as shown in Table 2.

Table 2. Suffixes and Prefixes that are removed by Light10

Prefixes	Suffixes
ال، وال، بال، كال، فال، لل، و	ها، ان، ات، ون، ين، يه، ية، ه، ة، ي

Although, light stemming can correctly conflate many variants of words into large stem classes, it can fail to conflate other forms that should go together. For example, broken (irregular) plurals for nouns and adjectives do not get conflated with their singular forms, and past tense verbs do not get conflated with their present tense forms, because they retain some affixes and internal differences. In spite of its simplicity and shortcomings, no more sophisticated approach has been shown to be more effective for AIR.

Stemming process is a very important process for language like Arabic [13]. Many searches are tackling stemming for Arabic language. Some of these researches concerns in light stemming while others tackle morphological analysis. However, the main prominent stemming in both searches is Light10 and Khoja which are dominant to Arabic language.

AIR research is still insufficient and lack of study from preprocessing until retrieval aspect. In preprocessing there is no standard in normalization, and no standard stopwords list for AIR system. The main factor is there is, no standard stemming to be used.

3   Experiment

This study explores the use of stop words, light stemmer and their effects on AIR. It compares the use of three different runs using TF.IDF term weighting scheme, and three stopwords lists. These stopwords lists and stemmer technique are examined using a large corpus that was not available before the introduction of Arabic Cross-Language Retrieval at TREC 2001.

The study evaluates these lists and technique using the standard recall and precision measures as the basis for comparison. It answers the following questions:

    a. What is the effect of the stopwords list on retrieval, i.e. how sensitive is retrieval to the use of stopwords; and which one of the lists, the general, the Khoja, or the combined list is superior to each other?

    b. What is the effect of the stopwords list with light stemming on retrieval?

First, performance of term weighting scheme with elimination of stopwords was compared, and then with light stemming with three different stopwords lists were run. Using precision and recall, the effectiveness of all stopwords lists and light stemming technique was evaluated to determine which combination achieves the optimal performance for Arabic language retrieval.

3.1   Data Set

This research used one Arabic test corpus, created in the Linguistic Data Consortium in Philadelphia, also used in the recent TREC experiments. The Arabic Newswire A corpus was created by David Graff and Kevin Walker at the Linguistic Data Consortium [13]. It is composed of articles from the Agence France Presse (AFP) Arabic Newswire. The source material was tagged using TIPSTER style SGML and was transcoded to Unicode (UTF-8). The corpus includes articles from 13 May 1994 to 20 December 2000. The data is in 2,337 compressed Arabic text data files. There are 209 Mbytes of compressed data (869 Mbytes uncompressed) with 383,872 documents containing 76 million tokens over approximately 666,094 unique words.

The query set associated with the LDC corpus was created for TREC 2001 and 2002 [14] and [15]. There were 25 topics with relevance judgments, available in Arabic, French, and English, with Title, Description, and Narrative fields. We used the Arabic titles and descriptions as queries in our monolingual experiment [16].

3.2   Stopwords Lists

The three stopwords list was tested in the research as we mentioned before are consist of the Khoja [4] stopwords list, Abu El-Khair [3] general stopwords list, and the combination of both in order to use the stopwords that are not appeared as part of Abu El-Khair general list.

3.2.1   Khoja Stopwords List

Khoja stopwords list that are used in many research mainly in [2], [4] and [17], and other researches. This stopwords list can be considered as the most frequent words in Arabic because it mainly consist of the Arabic preposition and conjunction but also is insufficient and cannot be considered enough for any AIR, natural language processing or any other system, due to its limitation.

3.2.2   The General Stopwords List

A general stopwords list was created by Abu El-Khair [3], based on the Arabic language structure and characteristics without any additions. The resulting list consisted of 1,377 words. The list was checked against Khoja [4] and Alshehri's [18] lists, and two standard English lists, the Okapi and SMART lists. But from our revision some stopwords of Khoja [4] stopwords list are not included.

3.2.3   Combined Stopwords List

The combined stopwords list consists of both previous stopwords lists Khoja and Abu El-Khair. In order to avoid forgotten the stopwords that are doesn't appeared in Abu ElKair general list we combined both of the two list together which has a little impact in performance.

3.3   Experiment Process

The data and the query set for the experiments were processed as follows.

    a.   The 383,872 files in the data set were converted from UTF-8 format to Windows.

    b.   Title and description for each of the 25 queries were extracted from the original query set.

    c.   The corpus and queries were normalized according to the following steps, same normalization used by [2]:

    d.   Remove punctuation

    e.   Remove diacritics (primarily weak vowels). Some entries contained weak vowels. Removal made everything consistent.

    f.   Remove non letters

    g.   Replace إ,أ , and آ with ا

    h.   Replace final ى with ي

    i.   Replace final ة with ه

    j.   Eliminating stop words for each run different stopwords list are examined

    k.   Tokenization as it will be described below.

    l.   Stemming using Light10 stemmer.

    m.   Evaluation using standard TFIDF weighting schema, precision and recall, the average un-interpolated precision for the 25 queries, and the compression rate for the system index for each run.

3.4   Tokenization

Tokenize the word in natural language processing is very important and not a trivial step. Tokenization process is responsible to split the text into tokens; determine word boundaries, abbreviations and numbers. Tokenizing Arabic text is a preliminary important step as part of preprocessing stage. Arabic token here in our research is a word delimited by white space as word boundaries which is a full form word. The next step after tokenization and stopwords removal is the stemming process.

3.5   Stemming

Light10 stemmer is used in stemming process and we followed the same steps by Larkey's:

    a.   Remove "و" ("and") for light2, light3, and light8, and light10 if the remainder of the word is 3 or more characters long. Although it is important to remove"و", it is also problematic, because many common Arabic words begin with this character, hence the stricter length criterion here than for the definite articles.

    b.   Remove any of the definite articles if this leaves two or more characters.

   c. Go through the list of suffixes once in the (right to left) order indicated in Table 2, removing any that are found at the end of the word, if this leaves two or more characters.

   d. The strings to be removed are listed in Table 2. The "prefixes" are actually definite articles and a conjunction. The light stemmers do not remove any strings that would be considered Arabic prefixes.

4   Evaluation

The performance of our experiment was evaluated using the standard measures of recall and precision, [6] A total of four runs were carried out where raw without stemming (2) Light10 with Khoja stopwords list, (3) Light10 with Abu El-Khair stopwords list and (4) a combination of Abu El-Khair and Khoja stopwords lists. The output results obtained from the system to give recall and precision. The same way as in Larkey's evaluation we follow the average non-interpolated precision.

Retrieval results were analyzed by calculating the differences between four runs by average non-interpolated precision and the compression rate score.

Table 3. Average Precision and compression rate for the three runs with Light10

	RAW	Khoja Stopwords list	Abu El-Khair Stopwords list	Combined Stopwords list
Average Precision	0.197	0.23	0.25	0.255
Compression Rate	0.37	0.43	0.45	0.46

5   Results of Running Comparisons

This study compared alternative stop word lists and their effect on retrieval effectiveness. Furthermore, the compression rate also stated her after stemming for three runs out of four. As it expected there is an effect of stopwords list removing on precision, for the combined stopwords list the percent of improving over light stemming with Khoja stopwords list is around 3% in un-interpolated average precision. Also the rate of data compression rate improved after light stemmer with combined stopwords list is very clear it's increased around 9% in data compression rate from the index data size which already increase the performance of the IR system.

The result are shown in Table 3 which shows the values of the three different runs for the three stopwords lists that already tested with light stemmer and the percent of change for the precision and compression rate for each run. According to the RAW run, we also achieved the same result for un-interpolated average precision achieved by Larkey's. But the comparison depend on the effects of stopwords list with light stemmer 'on the AIR system.

6   Conclusion and Future Work

In this research, we investigated the effect of stopwords list and stemming (Light stemming), and, on improving Arabic monolingual IR. We found that stopwords list size and contents with stemming has significant effect on improving retrieval effectiveness. The combined list with stemming are outperformed the Khoja stopwords list.

It would be better to compare these combined stopwords list with other published stopwords list(s) to find standard stopwords list to be used in AIR research. Finally, we concluded, the stopwords list has a clear impact on AIR effectiveness, also, with stemming. Testing the stopwords list over benchmark like we use here in our research, it will encourage the researchers to find the standard one.

For future work testing new stopwords list with different stemmers' light or morphological analysis or hybrid is our goal towards enhancing AIR effectiveness.

Acknowledgments. We would like to thank linguistic Data Consortium (LDC) for providing us with LDC2001T55 Arabic Newswire Part 1 at no-cost, as one of the student recipients of the Fall 2012 LDC Data Scholarship program.

References

1. Al-Maimani, M.R., Naamany, A.A., Bakar, A.Z.A.: Arabic Information Retrieval: Techniques, tools and challenges. In: 2011 IEEE GCC Conference and Exhibition (GCC), Dubai, pp. 541–544 (2011)
2. Larkey, L., Ballesteros, L., Connell, M.: Light stemming for Arabic information retrieval. Arabic Computational Morphology, 221–243 (2007)
3. El-Khair, I.A.: Effects of stop words elimination for Arabic information retrieval: a comparative study. International Journal of Computing & Information Sciences 4, 119–133 (2006)
4. Khoja, S.: APT: Arabic part-of-speech tagger, pp. 20–25 (2001)
5. Al-Shammari, E., Lin, J.: A novel Arabic lemmatization algorithm, pp. 113–118 (2008)
6. Croft, W.B., Metzler, D., Strohman, T.: Search engines: Information retrieval in practice. Addison-Wesley (2010)
7. Fox, C.: A stop list for general text. ACM SIGIR Forum, 19–21 (1989)
8. Larkey, L.S., Connell, M.E.: Arabic information retrieval at UMass in TREC-10. NIST Special Publication SP, pp. 562–570 (2002)
9. Chen, A., Gey, F.: Building an Arabic stemmer for information retrieval. In: Proceedings of TREC (2002)
10. Savoy, J., Rasolofo, Y.: Report on the TREC-11 experiment: Arabic, named page and topic distillation searches. In: TREC-11, pp. 765–774 (2003)
11. Darwish, K., Oard, D.W.: CLIR Experiments at Maryland for TREC-2002: Evidence combination for Arabic-English retrieval. DTIC Document (2003)

12. Aljlayl, M., Frieder, O.: On Arabic search: improving the retrieval effectiveness via a light stemming approach, pp. 340–347 (2002)
13. Manning, C.D., Raghavan, P., Schutze, H.: Introduction to information retrieval, vol. 1. Cambridge University Press, Cambridge (2008)
14. National Institute of Standards and Technology, TREC 2002 cross language topics in Arabic (2002), http://trec.nist.gov/data/topics_noneng/
15. National Institute of Standards and Technology, Data - Non-English Relevance Judgements File List (2001), http://trec.nist.gov/data/qrels_noneng/
16. Gey, F.C., Oard, D.W.: The TREC-2001 cross-language information retrieval track: Searching Arabic using English, French or Arabic queries. In: TREC, pp. 16–26 (2001)
17. Taghva, K., Elkhoury, R., Coombs, J.: Arabic stemming without a root dictionary. In: International Conference on Information Technology: Coding and Computing, ITCC 2005, pp. 152–157 (2005)
18. Alshehri, A.M.: Optimization and effectiveness of n-grams approach for indexing and retrieval in Arabic information retrieval systems (2002)

Multilingual Ontology Learning Algorithm for Emails

Majdi Beseiso[1], Abdul Rahim Ahmad[2], Roslan Ismail[2], and Mohammad Taher[3]

[1] Yanbu University College, Yanbu, Saudi Arabia
majdibsaiso@yahoo.com
[2] Universiti Tenaga Nasional (UNITEN), Selangor, Malaysia
{abdrahim,roslan}@uniten.edu.my
[3] FBK-IRST, Trento, Italy
tahirkhan@fbk.eu

Abstract. The e-mails used for personal or business proposals are becoming more and more difficult to manage. The abundant/overhead of e-mails and the lack of assistance from e-mail clients will certainly lead to losing tracks of information contained in e-mails. The multilingual nature of e-mails makes this situation even worse. In this research, we have explained the necessity for utilizing multilingual ontology learning processes to solve this problem. We have proposed the multilingual ontology learning algorithm for semantic email. The proposed algorithm is designed to handle the unstructured emails. This will lead to intelligent applications and information extraction for email users while keeping the semantic layer opaque to the end user.

Keywords: Semantic Web, Semantic Email, Multilingual Ontology Learning.

1 Introduction

Emails have become one of the most popular digital communication media. The Email service remains a crucial business communication tool and an important source of organization information and knowledge. There are enormous potential benefits that could be earned by using email for communication, although there is associated pitfall of information overload. It is really difficult to monitor the information contained in the relevant messages as those sandwiched between the irrelevant information [7].

These days organizations and large corporations are located in various countries employing and interacting with people of diverse cultural backgrounds and linguistic preferences. With the advent of internet and related technologies, the globalization and interaction among different cultures and countries is increasing at an exponential rate.

Due to this, e-mails used for personal or business proposals are becoming more and more difficult to manage. The abundant/overhead of e-mails and the lack of assistance from e-mail clients will certainly lead to losing tracks of information contained in e-mails. In addition to this, existence of organizations in different countries with using linguistic preference of that region for communication within the organization makes it difficult to build a single email archive for knowledge extraction.

S.A. Noah et al. (Eds.): M-CAIT 2013, CCIS 378, pp. 229–244, 2013.
© Springer-Verlag Berlin Heidelberg 2013

An effective and efficient knowledge management of Emails is critical for organizational growth. In this paper, we propose to extend the research on Semantic Web technologies into the realm of Emails and especially multilingual Emails. The aim is to propose and build a semantic Email application capable of filtering, categorizing and extracting important information from emails automatically without human intervention and based on the semantic email architecture proposed in [20]. An important step in this direction is to semantically annotate the emails and generate ontologies for multilingual emails. The research proposes an efficient algorithm followed by its implementation for automating the process of email ontology creation an and its implementation within existing email systems like Outlook.

A major issue faced is design of ontologies for multilingual concepts and the mapping between them is the linguistic breadth and diversity. Natural language processing techniques need to be applied in a specific linguistic environment and cultural context, for example, Arabic language structure is completely different from that of English. This poses a challenge in development of language agnostic ontologies. A particular natural language does not restrict knowledge and its representation, but representing this knowledge in a different language is a challenge. In order to address this issue, multilinguality is evident in ontologies.

More in detail, in this work we have investigated semantic emails and related ontologies in five broad domains like "Meeting & Schedule", "News", "Discussion and Reviews etc. We have also developed a detailed algorithm and its implementation to achieve the desired results. More importantly we have pondered over the multilingual aspect of emails, which is very important because of cultural diversity and linguistic challenges for emails in Arabic language taken along with English language emails. This entailed a mapping of ontologies in a language agnostic manner to achieve better results.

In the following section, we describe related work with respect to semantic web and semantic email. Section 3 describes our algorithm for ontology learning. In section 4 we present the implementation of our algorithm along with its evaluation in section 5. Finally in section 6 we conclude the paper and outline future work opportunities.

2   Related Work

The World Wide Web (WWW) has been proven to be a great success for its impact on how human beings access and share information among themselves. Nevertheless, it still has serious limitations to understand the knowledge as human beings. People use the computer to produce information and place the information on the Web. Enriching the existing Web with a layer of machine-interpretable metadata is the aim of the Semantic Web. When this aim is achieved, the computer program can draw conclusions and present new information from the one that exist[1]. The Semantic Web is an extension of the current web in which information is defined well in order to enable machines and people to work in a better co-operative way [2].

Over the last decade, the Semantic Web has evolved into several specific standards so as to support many semantic services. The "Semantic Web Layer Cake", making clear the important role played by the core technologies (RDF, OWL etc.) in the overall architecture of the Semantic Web[3]. The first and the fundamental technology of the Semantic Web was the Resource Description Framework (RDF)[4]. RDF is a language for asserting statements about any resources whether it is on the Web or not. The RDF provides primitives related to the semantic modeling, meanwhile some of the advanced requirements cannot be addressed, such as the cardinality restrictions. The Web Ontology Language [5, 6], OWL is considered to be a knowledge representation language for the Semantic Web with a formally defined meaning for creating ontologies. The OWL is a collection of basic "pieces of knowledge". There are some tools currently available for the RDF triple store, inference and reasoning engines, RDF generators, converters and validators, SPARQL data base search engines, content management systems, semantic web browsers, development environments and semantic wikis.

There is a work on the use of semantic emails to get the relevant information from emails is presented in l[8]. This works suggest to have email message consists of a semantic description or an update coupled with a corresponding explanatory text. The Semantic Email Process (SEP) have also been used for a wide range of useful interactions [8] and [9]. Another relevant work is the Information Lens [10], a tool that enables the user to generate a single email with a Semi-Structured content might assist the recipients with the act of filtering and prioritizing this message. Kalyanpur et al. [11] proposed that users semantically annotate messages for improving mail search, sorting and filtering. Many researchers have envisioned a Semantic Email philosophy during the last decade to deal with the information overload that hinders the workflow of the data that is to be handled.

Scerri et al. [12] describe a framework to develop ontology for speech acts and outline non-deterministic models to support the user in deciding the best course of action upon sending or receiving an email, it shows the classes and properties in the Semantic E-Mail ontology. The details in [13] show the most important aspect of the semantic E-Mail ontology "Speech Act Representation" There are several tools incorporatig the principle of Semantic Email.

Previous semantic email research focusing on use of templates where the template is send to users then get their responses [8]. To the best of our knowledge the research dealing with emails as a source of knowledge for different domains is very minimal.

There are many approaches proposed to solve the multilingual component, but most of them faced serious issues during the implementation process. There is one way for realizing this. It is by using the cross-lingual ontology mapping (CLOM). It is a process of establishing relationships among ontological resources from two or more ontologies where each ontology is labeled in a different natural language[14]. A good description of the available literature in the field of multilingual ontologies has been provided in [14].

3   Multilingual Ontology Learning Algorithm for Emails

In this section, we will discuss about the Multilingual Ontology Learning Algorithm. Algorithm is divided into three parts (Concepts extraction, Relation extraction,

Vocabularies for Ontology Extraction). The first part deals with Concepts extraction, part 2 deals with Relations extraction, the last part is for Vocabularies for Ontology Extraction. Fig. 1 shows an overview of goals that will be achieved through the email ontology learning algorithm.

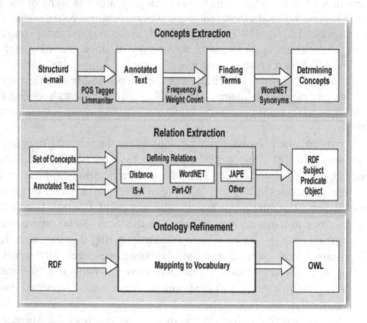

Fig. 1. Depiction of Concept Extraction, Relation Extraction and Ontology Refinemen

3.1   Concept Extraction

The concepts extraction part has two parts; first is Email Preprocessing and other is handling of body part of the email. Email preprocessing will provide the basic but important information or concepts of email. The body part handling will be needed for further analysis for Ontology Learning.

The first task is Concepts extraction in which a semi structured or unstructured email in html form or text is taken and terms are extracted. It reads and divides the header part and extracts terms from the Email. Then the body of the email, the message part which has sentences is split into phrases, and words are extracted and also Tagged by relative Part of Speech (POS).

The following steps describe this part in details:

Step 1: Email Preprocessing: The input for the algorithm will be semi-structured or unstructured text of emails. The goal of e-mail preprocessing is to prepare parts of emails for ontology extraction. The Email XML description generated by preprocessing step contains the basic information that could be obtained from the structure of an email, such as sender, receiver, date, etc. The sample of input & output is presented in Fig. 2:

Email	Preprocessed Email
From: Jack Smith	\<From\> Jack Smith \</From\>
To: Matine White martin.white@example.org	\<To\> Matine White martin.white@example.org \<\To\>
MessageId: 93B2345678C321D1	\<MessageId\> 93B2345678C321D1\</MessageId\>
Date: 22/10/2010 at 13:22:22	\<Date\> 22/10/2010 at 13:22:22 \</Date\>
Dear Mr.Matine,	\<Body\> Dear Mr.Matine,
The meeting has been scheduled on 02/11/2010 in Room32 at 15:00	The meeting has been scheduled on 02/11/2010 in Room32 at 15:00
We require your presence for the meeting... Regards	We require your presence for the meeting... Regards \</Body\>

Fig. 2. Example: Output of Pre-processing Implementation on Email

Step 2: NLP Annotation: NLP systems analyze the text to find phrase structures and group words according to their syntactic and semantic property. This implies that the text is pre-processed by a Part-of-Speech tagger, which groups the words in grammatical categories. The most important NLP components that will be used in this research are: Tokenizer, sentence splitter, POS tagger, Morphology, and Named Entity Recognition.

Step 3: Define email as the root (seed) ontology. ", six concepts are extracted using pattern matching by XML tags in preprocessing of an email viz. *Sender, Recipient, CC, BCC, Time, Date.*

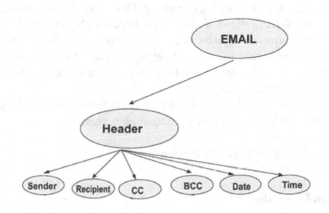

Fig. 3. Nodes of email Header and its child nodes

The following steps are required during this phase:

1. Iterate on Sentences: Parse the sentence and find the number of nouns and noun phrases in a sentence. Take a sentence E_{bm}<Sen_J>, if it has more than two POS E_{bm}<POS_I_J> as Nouns or Noun Phrases, then keep the sentence, otherwise discard it. This is required to extract the relations between concepts.
2. Frequency count needs to be check to find the Keywords or Concepts according to domain from the Email.

Step 4: Calculating the relevance of the words in email and their TFID occurrence. The following steps are required during this phase:

1. Take each token in the remaining sentences and if the token, E_{bm}<Tok_IJ>, has associated Noun POS, calculate the following:

    In email corpus E_{ITXT}

$$\left(NCOUNT_{I_J}\right)email = \frac{\left(Frequency\ Count\ of\ Ebm<Tok_{I_J}>\right)}{(Total\ words)} \tag{1}$$

    In general corpus like GATE corpus or Project Gutenberg, E_{GEN}

$$\left(NCOUNT_{I_J}\right)GEN = \frac{\left(Frequency\ Count\ of\ Ebm<Tok_{I_J}>\right)}{(Total\ words)} \tag{2}$$

    Calculate the ratio:

$$INDEX\_I\_J = \frac{(NCOUNT\_I\_J)email}{(NCOUNT\_I\_J)gen} \tag{3}$$

2. If the ratio $INDEX_{IJ} \geq 1.0$, keep that sentence for further analysis of domain specific ontology learning. Otherwise discard that sentence. The ratio $INDEX_{IJ}$ shows if the noun is appearing significantly enough in the email corpus to be taken as an email domain specific component.
3. An alternative is to use an $INDEX_{IJ}$ to count the match between subject tokens in the header part and tokens between email bodies to identify the domain of the email.
4. Use term weighing, TFID, with a predefined threshold limit to further limit down the noun tokens containing sentences.

$$tfidf(w) = tf(w).log\left(\frac{N}{df(w)}\right) \tag{4}$$

3.2   Relation Extraction

The relation extraction based on POS (part of speech, Noun verb, adverb, etc.) to find relations between these Tokens (terms) based on which POS they belong to like noun or verb or adverb. IS_A patterns are found using WORDNET and lexicon syntactic pattern is used to extract Part_Of relations. Semantic relations are also defined through JAPE rules. The output of which would be terms extracted with their semantic relations between them and their domain, these would form the subject predicate and object Or RDF model of data.

The following steps describe in details:

Step 1: Use the synsets obtained from WORDNET with most frequent sense. Then define " IS_A" relationship between Nouns in a sentence is found by checking for hypernyms in WordNet For example:

The output of hypernymically related synsets can be reconstructed by following the trail of hypernymically related synsets for example: {robin,redbreast}@→{bird} @→{animal,animate_being}@→{organism,life_form,living_thing}@→ is transitive, semantic relation that can be considered as IS-A or KIND OF and the direction of the sign indicates upward pointing.[15]

Step 2: Further relationships (part_of) are obtained using Lexicon-Syntactic pattern matching as defined in JAPE rules.

All the annotations are managed through the lookup lists from the Gazetteer. Where according to[16] the gazetteer list is the lookup list of entities. They are stored in various files which the GATE Gazetteer uses to detect, generally the initial phases of annotations. JAPE rules utilize these annotations along with annotations from other processing resources to further identify patterns/entities from the text. Following is an example of how JAPE rule can be written for our term Meeting to be under the category EVENT (like event meetings or Prayer meetings or group meetings [16].

Below the Fig. 4 is showing how to match and insert the pattern in JAPE for relation extraction from the sentence

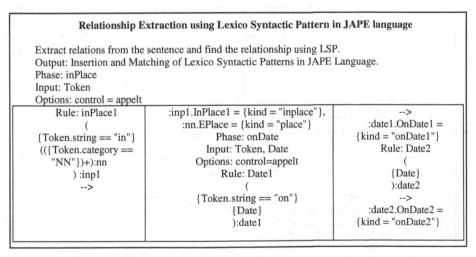

Relationship Extraction using Lexico Syntactic Pattern in JAPE language		
Extract relations from the sentence and find the relationship using LSP. Output: Insertion and Matching of Lexico Syntactic Patterns in JAPE Language. Phase: inPlace Input: Token Options: control = appelt		
Rule: inPlace1 ({Token.string == "in"} (({Token.category == "NN"})+):nn) :inp1 -->	:inp1.InPlace1 = {kind = "inplace"}, :nn.EPlace = {kind = "place"} Phase: onDate Input: Token, Date Options: control=appelt Rule: Date1 ({Token.string == "on"} {Date}):date1	--> :date1.OnDate1 = {kind = "onDate1"} Rule: Date2 ({Date}):date2 --> :date2.OnDate2 = {kind = "onDate2"}

Fig. 4. Lexico Syntactic Pattern of a Sentence

3.3   Vocabularies for Ontology Extraction

In the last stage, the obtained concepts and relationships are integrated and aligned to domain ontologies. User will select the email category for ontology extraction. The XML tags and structured information obtained in steps above are saved into an RDF store and will be used by GATE and ANNIE to obtain the adaptive ontology. Ontologies are identified and written in OWL format.

The extraction of ontologies refer to the one or more (in case that an e-mail contains more than one language) localizations of. Usually, the language agnostic ontology is not used directly in the extraction process unless we deliberately choose one localization as the "language-agnostic" ontology.

We also need some domain specific ontologies (such as Dublin Core, SKOS, FOAF, rNEWS etc.) to align concepts extracted from e-mails. Here, we just give several examples of reusing existing ontologies to model concepts in these domains. Four basic domain vocabularies are considered and defined for the emails: News, Discussion and Comments, Meetings & Schedule, and Collaborative and Technical Requests.

Fig. 5 demonstrates the use of LODE vocabulary for "Schedule and Meetings" domain.. An "Event" is a central concept in the LODE vocabulary which can be used to relate "ec: MeetingEmail" to "Event" by "ec: hasEvent" property. Various LODE properties can be used to relate various terms in e-mail related to "Schedule and Meetings" domain like "atPlace", "atTime" and "involdedAgent" to get the time & duration, location and persons required for the meeting.

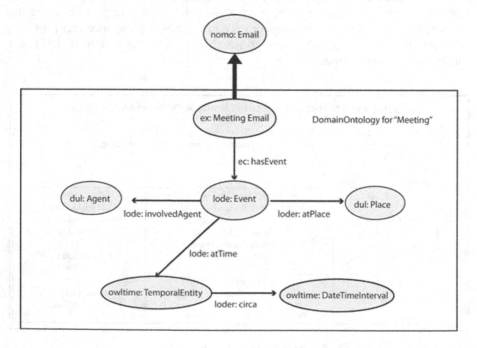

Fig. 5. Domain Ontology for Email "Meeting & Schedule"

3.4   Mapping Algorithm

In our ontology Algorithm we use both Linguistic Matching and Structure-Based Strategies. There are two parts required for mapping, the structure and labels of ontology for structure we use predefined rules based on vocabularies used for cross-lingual ontology matching. For label mapping, we use translation and transliteration tools available online. Fig. 6 describes the algorithm with the main required steps in each phase.

The following Fig. 6 shows the steps involved in ontology mapping. In this algorithm the input would be constructed ontology from RDF obtained from algorithm 1 which is concepts and their relationships that are extracted from semi structured and unstructured email text.

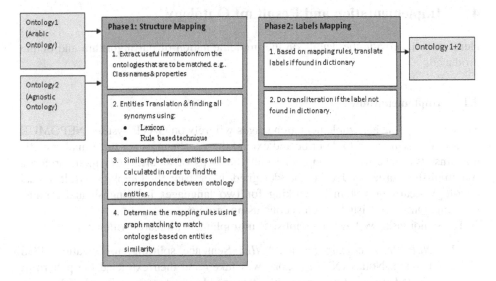

Fig. 6. Algorithm for ontology Mapping

In our algorithm for mapping, we have developed the agonistic ontology template based on the backbone ontologies. The mapping between the agonistic template graph and intermediate localized ontology is based on a matching relation between the graphs produced by two RDF documents

We developed a simple graph matching algorithm to be used for mapping. The following steps summarize how the algorithm functions:

1.   Translate the agonistic ontology template and intermediate ontology from their RDF format into graphs O1 and O2. The translation into graphs is done using the method RDF2Graph() that understands the definition of RDF schema.

2. Obtain an initial mapping, by using graphMapping() method, between O1 and O2 using a simple string matcher. The graphMapping() method defines some operators for string matching, such as, exact, like, and range matching.
3. If the graph O1 is found in O2, then the sub-graph in O2 is added to result in RDF format.
4. Iterate the mapping process until the complete graph O2 is traversed.

In phase two, the label mapping is done using resulted ontology structure from the previous phase and the set of rules to map the labels into specific classes or properties. Translation is used if the instance is found in dictionary otherwise transliteration tools available online are used if the instance is not found in the dictionary.

4   Implementation and Resultant Ontology

Here we will provide the details on the implementation of our algorithm and output produced.

4.1   Implementation

The preprocessing and implementation stages will rely on GATE system. NEPOMUK e-mail models are used to describe and extend the vocabularies for the e-mail specific domains. We have tried to implement all the steps of algorithm in modular form. Accomplished tasks in JAVA and designed User friendly Graphics. Multilingual Ontology Learning system is working for two languages i.e. English and Arabic. Dynamic Ontology Visualization is done using protégé.

The major tasks we have accomplishes in implementation are written below:

1. *NLP Preprocessing using ANNIE*: Sentence splitting, tokenization, POS tagging. Nouns (NN) category was taken and then extracted the pattern of Named entity, time, date, place using JAPE grammar in ANNIE using ANNIE Transducer.
2. *Concept Extraction*: The Nouns of sentences were taken and superclass was determined then relations were extracted. This integration was done using GATE and WORDNET
3. *Synonym Extraction:* Synsets were extracted using WORDNET.
4. *Domain Grammar integration*: Used own grammar through transducer in GATE. This was required to extract lexico-syntactic pattern.
5. *Placement of Prepositions:* Placement of prepositions like 'at place', 'in place' 'on date' etc was analyzed and integrated with the grammar in our own JAPE files.

The final goal is to achieve implementation in a way which makes it easy for end users to get an MS Outlook plugin which will help them to automatically annotate e-mails and track activities based on the content of the e-mails in a multilingual environment.

4.2   Resultant Ontology

The ontology for e-mails is a description of entities in the e-mail domain and their semantic relations. Ontological representations have taxonomic relations, extended vocabulary including words and synonyms using WordNet, conceptual models and knowledge structures. Well defined ontology must have a syntax which is unambiguous and an interpretation which is understood by the machines.

The following Fig. 7 shows the visualization of extracted ontologies with the entities and their relationships in email domain. We have taken the snapshot of extracted ontologies in Review, rNews and Event which consist according to figure common entities/classes such as "Agent", "Person" "Event". This whole ontology can be further optimized through mapping and matching techniques.

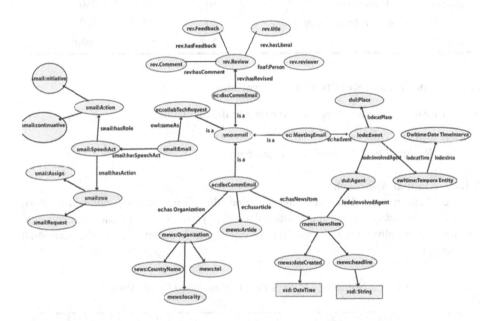

Fig. 7. Snapshot of extracted ontology

5   Evaluation

Ontology evaluation is a critical task when it is learned automatically by the automatic system. The Semantic E-Mail dataset corpus is a prerequisite and is essential for the evaluation of the ontology extraction technique proposed in this research. Three Email dataset corpuses have been considered for evaluation.

1-Enron Email Corpus with Categories,
2- British Columbia Conversation Corpus (BC3)
3-Custom E-Mail Corpus in Arabic language

We have tried to extract the knowledge based on some scenarios like meeting and schedule, Technical Support and Request, Collaboration, Business related information, news, legal documents, jokes, spam, attachments etc. These classified emails are being used for checking of accuracy of the system. In the end we classified all the emails from respective corpuses in four categories as shown in Table 1 below.

Table 1. Classification of Emails from the Three Email Corpus Dataset

Email Domain	Enron	BC3	Customized Arabic
Meeting and Schedule	345	24	225
News	230	18	125
Discussion and Comments	650	36	350
Collaborative and Technical Support	120	15	200
Total	**1345**	**93**	**900**

5.1   Experimental Results and Observation

Two evaluation methods were used to evaluate the algorithm.

a)   Manual Approach
b)   Recall, Precision and F1 measurements

5.1.1   Manual Approach

This evaluation was done keeping in mind the *Completeness, appropriateness, acquaintance, adequacy* of the extracted ontology, *relevance* of the concepts and relationship in ontologies and *usefulness* of ontologies in comparison to the traditional approach[17].

Table 2. Normalized Ratings from Experts based on questionnaire

Ontology Expert	Completeness	Consistency	Conciseness	Preciseness	Clarity	Average	Std. Deviation
1	0.70	0.67	0.73	0.74	0.72	0.71	0.025
2	0.75	0.71	0.73	0.70	0.73	0.72	0.017
3	0.92	0.79	0.77	0.75	0.76	0.80	0.062
4	0.65	0.61	0.67	0.65	0.70	0.66	0.029
5	0.82	0.77	0.81	0.79	0.78	0.79	0.019
6	0.90	0.69	0.78	0.74	0.75	0.77	0.070
7	0.60	0.61	0.63	0.59	0.62	0.61	0.014
8	0.58	0.63	0.67	0.64	0.71	0.65	0.043
Average	0.74	0.69	0.72	0.70	0.72		
Std. Deviation	0.122	0.065	0.059	0.063	0.046		

The normalization of the ordinal numbers provided by the experts is done by taking the average ratings given by experts for each category and dividing it by 5. The highest average rating by experts is given to the quality of completeness. Average rating for completeness is 0.74 (74%), which is quite high for first version of the implemented system for email ontology learning. Overall the ratings are close to 70% for all the other features which is also satisfactory.

The extracted concepts in terms of the numbers and the relations established among the concepts are shown in the Table 3 below. The obtained results shown are based on proposed architecture and used rule sets for ontology learning. We have taken emails of each domain and extracted the concepts and defined relationship among them.

Table 3. Summary of Results

Domain (English Emails)	Schedule & Meetings	News	Review	Technical support	Overall
No. of emails	78	54	120	48	300
No. of extracted concepts	1361	942	2094	840	5237
No. of extracted relations	830	574	1277	512	3193
Domain (Arabic Emails)	Schedule & Meetings	News	Review	Technical support	Overall
No. of emails	75	42	117	66	300
No. of extracted concepts	1163	651	2042	799	4655
No. of extracted relations	709	397	1106	626	2838

5.1.2 Precision, Recall and F Measure

The quality of the learned ontology can also be measured through Precision, Recall & F-measure metrics.

Precision is fraction of retrieved concepts and relations that are relevant and Recall is the fraction of relevant items that are retrieved by the algorithm and the automated system.

Let us say gold standard has given R results and the number of results and concepts extracted by the automated system are X of which N are correct after comparison with gold standard results.

Precision = N/X

Recall = N/R

We can also say that the correct matches are the intersection between R and X. N is equal to the common matches between the R and X. Table 4 shows the results for such an evaluation.

Precision and recall are respectively calculated using N/X and N/R. For meeting and schedule precision is 0.8955 and 0.8927 and recall is 0.7523 and 0.7084. The last column F-measures evaluates the accuracy of the system by calculating the harmonic mean of Precision and Recall. The F-measures for news domain are highest. For other three domains also the accuracy measures are reasonably good at more than 80%. The emails in news domain, in general, are more formal in nature. This minimizes the ambiguities related to natural language.

Table 4. Example of Precision, Recall and F Measure for domains

Domains (English Emails)	R	X	N	Precision	Recall	F-Measure
Meeting & Schedule	2608	2191	1962	0.8955	0.7523	0.8177
News	1805	1517	1358	0.8952	0.7524	0.8176
Discussion and Comments	4013	3372	3019	0.8953	0.7523	0.8176
Collaborative & Technical Support & Requests	1605	1348	1207	0.8954	0.7520	0.8175
Domains (Arabic Emails)	R	X	N	Precision	Recall	F-Measure
Meeting & Schedule	2455	1948	1739	0.8927	0.7084	0.7899
News	1699	1348	1204	0.8932	0.7087	0.7903
Discussion and Comments	3777	2997	2676	0.8929	0.7085	0.7901
Collaborative & Technical Support & Requests	1510	1198	1070	0.8932	0.7086	0.7903

6   Conclusion and Future Direction

Ontology Learning Algorithm is very efficient and can easily customized for any other language. Basic approach applied will be same. The basic idea of the IE in this research is that the multilingual entities are treated as first class objects and the ontology learning models are divided into different languages. Each language will have different extraction rules and lexicons for ontology learning. For the extraction ontology, we will define an agnostic ontology for each language; each language-specific ontology in different ontology learning models will have a mapping to the agnostic ontology.

Evaluation is done to perform a scientific study on efficacy of our ontology learning system. Ontology Learning is complex so it is more practical to do evaluation at different levels of ontology learning separately rather than doing it in totality. We tested the robustness of the algorithm in a large email dataset of different domains. It is giving decent results for English and Arabic both.

The current research work explores the possibility of ontology learning for email using the existing domain ontologies. The work then expands the scope to map the ontologies in multilingual framework.

We can explore other categories of e-mail domain e.g.: Business Letters and Documents, Legal Documents, Press release etc. With the same approach. We need to analyze the Lexico Syntactic patterns and write the JAPE rules then easily we can incorporate new category in the existing system. We can create exhaustive patterns from corpus in future for more accuracy. It involves more relation and sentence forms. Ontology extraction integration with mail servers for improved efficiency, information extraction from the attachments could be added.

References

1. Van Harmelen, F.: The semantic Web: what, why, how, and when. IEEE Distributed Systems Online 5(3) (2004)
2. Berners-Lee, T., Hendler, J., Lassila, O.: The Semantic Web. Scientific American, 34–43 (2001)
3. Kagal, L., et al.: Using Semantic Web Technologies for Policy Management on the Web. In: The Twenty-First National Conference on Artificial Intelligence (2006)
4. Klyne, G., Carol, J.: Resource Description Framework (RDF): Concepts and Abstract Syntax. In: McBride, B. (ed.) W3C Recommendation (2004)
5. W3COWL homepage
6. Smith, M.K., Welty, C., McGuiness, D.L.: OWL.: Web Ontology Language Primer. W3C Recommendation (2004)
7. Whittaker, S., Sidner, C.: Email overload: exploring personal information management of email. In: Proceedings of the SIGCHI Conference on Human Factors in Computing Systems: Common Ground, pp. 276–283. ACM, Vancouver (1996)
8. McDowell, L., et al.: Semantic email. In: Proceedings of the 13th International Conference on World Wide Web, pp. 244–254. ACM, New York (2004)
9. Pazienza, M.T., Stellato, A.: Linguistically motivated Ontology Mapping for the Semantic Web. In: Proceedings of SWAP 2005, the 2nd Italian Semantic Web Workshop, Trento, Italy, December 14-16. CEUR Workshop Preceedings (2005)
10. Malone, T.W., et al.: Intelligent information-sharing systems. Commun. ACM 30(5), 390–402 (1987)
11. Kalyanpur, A., et al.: SMORE -Semantic Markup, Ontology, and RDF Editor (1998)
12. Scerri, S., Davis, B., Handschuh, S.: Improving Email Conversation Efficiency through Semantically Enhanced Email. In: 18th International Workshop on Database and Expert Systems Applications, DEXA 2007 (2007)
13. Pazienza, M.T., Stellato, A.: Exploiting linguistic resources for building linguistically motivated ontologies in the semantic web. In: Second Workshop on Interfacing Ontologies and Lexical Resources for Semantic Web Technologies (OntoLex 2006), Held Jointly with LREC 2006, Genoa, Italy (2006)
14. Fu, B., Brennan, R.: Cross-lingual ontology mapping and its use on the multilingual semantic web. In: Proceedings of WWW Workshop on Multilingual Semantic Web (2010)
15. Fellbaum, C.: WordNet: An electronic lexical database (1998)WordNet is available from, http://www.cogsci.princeton.edu/wn
16. Thakker, D., Osman, T., Lakin, P.: GATE JAPE Grammar Tutorial (2009)

17. Dellschaft, K., Staab, S.: Strategies for the Evaluation of Ontology Learning. In: Proceedings of the 2008 Conference on Ontology Learning and Population: Bridging the Gap between Text and Knowledge, pp. 253–272. IOS Press (2008)
18. Wang, T., Hirst, G.: Extracting Synonyms from Dictionary Definitions. In: Recent Advances in Natural Language Processing (2009)
19. Hu, W., et al.: Gmo: A graph matching for ontologies. In: Proceedings of K-CAP Workshop on Integrating Ontologies (2005)
20. Beseiso, M., Ahmad, A.R., Ismail, R.: A New Architecture for Email Knowledge Extraction. International Journal of Web & Semantic Technology, IJWesT (2012)

Measuring the Compositionality of Arabic Multiword Expressions

Abdulgabbar Saif, Mohd Juzaiddin Ab Aziz, and Nazlia Omar

Center for Artificial Intelligence Technology, Faculty of Information Science and Technology,
Universiti Kebangsaan Malaysia, 43600 Bangi, Selangor, Malaysia
Gabore7@hotmail.com, {din,no}@ftsm.ukm.my

Abstract. This paper presents a method for measuring the compositionality score of multiword expressions (MWEs). Based on Wikipedia (WP) as a lexicon resource, the multiword expressions are identified using the title of Wikipedia articles that are made up of more than one word without further process. Through the semantic representation, this method exploits the hierarchical taxonomy in Wikipedia to represent the concept (single word or multiword) as a feature vector containing the WP articles that belong to concept of categories and sub-categories. The literality and the multiplicative function composition scores are used for measuring the compositionality score of an MWE utilizing the semantic similarity. The proposed method is evaluated by comparing the compositionality score against human judgments (dataset) containing 100 Arabic *noun-noun* compounds.

Keywords: multiword expression, semantic compositionality, wikipedia, semantic similarity.

1   Introduction

Recently, multiword expressions have received a wide attention in the linguistic and natural language processing due to the variety in their linguistic characteristics such as, semantic, syntactic, pragmatic or statistical properties. Handling the MWEs as the single units in different natural language processing tasks has been reported to increase the accuracy of the tasks, such as text mining [19], syntactic parsing [2] and [13], and Machine Translation [5]. One of the most important properties of MWE that plays a significant role in identification and extraction of MWEs is the semantic compositionality. In the semantic theory, according to Partee [14], the semantic compositionality is defined as a function in which the meaning of a complex expression is determined by the meanings of its constituents and its structure. Based on the semantic compositionality perspective, MWEs are classified into three levels [17]: high compositional such as *Nitrogen oxide* (all constituent words contribute in the meaning of MWE), intermediate compositional such as *night owl* and *zebra crossing* (the first or second word only contributes in the meaning of MWE), and non- compositional such as *smoking gun* and *gravy train* (Neither of constituent words contribute in the

S.A. Noah et al. (Eds.): M-CAIT 2013, CCIS 378, pp. 245–256, 2013.
© Springer-Verlag Berlin Heidelberg 2013

semantic of an MWE). Nevertheless, there are many of MWEs that fill between these levels of compositionality. The semantic compositionality measuring of MWE is the task to determine the degree of the interdependence between the meaning of MWE and its components.

The semantic representation of a given MWE and its component words constitute the main issue in estimating the compositionality degree of an MWE. Based on the source of background knowledge, semantic representation approaches can be classified into two main types: corpus-based approach and lexical database approach. The corpus-based approach [9], [8] and [18] depends on the distributional method that represents the semantic of term as a co-occurrence vector containing the patterns of word usage in large corpora. The co-occurrence vector of a word is populated by conflating all the corpus instances of the constituent that lead to the ambiguity for the words that have more than one meaning. These approaches also fixedly rely on the statistical measures for the contextual information extracted from large corpus that can be affected by several uncontrollable noises. Lexical database approach [16] is the knowledge-based method that uses the structured-information of predefined lexicon resources to represent the concept by the semantic relations in that resource.

In this paper, we propose the method based on Wikipedia for detecting the compositionality degree of an MWE depending on the semantic relatedness between MWE and its constituent words. The quantity and quality of Wikipedia are main motivations for exploiting Wikipedia in modeling the compositionality of MWEs. Wikipedia is defined as a goldmine containing a massive amount of highly structured information that is increasing and pruning continuously [10]. It includes many different types of links (redirect links, in-links and out-links) that can be used as the semantic relations (synonyms and associations) between Wikipedia concepts. The links between the concept and category can be utilized as the hierarchical relations (hyponymy; hypernymy), such as the concept *Chemistry is* hypernym of the category *Natural sciences*. The proposed method exploits the hierarchical taxonomy in Wikipedia to represent the concept (single word or multiword) as a feature vector containing the WP articles that belong to concept's categories and sub-categories. In measuring the compositionality of MWE, our hypothesis is that if an MWE is compositional, then its categories are similar or related to the categories of its parts.

The remainder of the paper is organized as follows: Section (2) presents details of previous related studies in measuring the compositionality score of MWE and Section (3) discusses the Wikipedia-based method for identification Arabic MWEs and detecting the compositionality score of MWE. Section (4) describes our experiment for evaluation method. Lastly, section (5) illustrates the conclusions and recommendations for future work.

2   Related Works

In recent years, various approaches have been proposed for analyzing the semantic compositionality of MWEs depending on the semantic representation. Baldwin *et al.* [3] applied the latent semantic analysis in distributional method to represent the meaning of the words and MWEs in vector space model. Two types of MWEs (the English verb-particle construction (VPCs) and noun compound) are modeled to detect the compositionality degree as the similarity between VPCs and their component verbs, and NCs and their component nouns. The traditional distribution method represents the word as a single vector without handling the polysemy for the ambiguous words that have more than one meaning. The collocation graph-based sense induction technique [8] is used in the distributional method to overcome the ambiguity for the polysemous words. The technique clusters the graph into different senses for the target concept to select the major sense of MWE and its semantic head. The Jaccard coefficient is used to measure the semantic similarity between the representation of MWE and its semantic head. The threshold parameter is introduced to decide whether the MWE is a compositional, or non-compositional. However, this approach uses only the head of MWE to test compositionality. Reddy et al. [18] proposed the Multi-Prototypes Based method for addressing the polysemy issue in vector space model of semantic representation of the concepts from the corpus. In this method, the word sense induction is used to cluster the contexts of the particular word to different groups according to the senses of the word. The method was also applied to measure the compositionality score for *noun-noun* expressions as the similarity between the expression and two nouns.

On the other hand, Piao *et al.* [16] proposed the knowledge-based approach that exploits the lexicon resource to measure the compositionality of MWEs using the relative similarity between the semantics of an MWE and its parts. The lexicon resource in the approach is the Lancaster English Semantic Lexicon [15] containing nearly 73,800 concepts that were categorized into 55,000 single-word entries and over 18,800 MWEs entries. Each concept (single or multiword) is tagged to the semantic field classes. The semantic distance between two concepts is measured by quantifying the similarity between their semantic field categories in the hand-tagged hierarchical semantic information. Then, the compositionality of a given MWE is calculated by multiplying the semantic distance between MWE and each its component words.

For the semantic representation vector, there are two models for measuring the compositionality score of an MWE: constituent based models []3, [17] and [18] and composition function based models [7] and [12]. The first model relies on the literality score of the constituent words for MWE that is defined for the word in compound as the semantic similarity between the compound and the word. The second model has been proposed to create the distributional semantic vector for the phrase from the distributional semantic vector of its component words and its structure [6] using the composition function. The simple addition ($a\vec{v_1} + b\vec{v_2}$) and multiplication function [12] are used to build the semantic vector for the phrase or sentences. Guevara [7] applied the partial least squares regression to form the composition function as a multivariate multiple regression.

For identifying Arabic non-compositional expressions, Attia et al. [1] proposed the cross-lingual correspondence asymmetry method that identifies the MWEs in Arabic Wikipedia. This method extracts the MWE candidates from Wikipedia titles that are made up of more than one word. The candidates that have the single word translation in other languages are considered as the non-compositional expressions. In spite of its simplicity and low cost, this method requires far less data and resources for classification the MWEs into two types: compositional and non-compositional expressions. Its binary decision about whether an expression is compositional or non- compositional is usually made based on the criterion of single word translatability.

3   Method

This section describes the method for measuring the compositionality score of Arabic multiword expressions using the Wikipedia. It includes four of four main steps: candidate identification, semantic similarity, alignment constituent words, and measuring the compositionality of MWEs.

3.1   Candidate Identification

This step is concerned with identifying the MWEs from the title of Wikipedia articles and classifying them into named entity and non-named entity. The title of article in Wikipedia can be classified into three types: single word, single word with description tags, and sequence of words. All articles that are titled "*word_ (description tag)*" are considered as a single word Wikipedia concepts. The set of Wikipedia *article titles* that are made up of more than one word are selected as MWE candidates. There are some of candidates that are not constituted the MWEs such as the articles that are tiled "***** قائمة" '*List of ****'* or "*city name, country name*" that represent the name of single word location with its country. All the candidates that belong to this kind are filtered out from the MWE candidates. The set of selected candidates contains many different types of multiword expressions: named entities (personal names, movie names, location names, and organizations), technical terms and general-language terms of varying compositionality (including non-compositional multiword expressions and collocations).

Table 1. Patterens for classification NEs

NE types	Patterns	English patterns
	اشخاص على قيد الحياة	Living people
Personal names	مواليد ****	**** births
	وفيات ****	**** deaths
	مدن ****	cities ****
Locations	عواصم ****	Capitals ****
	بلدان ****	Counties ****
Organizations	منظمات ****	Organizations ****
	شركات *****	Companies ****
Movie names	أفلام *****	***** films
	مسلسل تلفزيوني *****	Television series *****

In this paper, heuristic rules are used to classify the MWEs into two types: named entity MWE and non-NE MWEs. The heuristic rules based on the Wikipedia categories are introduced for each type of named entity: personal name, movie name, location, and organization. The multiword/single Wikipedia articles that have the category matches one personal category pattern are considered as personal names. The same can count for the movie names, locations and organizations. Table 1 presents the category patterns for the named entities (personal names, movie names, locations and organizations). For each MWE candidate, if its categories match at least one of the patterns, it is marked as a NE-MWE otherwise Non-NE-MWE (technical terms).

3.2   Semantic Similarity of Concepts

The basic unit in Wikipedia as a lexical resource is the article that describes the single concept. Wikipedia concepts are categorized as hierarchical organizations ranking from highly specific to highly generic classes. Each Wikipedia concept belongs to one or more than one category. For example, the concept *'chemistry'* belongs to three categories *'chemistry'*, *'Natural sciences'*, and *'Physical sciences'*. Each category contains the set of Wikipedia concepts that are related to each other. In this method, our hypothesis is that the semantic meaning of the concept can be represented by the sibling concepts in hierarchal taxonomy of Wikipedia. In order to measure the semantic similarity between two concepts, we proposed the vector-based method that relies on representation of the concept as the features derived from a structured knowledge in Wikipedia, instead of using co-occurrence counts or contexts. The core idea of this representation is that the concept in Wikipedia belongs to two categories. This means there is a semantic association between these categories regardless of their positions in the hierarchical tree. However, there are some of categories in Wikipedia that do not have any semantic association. For example, *category: page need revise, Articles with dead external links,* and *Articles with inconsistent citation formats* are those categories which are not semantically associated. For this reason, we focus only on the *main category* 'التصنيف الرئيسي'in Wikipedia and other categories are filtered out.

For a given Wikipedia concept x, the concept is represented as the feature vector containing the set of concept (Wikipedia articles) that belongs to the categories of the concept x in hierarchal taxonomy of Wikipedia. The feature vector of the Wikipedia concept is represented as the following:

$$\vec{v_x} = \{(c_1, w_1), (c_2, w_2), \dots \dots, (c_m, w_m)\}$$

In this feature vector representation, each concept is assigned to a weight w_i that is measured to rank the concepts in the vector as the following:

$$w(c_i) = f_i * (WLM(c_i, x) + \beta) \tag{1}$$

Where, f_i is the number of categories (frequency of concept) in which the concept c_i belongs to in. The concept with high frequency is more related to x than the concept with low frequency. However, this is not enough to weigh the concept in vector with respect to the Wikipedia concept. This is because some Wikipedia articles belong to only one category, leading to the same frequency of all concepts in the feature vector. The WLM is the Wikipedia-link measure [11] that computes the semantic relatedness between two Wikipedia articles based on the hyperlinks (in-links). This measure utilizes the *Normalized Google Distance* [4] to apply in Wikipedia as the following:

$$NGD(C_1, C_2) = \begin{cases} \frac{\log(Max(|L_{c1}|,|L_{c2}|)) - \log(|L_{c1} \cap L_{c2}|)}{\log(|W|) - \log(Min(|L_{c1}|,|L_{c2}|))} & , \ |L_{c1} \cap L_{c2}| > 0 \\ 1 & otherwise \end{cases} \quad (2)$$

Where L_{c1} is the set of hyperlinks (in-links) of the concept c_1 and $|w|$ is the number of Wikipedia concepts. The Wikipedia-link measure is defined to measure the semantic relatedness between two Wikipedia concepts as the following:

$$WLM(C_1, C_2) = \begin{cases} 1 - NGD(C_1, C_2) & NGD(C_1, C_2) \leq 1 \\ 0 & otherwise \end{cases} \quad (3)$$

Finally, this measure is smoothed by adding a small quantity $\beta=0.01$ to avoid zeroing the weight of concepts. The semantic similarity between two Wikipedia concepts x and y is calculated as the cosine function between the vector representation of two concepts.

$$SS(x, y) = \cos(\vec{x}, \vec{y}) = \frac{|\vec{x}||\vec{y}|}{|\vec{x}^2||\vec{y}^2|} \quad (4)$$

3.3   Alignment Constituent Word to Wikipedia Concept

This step is to link the constituent words of MWE to the corresponding concepts in Wikipedia by matching the word to the title of articles. For example, the MWE in Wikipedia "أكسيد النتروجين" *'Nitrogen oxide'* consists of two words "النتروجين"*'Nitrogen'* and "أكسيد"*'oxide'*; each of them corresponds to concept as a Wikipedia article title. There are two cases for alignment of the constituent word of MWE to the Wikipedia concept: the word has only one concept (unambiguous word) and an ambiguous word. For the unambiguous word, the Wikipedia concept that is titled by the word is selected without further processing.

The ambiguous word can be aligned to more than one Wikipedia concept according to the number of senses for the word. For example, the MWE "شعر الإبط" *'Underarm hair'* contains the ambiguous word 'شعر' that can be aligned to two Wikipedia concepts "شعر (أدب)" *'poetry'* and "شعر (تشريح)" *'hair'*. In order to solve the ambiguity, let S1 be the set of potential Wikipedia concepts that correspond to the particular word, and the most related concept to MWE is selected.

$$S_1 = \{\overrightarrow{W_{11}}, \overrightarrow{W_{12}}, \dots, \overrightarrow{W_{1n}}\}$$

To select the most related concept, we use the semantic similarity between MWE and each concept candidates. The concept with the highest similarity score is selected as the sense of the given word.

$$mrc(MWE, S_1) = \arg\max_{x \in S_1} SS(MWE, x) \quad (5)$$

3.4   Measuring the Compositionality

For a given MWE with its component words w_1 and w_2, we use two functions for measuring the compositionality score based on constituent based models [3] and composition based models [12]. The first function relies on the literality score of the constituent words for MWE which are defined for the word in compound as the semantic similarity between the compound and the word. The first function is defined as the following:

$$Cmp(MWE) = \frac{1}{2}(SS(MWE, w_1) + SS(MWE, w_2))$$ (6)

This equation can be extended to capture the compositionality of MWE that contains more than two words $\{w_1, w_2, ..., w_n\}$, the compositionality score is defined as the following:

$$Cmp(MWE) = \frac{1}{n}\sum_{i=1}^{n} SS(MWE, w_i)$$ (7)

The second function depends on the composition based model that builds the semantic representation vector of MWE by utilizing the mathematic operation (point-wise multiplication) between the semantic vectors of its parts. In this function, the composed semantic vector of MWE is created by multiplying the feature representation vector of its constituent words. It is defined as the following:

$$Cmp(MWE) = SS(MWE, w_1 \times w_2)$$ (8)

4   Results and Evaluation

4.1   Results

We applied the proposed method on Arabic Wikipedia that has been released on March 15, 2012. The JWPL system [18] has been used for parsing AWP to transform the Wikipedia dump into organized database including the following tables: page, page-in links, page redirect, category, category-pages and category-in links. Table 2 shows the statistics for each table in the organized database.

Table 2. Statistics for Arabic Wikipedia

Tables	Statistics	Table	Statistics
Pages	164,760	Category	50,377
Redirects	150,370	Category-page	665,953
In-links	5,662,744	Category-inlinks	95,175

As previously mentioned, the titles of Wikipedia articles include the single words and multi-words that can be classified into NE-MWE and non-NE-MWE. The results after applying the heuristic rules for identification of the Wikipedia concepts are shown in Table 3.

Table 3. Number of Wikipedia concepts

Types	Multiwords	Single words
Personal names	37,060	1,080
Locations	9,106	13,489
Organizations	8,959	795
Movie names	4,060	964
Technical terms	62,445	19,215
Total	121,630	35,543

4.2 Evaluation

The Methods for measuring the compositionality degree of MWE were typically evaluated by the in-vitro experiments. In these experiments, the compositionality score of MWE was compared directly against a gold standard dataset. Baldwin *et al.* [3] used the WordNet as a gold standard of MWEs and the WordNet-based similarities between MWE and its parts as the compositionality score of MWE. Based on other experiments [17], the compositionality score of MWE was compared directly against the human score in a gold standard. In this paper, we used three experiments to evaluate the proposed method for measuring the compositionality degree of MWE.

In-vitro Experiment. The first experiment aimed to evaluate the performance of the compositionality scores of MWE against human judgment (golden standard dataset). To our knowledge, there is no golden standard (available evaluation datasets) for modeling the compositionality of Arabic MWEs. In this experiment, we describe how to manage building the golden standard dataset for Arabic. Since the single word concepts in Wikipedia are placed under one part of speech (noun), we collected 100 (noun-noun) concepts from Wikipedia that vary in their compositionality degree varying from high compositional phrases to non-compositional compounds. In this sample, the manual measuring compositionality of MWE was performed by ten human annotators. The compositionality score of a MWE was estimated by linguistic experts as a measure of how literal the compound is. For each *N-N* MWE, the human annotators were asked to give the compositionality score for the first noun, second noun and the compound. The compositionality score ranges between 0 (totally different from the meaning of its parts) and 10 (fully compositional).

The proposed method was applied to the golden standard dataset of the compositionality analysis. For each noun compound in the dataset, the compositionality score was calculated by using two functions in the semantic representation of the noun compound and its constituent words. To compute the correlation with human judgments in the golden dataset, we used the Spearman's rank-order correlation coefficient. Table 4 presents the Spearman's correlation of the compositionality scores with the human scores. The results of this experiment indicate that the literality score of the constituent words of MWEs outperforms the multiplicative function composition score.

Table 4. Correlations of compositional functions

Functions	Word 1	Word 2	Noun compound
Constituent	0.65	0.57	0.71
Multiplicative	-	-	0.67

Non-compositional Experiment. Unlike the first experiment, the main aim of conducting the second experiment was test the proposed method that classifies the noun compounds into two types: compositional and non-compositional. The movie name compounds are considered the high non-compositional compounds in which the semantics of its constituent words are totally different from the semantic of compounds. For example, the semantic representation of the movie name compound "ملح الأرض" lit. 'Salt of the Earth' contains the following top concepts: 'سينما' *cinema*, 'مسرح' *Theatre*, 'ممثل' *actor*, 'تلفاز' *Television*, and 'جمهور' *audience*; but the semantic representations of its parts are different.   This experiment also aimed to evaluate the proposed method by measuring the compositionality scores of compounds of the movie name that have been identified from Wikipedia using the heuristic rules. From the movie list, only the compound that each of its constituent word can be aligned to at least one Wikipedia concept was selected. The compound is non-compositional when the compositionality score is below a parameter *threshold ($\lambda=0.01$ in this experiment)*, otherwise it is a compositional. To assess the performance of the method in this experiment, its accuracy was computed as the percentage of non- compositional movie name compounds over the size of dataset. Table 5 shows the accuracy of the method on the dataset of the movie name compounds.

Table 5. Accuracy of non-compositional movie name compounds

Patterns of compound	No. of compounds	Accuracy
N-N	746	0.84
N-N-N	193	0.85
N-N-N-N	47	0.95

Cross-lingual Experiment. The third experiment tested the cross-lingual hypothesis proposed for identifying the Arabic MWEs identification by [1]. According to this, it was hypothesized if the MWE is translated as a single word in other languages, it is considered a non-compositional expression. Based on the compositionality detection, the many-to-one (*M1*) correspondence relationships can be used as the measure for weighting the MWE candidates depending on the translations.   This measure is the percentage of number of languages with a single word translation of MWE over the number of languages under testing. For example, the MWE "فقر الدم" *'Anemia'* in Arabic that has the translation as a single word in all languages is considered a high non-compositional phrase with the M1 measure value (100%) according to the hypothesis. On the other hand, the MWE "دائرة كهربائية" *'electrical network'* in Arabic that has the translation as a single word in only one language among 20 languages in

Wikipedia is considered a low non-compositional phrase with the M1 measure value (5%). Obviously, the low non-compositional phrase is the same high compositional phrase and the (1-M1) can be used to measure the degree of compositionality for the MWEs. In order to compare between the cross-lingual measure and our measure, the MWEs that have single word translation in at least one language among 22 languages were extracted as the non-compositional MWE candidates and the (1-*M1*) measure is calculated for each MWE. For each MWE candidate, the compositionality score was measured by using the constituent function model. In this experiment, 1,405 noun compounds were selected as the dataset for assessing the cross-lingual hypothesis against the compositionality score of the proposed method. Fig. 1 presents the results of the compositionality score and M1 measure.

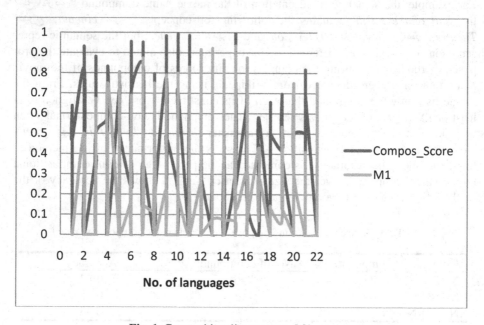

Fig. 1. Compositionality score vs. M1 measure

Based on the results displayed in Fig. 1, it is obvious that there is a clear disparity between the compositionality score and the many-to-one correspondence measure. The Spearman rank-order correlation coefficient between the compositionality score and the many-to-one correspondence measure is 0.135. This means that the many-to-one correspondence relationships can be used for identifying the multiword expressions, but it cannot be used for measuring the compositionality degree of an MWE. For example, the compound "ثعبان البحر" '*Eel*' has a single word translation in 22 languages, but it is a compositional phrase in which the semantic representation can be extracted from the semantic of its words "ثعبان" 'snake' and "البحر" 'sea'.

5   Conclusion

In this paper, we presented a novel method for measuring the compositionality score of an MWE. We hypothesized that the semantic of the concept can be represented by the sibling concepts in the hierarchal taxonomy of Wikipedia instead of using co-occurrence counts or contexts. In applying the proposed method to the Arabic noun compounds found as Wikipedia concepts, three experiments were conducted to evaluate the compositionality score of Arabic noun compounds. In the first experiment, the literality score and the multiplicative function composition score were compared against the human judgment (golden standard dataset). This experiment showed an encouraging correlation of 71% for literality score and 67% for the multiplicative function. The second experiment carried out on the compounds of the Arabic movie name to classify them into compositional and non-compositional phrases. The average accuracy of the proposed method obtained in this experiment was 88%. The final experiment indicated that the many-to-one correspondence relationships used for extracting the MWEs cannot be used for measuring the compositionality score of MWEs. Therefore, the proposed method can be extended to measure the compositionality score for other types of MWEs.

Acknowledgments. The authors would like to express their gratitude to the Ministry of Higher Education for funding this project under the grant no. (FRGS/1/2012/SG0S/UKM/02/14).

References

1. Attia, M.A., Tounsi, L., Pecina, P., van Genabith, J., Toral, A.: Automatic extraction of arabic multiword expressions. In: Workshop on Multiword Expressions: from Theory to Applications (MWE 2010), pp. 19–27 (2010)
2. Attia, M.A.: Accommodating Multiword Expressions in an Arabic LFG Grammar. In: Salakoski, T., Ginter, F., Pyysalo, S., Pahikkala, T. (eds.) FinTAL 2006. LNCS (LNAI), vol. 4139, pp. 87–98. Springer, Heidelberg (2006)
3. Baldwin, T., Bannard, C., Tanaka, T., Widdows, D.: An empirical model of multiword expression decomposability. In: Proceedings of the ACL 2003 Workshop on Multiword Expressions: Analysis, Acquisition and Treatment (2003)
4. Cilibrasi, R.L., Vitanyi, P.M.B.: The Google Similarity Distance. IEEE TKDE 19(3), 370–383 (2007)
5. Deksne, D., Skadiņš, R., Skadiņa, I.: Dictionary of multiword expressions for translation into highly inflected languages. In: Proceedings LREC, Marrakech, pp. 1401–1405 (2008)
6. Erk, K., Padó, S.: A structured vector space model for word meaning in context. In: Proceedings of the 2008 Conference on Empirical Methods in Natural Language Processing (EMNLP), Edinburgh, UK, pp. 897–906 (2008)
7. Guevara, E.: Computing Semantic Compositionality in Distributional Semantics. In: Proceedings of the Ninth International Conference on Computational Semantics (IWCS 2011), Oxford, England, UK, pp. 135–144 (2011)

8. Korkontzelos, I., Manandhar, S.: Detecting Compositionality in Multi-Word Expressions. In: Proceedings of the ACL-IJCNLP 2009 Conference Short Papers, pp. 65–68. Association for Computational Linguistics (2009)
9. Landauer, T., Dumais, S.: A solution to plato's problem: The latent semantic analysis theory of acquisition, induction, and representation of knowledge. Psychological Review 104(2), 211 (1997)
10. Medelyan, O., Milne, D., Legg, C., Witten, I.H.: Mining meaning from wikipedia. Int. J. Hum.-Comput. Stud. 67(9), 716–754 (2009)
11. Milne, D., Witten, I.H.: An Effective, Low-Cost Measure of Semantic Relatedness obtained from Wikipedia Links. In: Proceedings of AAAI Workshop on Wikipedia and Artificial Intelligence (WIKIAI), pp. 25–30. AAAI, Menlo Park (2008)
12. Mitchell, J., Lapata, M.: Vector-based models of semantic composition. In: Proceedings of the 46th Annual Meeting of the Association for Computational Linguistics: Human Language Technologies, pp. 236–244 (2008)
13. Nivre, J., Nilson, J.: Multiword Units in Syntactic Parsing. In: Proceedings of MEMURA 2004 Workshop, Lisbon, pp. 39–46 (2004)
14. Partee, B.H.: Lexical semantics and compositionality. In: Osherson, D. (series ed.), Gleitman, L., Liberman, M. (volume eds.) Invitation to Cognitive Science. Part I: Language, pp. 311–360. MIT Press, Cambridge (1995)
15. Piao, S.L., Archer, D., Mudrayam, O., Rayson, P., Garside, R., McEnery, T., Wilson, A.: A Large Semantic Lexicon for Corpus Annotation. In: Proceedings of the Corpus Linguistics Conference, Birmingham, UK (2005)
16. Piao, S., Rayson, P., Mudraya, O., Wilson, A., Garside, R.: Measuring MWE compositionality using semantic annotation. In: Proceedings of the Workshop on Multiword Expressions: Identifying and Exploiting Underlying Properties, Sydney, Australia, pp. 2–11 (2006)
17. Reddy, S., Klapaftis, I., McCarthy, D., Manandhar, S.: An Empirical Study on Compositionality in Compound Nouns. In: Proceedings of the 5th International Joint Conference on Natural Language Processing, Chiang Mai, Thailand, pp. 210–218 (2011)
18. Reddy, S., Klapaftis, I., McCarthy, D., Manandhar, S.: Dynamic and static prototype vectors for semantic composition. In: Proceedings of the 5th International Joint Conference on Natural Language Processing, Chiang Mai, Thailand, pp. 705–713 (2011)
19. SanJuan, E., Ibekwe-SanJuan, F.: Text mining without document context. Inf. Process. Manage. 42(6), 1532–1552 (2006)
20. Zesch, T., Müller, C., Gurevych, I.: Extracting Lexical Semantic Knowledge from Wikipedia and Wiktionary. In: Proceedings of the 6th International Conference on Language Resources and Evaluation, Marrakech, Morocco, pp. 1646–1652 (2008)

Joint Distance and Information Content Word Similarity Measure

Issa Atoum and Chih How Bong

Faculty of Computer Science and Information Technology,
University of Malaysia Sarawak, 94300 Kota Samarahan,Sarawak,Malaysia
Issa.Atoum@gmail.com, chbong@fit.unimas.my

Abstract. Measuring semantic similarity between words is very important to many applications related to information retrieval and natural language processing. In the paper, we have discovered that word similarity metrics suffer from the drawback of obtaining equal similarities of two words, if they have the same path and depth values in WordNet. Likewise information content methods which depend on word probability of a corpus tends to posture the same drawback. This paper proposes a new hybrid semantic similarity to overcome the drawbacks by exploiting advantages of Li and Lin methods. On a benchmark set of human judgments on Miller Charles and Rubenstein Goodenough data sets, the proposed approach outperforms existing methods in distance and information content based methods.

Keywords: semantic similarity, similarity measures, edge counting, information content, word similarity, WordNet.

1   Introduction

Semantic similarity encompasses the semantic likeness between compared words. For example, *fork* and *food* are related but they are not similar, whereas *food* and *salad* are more similar in semantics. Resnik illustrated that word similarity is a subcase of word relatedness [1]. However, many existing word similarity measures do not clearly distinguish between similarity and relatedness but instead they use a score on a scale between 0 and 1 to indicate the degree of semantic relatedness. Semantic similarity measures have been seen widely used in different applications such as: word sense disambiguation, information retrieval, question answering, summarization, machine translation and automatic essay grading[2].

Although many semantic similarity measures exist in past literatures, most of them are either distance or Information Content (IC) based. The distance based (also called edge counting method) usually uses a thesaurus (such as WordNet) to find words' pair shortest path or Least Common Subsumer (LCS) depth length (shortly referring to it depth) to derive a semantic score. Although it seems the methods are reported to work well, one intuitive problem is that words having the same path and same depth will yield an identical score, even though they postulate semantic differences. On the other hand, the fine-grained IC methods not only relying on the structure of thesaurus, but it

S.A. Noah et al. (Eds.): M-CAIT 2013, CCIS 378, pp. 257–267, 2013.

also depend on the probability of words used in a dictionary. The significant fact is that both methods depend on the LCS, which is the shared ancestor of the two concept words to determine semantic similarity. For example, the LCS of *school bus* and *train* is *public transport*, which is depicted in Fig 1.

Fig. 1 illustrates a fragment of WordNet hierarchy. The numbers indicate the IC values of a node extracted from Brown Dictionary. Noted here that using the distance based methods, the word pairs *wheeled vehicle–rocket* and *bus–train* yield identical score as they have the same shortest path to each other and the same depth in the tree, which is (2,8)[1] respectively. Similarly, the words *car– public transport* will have the same similarity as *rocket– train* where intuitively the later should yield lower similarity score.

By obtaining the IC, we will get that the words pairs *bus* and *train* are more similar than the word pairs *rocket* and *wheeled vehicle*. Likewise, using a distance based method we will get a similarity for *car* and *scooter* rather than getting zero using an information content approach due to unavailability of information for *scooter* in Brown Dictionary.

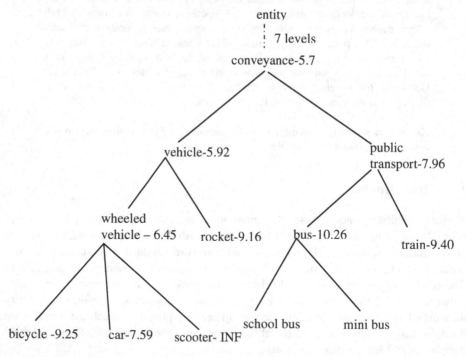

Fig. 1. Fragment of WordNet showing nodes in an *"is-a"* relationship and the information content of each node. The figure shows that word pairs such as *train - wheeled vehicle* and *bus-rocket* can have same similarity value using a distance based method because they have the same path and depth values (4,7).

---

[1] (path, depth).

When the information content ratios between words commonality (i.e LCS) and their individual information content are almost identical will be illustrated in Table 1.

Table 1 illustrates the above-mentioned drawbacks using one example of distance based measures Li et al.[3] (shortly Li) and information content method of Lin[4]. In Table 1.a when the word pairs are been measured by Li and Lin separately, their similarity scores in Li are identical (0.818 for the first two pairs and 0.448 to the second two pairs) but according to the human means these word pairs somehow show semantic differences. In Table 1.b when the word pairs are been measured by same measures separately, their similarity scores in Lin are identical (1.00 for the first two pairs and 0.123 the second two pairs) but their human means are relatively semantically different. Close examination of in the first case shows that this occurs because the word pairs have same path and their LCS have the same depth length while in the second case the word pairs contain equal information content values through their LCS.

Table 1. a) Comparison of word pairs similarity given same path and depth of (1:10, 4:7) respectively. All these word pairs have the same similarity using Li method (0.818 for the first two pairs and 0.448 to the second two pairs). b) Comparison of word pairs similarity given equal IC values in Brown Corpus. All pairs have the same similarity measure using Lin method (1.00 for the first two pairs and 0.123 the second two pairs) because the ratio between commonality to individual IC is equal. The calculation of these values can be found in section 3, equations (1) and (4).

a)   Words that have the same Li similarity due to equal path and depth.

word pairs(w1,w2)	human means	Path length	LCS depth	Li Sim.
journey - voyage	3.84	1	10	0.82
asylum - madhouse	3.61	1	10	0.82
lad-brother	1.66	4	7	0.45
monk-slave	0.55	4	7	0.45

b)   Words that have the same Lin similarity due to equal IC ratios.

word pairs(w1,w2)	human means	IC LCS(w1,w2)	IC(w1)	IC(w2)	Lin Sim.
car - automobile	3.92	7.59	7.59	7.59	1.00
gem-jewel	3.94	12.07	12.07	12.07	1.00
cemetery-woodland	0.95	11.29	9.61	1.29	0.12
forest-graveyard	0.84	9.61	11.29	1.29	0.12

Presenting the discovered drawbacks, this paper is to suggest an approach to resolve these issues by utilizing a hybrid approach over edge counting approach. We propose a new weighting factor on the aggregated score of Li and Lin to resolve the equal values for IC under the LCS.

First, the related works are summarized. Next the proposed approach is presented and explained. Then, the proposed approach is evaluated and finally, the paper is concluded.

2   Related Work

Since many semantic similarity measures are built over WordNet[5], here, we give a brief about it. WordNet is a lexical knowledge of English that contains more than 155,000 words organized into a taxonomic ontology of terms[2] (see Fig. 1). It contains nouns, verbs, adjectives and adverbs that are clustered into synonym sets (called synsets). Each synset can be linked to different synsets via a specific relationship entailed between concepts. Hyponym and hypernym represent *is-a* relationship, and the meronym and holonym which represent *part-of* relationship are some of the common relationships that can be found in WordNet. The norm of the relationship is *is-a* relationship which covers more than 80% of WordNet taxonomy.

Below is some related works which we have categorized into interrelated groups: distance based, IC based, and hybrid methods. Our approach falls on the hybrid methods that we defer to the last subsection of related works to provide a readability sequence. An extensive literature review is outside the scope of this paper.

2.1   Distance Based Methods

Distance based methods, as its name implies, depend on hierarchy structure of dictionaries to determine the word similarity between word senses. Classical methods use shortest path alone [6] , while others extend the path measure with depth of the LCS of compared words [3][7] . Leacock Chodorow [8] proposed a similarity measure based on number of nodes and depth in which terms occur. Hirst and St-Onge [9] considered all types of WordNet relations that represents the path length and its change in direction.

2.2   IC Based Methods

Information content methods are methods that employ the thesaurus structure and with probability of words in a corpus. Resnik [1], Jiang and Conrath method [10] (known as Jcn) proposed a similarity measures based on IC of the LCS of compared

---

[2] Version 3.0 as of Feb 2013.

term synsets. The difference between these measures is the used formulas. These methods can use another approach to get similarity using intrinsic information in thesauri such as the approach of Seco et al.[11], that depends on hyponyms count and maximal number of concepts.

Also, dictionary based methods (known as Lesk family of methods) depend on measuring the overlapping between glosses of related word pairs. In the original Lesk algorithm[12], the target word signature is compared with signatures of each of the context words. The extended Lesk measure of Patwardhan et al. [13] identifies that two concepts are similar if their glosses contain similar overlapping words based on phrases length and different thesaurus relation types.

2.3   Hybrid Methods

These methods combine any of the previous presented methods. Zhou et al. in his paper [14] proposed a similarity measure as a function of IC and path length. On the other hand, Rodriguez and Egenhofer [15] used weighted sum between synsets, features and neighbor concepts of evaluated concepts. Dong et al. [16] propose a weighted edge approach to give different weights for words that share the same LCS and have the same graph distance; words with lower edge weights are more similar than words with higher edge weights.

3   Proposed Joint Distance Information Content (JDIC) Word Similarity Measure

Li proposed that the similarity of two words can be measured using a function of the attributes: path length and depth in the WordNet. The similarity measure has the benefit of balancing the word depth and the shortest path in WordNet. The similarity is a non-linear function where it is monotonically decreasing with path length and monotonically increasing with depth length. In other words, the function projects low similarity for word pairs that has long path and shallow depth in the WordNet taxonomy which indicates its general meaning. Li similarity metric is shown in (1) .

$$Sim_{Li}(w_i, w_j) = e^{-\alpha l} \cdot \frac{e^{\beta h} - e^{-\beta h}}{e^{\beta h} + e^{-\beta h}} \tag{1}$$

w_i and w_j are given words, l is the shortest path between the word pairs, where h is the depth of the LCS. Both α and β are the smoothing factors for WordNet thesaurus that has been empirically found to be 0.2 and 0.45 respectively [17].

On the other hand, Lin's similarity measure can be described as the ratio of the information shared in common to the total amount of information possessed by the two words. It was derived based on Resnik's [1] similarity shown in (2) where it intakes are depended on the LCS of the two words.

$$Sim_{Resnik}(w_i, w_j) = IC\left(LCS(w_i, w_j)\right) \tag{2}$$

IC refers to the Information Content that can be computed using (3), which can be calculated by taking the probability of a word in a corpus over the logarithmic function that have the same part of speech tag (POS).

$$IC(w) = -\log p(w) \tag{3}$$

$p(w)$ is the probability of the word w based on the observed frequency counts in the information content of Brown Corpus. The equation below show Lin's word similarity measure,

$$Sim_{Lin}(w_i, w_j) = \frac{2 * Sim_{Resnik}\left((w_i, w_j)\right)}{IC(w_i) + IC(w_j)} \tag{4}$$

Our proposed measure borrows Li and Lin similarity measures. Li method has provided good performance by taking into consideration path and depth together. On the other hand, Lin method can help in getting similarity of words that are commonly used.

We proposed a similarity metric that is able find a more accurate similarity for word pairs that have the same path and depth by utilizing also the words' information content. It can also produce a similarity score if the words have the same value of commonality to total information content of compared words. The proposed similarity measure is illustrated in (5). For simplicity of reference, we will refer to the proposed measure as JDIC, which stand for Joint Distance and Information Content word similarity measure. A special case of (5) is that when the Sim_{Lin} equals to zero then we take only the Sim_{Li} as in (1) so not to lose the whole value of similarity, likewise for vice versa.

$$Sim_{JDIC} = \psi * log_2\big(Sim_{Li}(w_i, w_j) + 1\big) * log_2\big(Sim_{Lin}(w_i, w_j) + 1\big) \tag{5}$$

$$\psi = 1 - e^{-(log2(IC(w1)*IC(w2)+1)+e^{-\alpha l})} \tag{6}$$

$\psi \in [0,1]$ is a weighting factor that combines the IC of the pairs and its shortest path contribution as shown in (6). This factor gives more similarity to words that have higher IC given that they have the same path. Our experiment shows that the path weight is more influential than the depth of LCS. This weighting factor also considers the case when a word has no IC, or if the compared words' ICs are almost identical.

In other words, when IC is zero, then the path weight will decide the value of this equation. Contrariwise, when path length weight value is very low, then the IC will decide the value of this equation. The lowest value of this weighting factor is zero when the one or both of the compared words are not in the scope. The highest value will be 1 when there is sufficient information about the compared words in a corpus.

4   Evaluation and Discussion

4.1   Data Sets

To evaluate the performance of the proposed algorithm, we have selected Rubenstein Goodenough[18] and Miller Charles[19] word pairs, which are widely used for benchmarking semantic similarity. Rubenstein and Goodenough investigate synonymy judgements of 65 noun pairs categorized by human experts on the scale of 0.0 to 4.0. Whereas Miller and Charles selected 30 word pairs out of the 65 pairs of nouns, organized them under three similarity levels, and study the semantic similarity in the context of how these words are used.

4.2   Similarity Measures Comparison

All our experiments are run with WordNet 3.0. Brown Dictionary [20] was used as a base to obtain IC. Rada [6], Wup [7], Resnik [1], Jcn [10], Lch [8], Li et al. [3], Lin[4] are implemented in Python to obtain similarity score. Table 2 shows similarity results on Miller data set. Fig. 2 and Fig. 3 respectively summarize the correlation of different similarity measures against human means on the Miller and Goodenough data set. Similarities may have a minor difference from Pedersen[3] Word Similarity due to difference in implementation and used dictionary (Brown in our case).

Among the distance based measures (Hso, Rada, Wup, Lch, and Li), those consider only path length (Rada and Hso) yields the lowest correlation. Other measures in the same category such as Wup, Lch and Li are performed slightly better due to inclusion of additional information LCS, besides depth.

For the IC based measures (Jcn, Res, and Lin), Lin achieves the highest correlation, outperform the others.

In Table 2, it is obvious that the proposed word similarity measure, JDIC outperforms the others when correlates the similarity scores to the human means on both Miller and Rubenstein data sets. JDIC measure is able to detect semantic differences even the two word has the length and depth. The reason behind this performance is that the new measure incorporates IC from the corpus and at the same time utilizes WordNet taxonomy information, which yield different similarity score even words have identical path and length. For example, using JDIC, the word pairs *car - automobile* and *gem – jewel* in Table 2 will get the similarity of (0.997, 0.957) respectively that correlates better than Li and Lin measures with respect to human means.

---

[3] http://marimba.d.umn.edu/cgi-bin/similarity/similarity.cgi

Table 2. A comparisons between JDIC to other word similarity against Miller Charles Data set

Word Pairs	HM	Jcn	Hso	Rada	Lch	Wup	Res	Li	Lin	JDIC
car-automobile	3.92	1.000	16	1.00	3.638	1.000	7.59	0.999	1.000	0.997
gem-jewel	3.84	1.000	16	1.00	3.638	1.000	12.07	0.947	1.000	0.957
journey-voyage	3.84	0.298	4	0.50	2.944	0.952	7.11	0.819	0.809	0.737
boy-lad	3.76	0.292	5	0.50	2.944	0.947	8.40	0.818	0.831	0.751
coast-shore	3.70	1.000	4	0.50	2.944	0.909	9.42	0.801	0.963	0.825
asylum-madhouse	3.61	0.313	4	0.50	2.944	0.952	9.48	0.819	0.856	0.769
magician-wizard	3.50	1.000	16	1.00	3.638	1.000	11.98	0.818	1.000	0.862
midday-noon	3.42	1.000	16	1.00	3.638	1.000	11.06	1.000	1.000	0.999
furnace-stove	3.11	0.064	5	0.10	1.335	0.526	2.31	0.162	0.228	0.063
food-fruit	3.08	0.092	0	0.10	1.335	0.400	1.59	0.118	0.161	0.033
bird-cock	3.05	0.223	6	0.50	2.944	0.952	7.68	0.819	0.774	0.712
bird-crane	2.97	0.193	5	0.25	2.251	0.870	7.68	0.549	0.748	0.507
tool-implement	2.95	1.000	4	0.50	2.944	0.933	5.88	0.816	0.947	0.824
brother-monk	2.82	1.000	4	0.50	2.944	0.952	9.26	0.819	0.986	0.853
lad-brother	1.66	0.073	3	0.20	2.028	0.667	2.33	0.448	0.255	0.173
crane-implement	1.68	0.086	3	0.20	2.028	0.750	3.26	0.445	0.359	0.233
journey-car	1.16	0.069	0	0.06	0.747	0.105	0.00	0.014	0.000	0.003
monk-oracle	1.10	0.062	0	0.13	1.558	0.571	2.33	0.246	0.226	0.092
cemetery-woodland	0.95	0.055	0	0.11	1.440	0.429	1.29	0.176	0.123	0.038
food-rooster	0.89	0.063	0	0.06	0.865	0.211	0.80	0.036	0.092	0.006
coast-hill	0.87	0.127	4	0.20	2.028	0.667	5.88	0.425	0.599	0.345
forest-graveyard	0.84	0.055	0	0.11	1.440	0.429	1.29	0.176	0.123	0.039
shore-woodland	0.63	0.061	3	0.20	2.028	0.600	1.29	0.393	0.136	0.086
monk-slave	0.55	0.073	3	0.20	2.028	0.667	2.33	0.448	0.254	0.173
coast-forest	0.42	0.058	2	0.17	1.846	0.545	1.29	0.322	0.131	0.070
lad-wizard	0.42	0.073	3	0.20	2.028	0.667	2.33	0.448	0.255	0.173
chord-smile	0.13	0.062	0	0.13	1.240	0.375	2.62	0.097	0.246	0.040
glass-magician	0.11	0.060	0	0.13	1.558	0.471	2.28	0.177	0.214	0.064
rooster-voyage	0.08	0.044	0	0.04	0.460	0.080	0.00	0.004	0.000	0.000
noon-string	0.08	0.059	0	0.08	1.153	0.267	0.60	0.079	0.066	0.010
Pearson Correlation	1.0	0.691	0.667	0.752	0.779	0.782	0.813	0.794	0.835	**0.841**

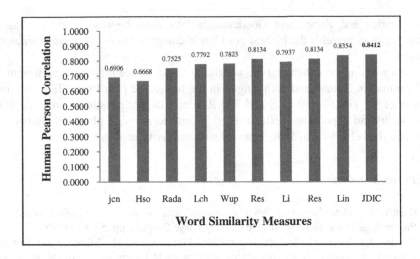

Fig. 2. Pearson correlations with human means on Miller Charles data set

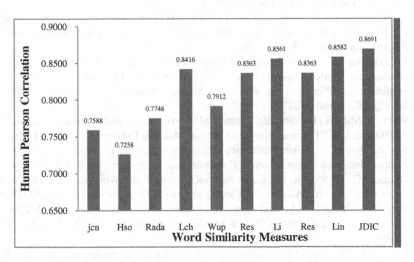

Fig. 3. Pearson correlations with human means on Rubenstein and Goodenough data set

5   Conclusion and Future Work

This paper derives a word similarity measure to address the identical similarity score due to words that have the same path length, depth and information content. We proposed an approach to find the similarity between two word pairs based on distance based approach of Li and the information content approach of Lin. The proposed approach has the flexibility of the weighting factor which allows discriminating equal words similarities. Benchmarking the proposed approach with human judgments on

Miller Charles and Rubenstein Goodenough data sets, the results shows that the proposed approach yields the highest correlation compared to existing word similarity measures.

We are working on enhancing the weighting factor to identify the best weight for both Information content and path length. In the future, we plan to test JDIC on other data sets of 80 TOEFL, 50 ELS and 300 Reader's Digest questions, compare JDIC with other hybrid approaches, extend the similarity to measure phrase similarity, and to measure the behaviour of JDIC in a sentence similarity measure.

References

[1] Resnik, P.: Disambiguating noun groupings with respect to WordNet senses. In: Proceedings of the Third Workshop on Very Large Corpora, pp. 54–68 (1995)

[2] Varelas, G., Voutsakis, E., Raftopoulou, P., Petrakis, E.G.M., Milios, E.E.: Semantic similarity methods in wordNet and their application to information retrieval on the web. In: Proceedings of the 7th Annual ACM International Workshop on Web Information and Data Management, pp. 10–16 (2005)

[3] Li, Y., McLean, D., Bandar, Z.A., O'Shea, J.D.D., Crockett, K.: Sentence similarity based on semantic nets and corpus statistics. IEEE Transactions on Knowledge and Data Engineering 18(8), 1138–1150 (2006)

[4] Lin, D.: An information-theoretic definition of similarity. In: Proceedings of the 15th International Conference on Machine Learning, vol. 1, pp. 296–304 (1998)

[5] Fellbaum, C.: WordNet: An Electronic Lexical Database, vol. (2000), pp. 231–243. Springer, Cambridge (2010)

[6] Rada, R., Mili, H., Bicknell, E., Blettner, M.: Development and application of a metric on semantic nets. IEEE Transactions on Systems, Man and Cybernetics 19(1), 17–30 (1989)

[7] Wu, Z., Palmer, M.: Verbs semantics and lexical selection. In: Proceedings of the 32nd Annual Meeting on Association for Computational Linguistics, pp. 133–138 (1994)

[8] Leacock, C., Chodorow, M.: Combining local context and WordNet similarity for word sense identification. WordNet: An Electronic Lexical Database 49(2), 265–283 (1998)

[9] Hirst, G., St-Onge, D.: Lexical chains as representations of context for the detection and correction of malapropisms. WordNet: An Electronic Lexical Database 5, 305–332 (1998)

[10] Jiang, J.J., Conrath, D.W.: Semantic similarity based on corpus statistics and lexical taxonomy. arXiv preprint cmp-lg/9709008 (1997)

[11] Seco, N., Veale, T., Hayes, J.: An Intrinsic Information Content Metric for Semantic Similarity in WordNet. In: ECAI 2004, the 16th European Conference on Artificial Intelligence, vol. (Ic), pp. 1–5 (2004)

[12] Lesk, M.: Automatic sense disambiguation using machine readable dictionaries: how to tell a pine cone from an ice cream cone. In: Proceedings of the 5th Annual International Conference on Systems Documentation, pp. 24–26 (1986)

[13] Patwardhan, S., Banerjee, S., Pedersen, T.: Using Measures of Semantic Relatedness for Word Sense Disambiguation. In: Gelbukh, A. (ed.) CICLing 2003. LNCS, vol. 2588, pp. 241–257. Springer, Heidelberg (2003)

[14] Zhou, Z., Wang, Y., Gu, J.: New model of semantic similarity measuring in wordnet. In: 3rd International Conference on Intelligent System and Knowledge Engineering, ISKE 2008, vol. 1, pp. 256–261 (2008)

[15] Rodriguez, M.A., Egenhofer, M.J.: Determining semantic similarity among entity classes from different ontologies. IEEE Transactions on Knowledge and Data Engineering 15(2), 442–456 (2003)
[16] Dong, L., Srimani, P.K., Wang, J.Z.: WEST: Weighted-Edge Based Similarity Measurement Tools for Word Semantics. In: 2010 IEEE/WIC/ACM International Conference on Web Intelligence and Intelligent Agent Technology (WI-IAT), vol. 1, pp. 216–223 (2010)
[17] Li, Y., Bandar, Z.A., McLean, D.: An approach for measuring semantic similarity between words using multiple information sources. IEEE Transactions on Knowledge and Data Engineering 15(4), 871–882 (2003)
[18] Rubenstein, H., Goodenough, J.B.: Contextual correlates of synonymy. Commun. ACM 8(10), 627–633 (1965)
[19] Miller, G.A., Charles, W.G.: Contextual correlates of semantic similarity. Language and Cognitive Processes 6(1), 1–28 (1991)
[20] Francis, W.N., Kucera, H.: Brown corpus manual. Letters to the Editor 5(2), 7 (1979)

Definition-based Information Content Vectors
for Semantic Similarity Measurement

Ahmad Pesaranghader and Saravanan Muthaiyah

Multimedia University, Jalan Multimedia,
63100 Cyberjaya, Malaysia
ahmad.pgh@sfmd.ir, saravanan.muthaiyah@mmu.edu.my

Abstract. Ontologies, as representation of shared conceptualization for variety of specific domains, are the heart of the Semantic Web. In order to facilitate interoperability across multiple ontologies, we need an automatic mechanism to align ontologies. Therefore, many methods to measure similarity between concepts existing in two different ontologies are proposed. In this paper, we will enumerate these methods along with their shortcomings in each case. In information content (IC) based similarity measures, the process of IC computation for concepts is so challenging and in many cases with failing. We will propose our new approach that is based on concepts' definitions. These definitions would help us to compute reliable and easy to calculate information contents for concepts. Applying these methods to the biomedical domain, using MEDLINE as corpus, International Classification of Diseases, Ninth Revision, Clinical Modification (ICD9CM) as thesaurus, and available reference standard, we will find our method outperforms other similarity measures.

Keywords: Semantic Similarity, Semantic Web, Bioinformatics, MEDLINE, UMLS, ICD9CM, Biomedical Text Mining, Ontology Alignment.

1 Introduction

Considering human ability for understanding different levels of similarity and relatedness among concepts or documents, semantic similarity and relatedness functions attempt to estimate these levels computationally. One of the main applications of these functions is ontology matching and ontology alignment [1]. This will establish interoperability among different ontologies whether within a specific domain or from different domains of knowledge. These functions have also wide usage in machine translation [2], automatic speech recognition [3], and text categorizing. The output of a similarity or relatedness measure is a value, ideally normalized between 0 and 1 inclusive, indicating how much two given concepts (or documents) are semantically similar (or related).

We will list existing measures of semantic similarity and relatedness in Section 2, with explanation of concomitant problems of each measure. In Section 4, we will represent our new approach in order to overcome these shortcomings and meet benefits of previously proposed measures in one place. Our new similarity measure is

S.A. Noah et al. (Eds.): M-CAIT 2013, CCIS 378, pp. 268–282, 2013.

based on a novel way of information content computation for a concept. In this regard, by using concepts' definitions from a thesaurus and applying information extracted from an external corpus, we will represent information content of a concept as a vector (instead of a scalar value). After information content vectors computation, we will employ them for similarity estimation between concepts. In order to be certain about the reliability of these estimations, mainly in specific areas of knowledge, through conducting our experiments in the biomedical domain we examined this new approach. The resources used in the experiments are discussed in Section 3. Regarding to the experiments results, in Section 5, we have evaluated the similarity estimations of different similarity functions for a reference standard of concept pairs. This reference standard is already scored by medical residents.

In the following section we will discuss about two main computational classes which semantic similarity and relatedness methods get mostly divided into.

2   Similarity Estimation with Distributional vs. Thesaurus-Based Algorithms

Different proposed approaches for measuring semantic similarity and semantic relatedness get largely categorized under two classes of computational techniques. These two classes are known for their distributional versus thesaurus-based attitudes for calculation of their estimations. By applying a technique belonging to any of these classes, it would be possible to construct a collection of similar or related terms automatically. This would be achievable by either using information extracted from a large corpus in distributional methods or considering concepts' positional information exploited from an ontology (or thesaurus) in thesaurus-based approaches. In the following subsections, we will explain each of these classes and functions of their kind by details along with the challenges coming in any case. In Section 4, we will present our method for overcoming these difficulties.

2.1   Distributional Model

The notion behind methods of this kind comes from Firth idea (1957) [4] indicating "a word is characterized by the company it keeps". In practical terms, such methods learn the meanings of words by examining the contexts of their co-occurrences in a corpus. While these co-occurred features (of words or documents) get represented in a vector space the distributional model is also recognized as the vector space model (VSM). The values of a vector belonging to this vector space get calculated from applied function in the model. The final intention of this function is to measure the relatedness of term-term, document-document or term-document based on their corresponding vectors. The result would be achieved through computing the cosine of the angle between two inputs' vectors.

Term Frequency/Inverse Document Frequency - The TF-IDF weighting is commonly applied to calculate values of each dimension for each vector in the VSM. For storing these weightings the term-document matrix gets generated. Mainly, the

TF-IDF value for a term in a document arises in proportion to the number of times the term occurs in that document, but gets neutralized by the frequency of that term in the corpus. The TF-IDF weighting scheme assigns to term t in document d a weight given by:

$$TF - IDF_{t,d} = TF_{t,d} \times IDF_t \tag{1}$$

Where $TF_{t,d}$ denoting term frequency is calculable by:

$$TF_{t,d} = \begin{cases} 1 + \log(f(t,d)) & \text{if } f(t,d) \neq 0 \\ 0 & \text{if } f(t,d) = 0 \end{cases} \tag{2}$$

$f(t,d)$ represents frequency of term t in document d

and, IDF_t expressing inverse document frequency is reachable through:

$$IDF_t = \log(\frac{D}{df_t}) \tag{3}$$

D represents total number of documents,
df_t is number of documents which contain term t

While many researches use uniform TF weighting instead of TF-IDF just for the sake of simplicity, numerous studies have tried to devise new formulae to enrich the vector space model by means of new weighting schemes. Some of these approaches like point-wise mutual information (PMI) take advantage of statistical techniques for this purpose. Other methods like latent semantic analysis (LSA) try to apply algebraic rules in order to optimize the VSM. Some other studies attempt to augment concept's (a specific sense of a word) vector by using definition of the concept and terms involved in it, instead of just looking for the word representing that concept.

Point-wise Mutual Information - PMI is a measure of association used in information theory and statistics. In computational linguistics, PMI for two given concepts indicates the likelihood of finding one concept in a text document that includes the other concept. The results for PMI measure illustrate acceptable approximating of human semantics which will specify the VSM's values in term-term matrices. The following formula represents the measure:

$$PMI(c_1, c_2) = \log \frac{P(c_1, c_2)}{P(c_1) \times P(c_2)} \tag{4}$$

$P(c_1, c_2)$ is the probability that concepts c_1 and c_2 co-occur in the same document, and $P(c_1)$ and $P(c_2)$ are respectively the probabilities for occurrence of c_1 and c_2 in a document

Latent Semantic Analysis - LSA was introduced by Landauer and Dumais in 1997 [5]. The method works based on parsing of a collection of different documents. When a set of documents are given to the LSA, for a certain term, the algorithm first compute the number of occurrence of that term in each document and then populate the well known term-document matrix. The singular value decomposition (SVD) is then applied to find principal components of this matrix. Considering A as our input term-document matrix the result of SVD is as follows:

$$A_{t\times d} = SVD(A) = U_{t\times f}S_{f\times f}V^t_{f\times d} \tag{5}$$

The SVD function applied on a matrix, first computes singular values of the matrix (values on S's diagonal) by considering a new latent feature (apart from terms and documents in our case), then produces three matrices based on input matrix and this new feature. The S is a rectangular diagonal matrix in which non-negative singular values are sorted downward on the diagonal. By selecting the first k indices out of the f indices from the produced matrices we will select the k principal components. Using k for creating new matrices as (6) shows, their multiplication will construct a new matrix that is closely similar to A.

$$A_{t\times d} \approx \overline{A}_{t\times d} = U_{t\times k}S_{k\times k}V^t_{k\times d} \tag{6}$$

LSA promises by selecting an appropriate k the reproduced matrix is highly optimized. Therefore, using cosine approach, this new matrix can be used for any term-term, document-document or term-document relatedness estimation. For the sake of dimensionality reduction, rows of the matrix produced from $U_{txk}S_{kxk}$ can be used for term-term, and columns of the matrix produced from $S_{kxk}V^t_{txk}$ can be used for document-document measurement of relatedness without any need to store whole new produced matrix.

Definition-based Relatedness Measures - A number of studies try to augment the concept's vectors by taking the definition of input term into consideration. In this regard, a concept will be indicative of one sense for the input term and its definition can be accessible from an external resource (or resources) such as a dictionary or thesaurus. Be aware that employing thesaurus here does not imply these methods get categorized under thesaurus-based model. This is because in these methods the full positional and relational information for a concept from the thesaurus is still unexploited and thesaurus here performs just as an advanced dictionary.

WordNet is the known thesaurus which often gets used in non-specific contexts for this type of relatedness measures. In WordNet, terms are characterized by a synonym sets called synsets while each has its own associate definition or gloss. Synsets are connected to each other through semantic relations such as hypernym, hyponym, meronym and holonym. Having these definitions available, the Lesk algorithm [6] is to compare the thesaurus glosses of different senses for an ambiguous term in a context (target word) with the terms contained in its neighborhood in the context. Banerjee and Pedersen [7] in the adapted Lesk algorithm extended the Lesk idea by also including the definition of its related synsets, referring to it as the extended gloss. In another study Patwardhan and Pedersen [8] use the co-occurrence information derived from an external corpus along with the WordNet definitions to build gloss vectors corresponding to each concept in WordNet. For two input terms (concepts), the cosine of the angle between their respective gloss vectors would be indicative of the level of similarity (relatedness to be exact).

The Unified Medical Language System (UMLS) is a mega-thesaurus specifically designed for medical and biomedical purposes. Applying forgoing techniques on this specific domain of knowledge, the scientific definition of medical concepts can be derived from the resources (thesauruses) included in the UMLS.

The functions listed above are just a few samples to represent main specifications of methods belonging to the distributional model. Second order co-occurrence PMI (SOC-PMI) [9], generalized latent semantic analysis (GLSA) [10] and explicit semantic analysis (ESA) [11] are other functions which take into account the VSM and operate based on the distributional model to a great extent.

Distributional Model Challenges - The functions categorized into the distributional model are widely useful for measuring semantic relatedness between two input terms or documents. The majority of approaches in this model are incapable for estimating the true semantic similarity for many target pairs. The results of similarity measures are not always comparable with the relatedness functions estimations. For example, the result of semantic relatedness for the terms *computation* and *computer* would be very high when low level of similarity between these terms is actually expected. This problem is more noticeable when two inputs are negation words (e.g. *honesty* and *dishonesty*). Even though semantic relatedness and semantic similarity can be used in many cases interchangeably, we will here distinguish semantic relatedness from semantic similarity as a result of abovementioned inconsistency as well as specific functionality of each one in different cases.

In order to measure the amount of semantic similarity between to input concepts the least common subsumer (LCS) of those concepts are intuitively appealing. In other words, we should know how far their nearest shared super-parent is from them. Therefore, for similarity measurement we need to rely on external sources such as thesaurus or ontologies to find this type of information. In the following subsection, the proposed thesaurus-based functions and their weaknesses are discussed in details. To resolve some restraining difficulties of this measure type we will present our new semantic similarity approach in Section 4.

2.2   Thesaurus-Based Model

The main factor showing the unique characteristic of measures belonging to this class is their capability for estimating semantic similarity (rather than relatedness). The functions of this kind, as the model's name implies, rely on positional information of concepts derived from a thesaurus or ontology. Depending on different degrees of information extraction from thesaurus and option of employing a corpus for enriching it, the methods for similarity measurement get categorized into three main groups.

Path-based Similarity Measures - For these methods distance between concepts in the thesaurus is generally important. It means the only issue for similarity measurement of input concept pair is the shortest number of jumps from one concept to the other one. The semantic similarity measures based on this approach are:

- Rada et al., 1989 [12]

$$\text{sim}_{\text{path}}(c_1, c_2) = \frac{1}{\text{shortest is-a path}(c_1, c_2)} \tag{7}$$

In path-based measures we count nodes (not paths)

- Caviedes and Cimino, 2004 [13]

It is sound to distinguish higher concepts in the thesaurus hierarchy, which are abstract in their meaning and so less likely to be similar, from those lower concepts, which are more concrete in their meanings (e.g. *physical_entity/matter* two high level concepts vs. *skull/cranium* two low level concepts). This is the main challenge in path-based methods as they do not consider this growth in specificity when we go down from the root towards leaves.

Path-based and Depth-based Similarity Measures - These methods are proposed to overcome the abovementioned drawback of path-based methods. They are generally based on both path and depth of concepts. These measures are:

- Wu and Palmer, 1994 [14]

$$\text{sim}_{\text{wup}}(c_1, c_2) = \frac{2 \times \text{depth}(\text{LCS}(c_1, c_2))}{\text{depth}(c_1) + \text{depth}(c_2)} \tag{8}$$

LCS is the least common subsumer of the two concepts

- Leacock and Chodorow, 1998 [15]
- Zhong et al., 2002 [16]
- Nguyen and Al-Mubaid, 2006 [17]

When one concept is less frequent than the other concepts (especially its siblings) we should consider it more informative as well as concrete in meaning than the other ones. It is a notable challenge in path-based and depth-based similarity functions. For example when two siblings get compared with another concept in these measures, they end up with the same similarity estimations while it is highly possible one of the siblings (more frequent/general one) has greater similarity to the compared concept.

Path-based and Information Content Based Similarity Measures - As mentioned, depth shows specificity but not frequency, meaning low frequent concepts often are much more informative so concrete than high frequent ones. This feature for one concept is known as information content (IC) [18]. IC value for a concept in the thesaurus is defined as the negative log of the probability of that concept in a corpus.

$$\text{IC}(c) = -\log(P(c)) \tag{9}$$

P(c) is probability of the concept *c* in the corpus

$$P(c) = \frac{tf + if}{N} \tag{10}$$

tf is term frequency of concept *c*, *if* is inherited frequency or frequency of concept's descendants summed together, and *N* is sum of all concepts' frequencies in the ontology

Different methods based on information contents of concepts are proposed:

- Resnik, 1995 [18]

$$\text{sim}_{res}(c_1, c_2) = IC(LCS(c_1, c_2)) \tag{11}$$

- Jiang and Conrath, 1997 [19]
- Lin, 1998 [20]

$$\text{sim}_{lin}(c_1, c_2) = \frac{2 \times IC(LCS(c_1, c_2))}{IC(c_1) + IC(c_2)} \tag{12}$$

Typically path-based and IC-based Similarity measures must yield better result comparing with path-based and path and depth-based methods. Nonetheless in order to produce accurate IC measures to achieve a reliable result from IC-based similarities many challenges are involved.

The first challenge is that IC-based similarities are highly reliable on an external corpus, so any weakness or imperfection in the corpus can affect the result of these similarity measures. This issue is more intense when we work on a specific domain of knowledge (e.g. healthcare and medicine or disaster management response) when accordingly a corpus related to the domain of interest is needed. Even when the corpus is meticulously chosen, absence of corresponding terms in the corpus for the concepts in the thesaurus is highly likely. This would hinder to compute IC for many concepts (when the term frequencies of their descendants are also unknown) and consequently this will avoid semantic similarity calculation for those concepts.

The second challenge is associated with the ambiguity of terms (terms with multiple senses) in their context during the parsing phase. As we really need concept frequency rather than term frequency for IC computation, this type of ambiguity for numerous encountered ambiguous terms in the corpus will lead to miscounting and consequently noisiness and impurity of produces ICs for concepts.

The third problem arises for concepts having equivalent terms that are multi-word rather than just one word (e.g. *semantic web*). In this case the parser should be so smart to distinguish these terms (multi-word terms) in corpus from times when they appear alone; otherwise the frequency count for this type of concepts will be also unavailable. As a result, there will be no chance for IC measurement and similarity estimation for the concept.

The last setback regarding to the usage of IC-based measure is extremely associated with those specific domains of knowledge which are in the state of rapid evolution and new concepts and ideas get added to their vocabularies and ontologies every day. Unfortunately, the speed of growth for documentation regarding to these new topics is not the same as their evolution speed. Therefore, it will be another hindrance for calculating required ICs for these new concepts.

In this research, by presenting new approach for computing information content for concepts, we will try to avoid the foregoing problems linked to the IC-based functions. By considering concepts' definitions, the conventional numeric IC will be replaced by information content vectors. Therefore, in our proposed method for semantic similarity estimation all benefits of distributional and thesaurus-based model will be gathered at one point. Our approach for computation of information content vectors is fully discussed in Section 4. Through experiments of our study on the specific domain of biomedical, in Section 5, the result of our similarity estimation gets compared with other similarity measures. For this purpose, we have applied these measures on the biomedical ontology of *International Classification of Diseases, Ninth Revision, Clinical Modification* (ICD9CM) from the Unified Medical Language System (UMLS) by using MEDLINE as the biomedical corpus for IC computations. The produce similarity results get evaluated against subset of concept pairs as reference standard which are previously judged by medical residents. Following section will explain each of abovementioned resources employed in our experiments.

3   Experimental Data

Resources existing for the study are a thesaurus to find positional and relational information among concepts, a corpus to extract required information feeding to the semantic similarity measures, and a dataset used for final evaluation against human judgment. While MEDLINE abstract is used for building term-term matrix of parsed bi-grams, the ICD9CM is used for constructing definitions for concepts.

3.1   International Classification of Diseases, Ninth Revision, Clinical Modification (ICD9CM)

The ICD9CM is a biomedical ontology for coding and classifying morbidity data from the inpatient and outpatient records, physician offices, and most National Center for Health Statistics (NCHS) surveys. We will use the UMLS which contains ICD9CM. The UMLS is a knowledge representation framework designed to support biomedical and clinical research. Its fundamental usage is provision of a database of biomedical terminologies for encoding information contained in electronic medical records and medical decision support. It comprises over 160 terminologies and classification systems that ICD9CM is one of them. In this research we have employed 2012AB release of the UMLS containing the latest update of ICD9CM.

3.2   MEDLINE Abstract

MEDLINE is a bibliographic database of life sciences and biomedical information. It contains bibliographic information for articles coming from academic journals which cover medicine, nursing, pharmacy, dentistry, veterinary medicine, and healthcare. MEDLINE also covers much of the literature in biology and biochemistry, as well as fields such as molecular evolution. The database contains more than 21.6 million records from 5,582 selected publications covering biomedicine and health from 1950 to the present.

For the current study we used MEDLINE article abstracts as the corpus to build a term-term co-occurrences matrix for subsequent computation of IC vectors and semantic similarity. Term frequencies for IC-based similarity measures are also obtained from this corpus. We used the 2013 MEDLINE abstract.

3.3   Reference Standard

The reference standard employed in our experiments was a subset of 566 medical concept pairs provided particularly for testing automated measures of semantic similarity, freely available from University of Minnesota Medical School created in their experimental study [21]. Eight medical residents were invited for participation in scoring similarity of these concept pairs in order to have them voted based on human judgment. Many of concepts from original reference standard are not included in the ICD9CM ontology. Therefore, after removing them from the original dataset, a subset of 64 concept pairs for testing on different semantic similarity functions including our measure was available. The main reason for ICD9CM selection with this rather small coverage of concept pairs was its potential to represent the weakness of conventional IC computation approach in the most understandable way, by emphasizing on a need for an alternation. Moreover, we considered that the ontology under study should have still its own application, which is reviewed to date and gets updated regularly.

4   Methods

The proposed approach for having information content vectors instead of conventional numeric ICs is to avoid problems regarding to the previous IC-based measures. These IC vectors will be stored in a matrix where each IC vector belonging to a specific concept in the ontology is represented as a row of that matrix. Therefore, the total number of matrix rows is equal to the quantity of concepts in the ontology indicating first dimension of the matrix. The second dimension will be words included in the MEDLINE and the total quantity of them will be different, as it depends on whether a list of stop words and a threshold of low frequent cutting-off point are considered to remove some of them. This removing step at times would help to find more valuable words. In order to construct this information content matrix we need to go through four steps. The sequence of these steps is illustrated in Fig. 1.

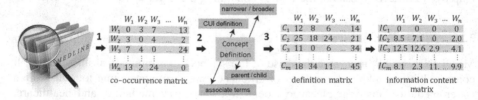

Fig. 1. The 4 steps of definition-based information content vectors construction

In the first step, we would record the frequency of every word co-occurrence with other words in its immediate context from MEDLINE. After gathering all of them we can construct a matrix of these co-occurrences. As a result, we will have a word-word matrix which is both square and symmetric. The removal phase for unwanted words can be enforced in this phase.

In the second step, we will retrieve assigned definition for each concept from ICD9CM. As all the concepts on the ontology do not have definition already assigned internally, we will build a definition for them by concatenating definition of their immediate parents and children (if their definitions exist) and their definitions from other vocabularies included in the UMLS. By applying this extension strategy for all concepts we will try to enrich concepts definition as well.

In the third step, we will construct definition-word matrix. For building this matrix the definition of a concept (a row of the matrix) is represented as a vector calculated through summation of all word vectors of constituent words in the definition and then normalized by their quantity. These words' vectors are available from our first word-word co-occurrence matrix.

In the forth step, we will calculate information content vectors for concepts from the definition-word matrix constructed in the previous step. The information content vector for a concept is negative log of probably vector of that concept (element-wise). The probability vector for a concept is computable having its definition vector from definition-word matrix summed with definition vectors of the descendants of that concept on the ontology and then divided by summation of all concepts' definition vectors on the ontology (element-wise).

Now, as we have information content vectors for all the concepts in the ontology, we can compute semantic similarity between two input concepts. For the similarity measurement, the information content of two concepts combined together would be compared with the information content vector of their least common subsumer (LCS). In other words, the similarity of the two concepts can be defined as cosine (θ) between their information content vectors united together and information content vector of their LCS. When the similarity is 1, the two concepts are exactly the same, and when the similarity is 0, they are strongly unlike. Other values in between indicate relative degrees of similarity.

Different ways for combining two IC vectors of input concepts are possible. After this combination we can calculate the cosine of this new vector with the LCS's IC vector. In our experiments we heuristically reached to the following formula which yields the best result.

$$\text{sim}_{vec}(c_1, c_2) = \text{cosine}(\text{IC}_{vec}(\text{LCS}(c_1, c_2)), \text{sqrt}(\text{IC}_{vec}(c_1) . \times \text{IC}_{vec}(c_2))) \qquad (13)$$

In above, the sign $. \times$ indicates dot multiplication or element-wise multiplication and *sqrt* function returns the square root of each element of a vector.

5   Experiments

Having our reference standard, the experiments are developed from two aspects: At first, without applying add-one technique, by computing the similarity results for all similarity measures, we will draw a comparison among the produced results. After that, as many of our concepts from reference standard do not have IC due to absence of their equivalent terms in the MEDLINE, by considering the technique of add-one for having IC for all the concepts, we will again compare the results of IC-based measures with our new approach. For the comparisons, the correlations of measures' similarity estimations against humane judgment of them are computed. It is done by using Spearman's correlation coefficients assuming there is no relationship between the two sets of data (produced estimation results and humane scores).

5.1   New Approach vs. Other Measures without Add-one Technique

We already mentioned that one of the drawbacks for IC-based measures is their demands for a very rich corpus which covers all the terms (concepts) in the thesaurus; otherwise ICs of many concepts for computing their similarities with other concepts would not be available. To solve this problem we proposed our approach for IC calculation which is based on the concepts' definitions representing ICs as vectors. Table 1 shows out of 20991 concepts (including root) in ICD9CM ontology how many of them have theirs IC calculated.

Table 1. Information contents and concept pairs available for testing

IC Calculation Method	Number of ICs available	Pairs for Testing
Conventional IC (scalar)	1186 / 20991	24 / 64
Definition-based IC (vector)	11578 / 20991	64 / 64

From above we can see even though we have chosen MEDLINE, one of the most comprehensive corpora for the biomedical domain, many ICs for concepts in conventional approach are still unavailable (94%). As a result, for 64 concept pairs provided for semantic similarity testing we can measure similarity for 24 of them. Moreover, with the absence of definitions of some concepts (and their descendants) in the UMLS, the definition-based IC vector of them are incalculable (44%).

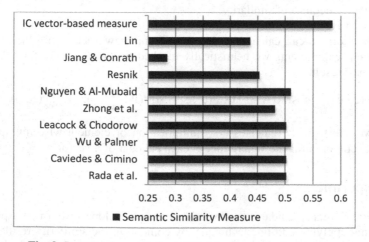

Fig. 2. Spearman's correlation of 24 pairs for similarity measures

Fig. 2 represents the result of the Spearman's correlation on 24 concept pairs (pairs with available ICs) for nine similarity measures proposed already and our new approach. Table 2 shows precisely the Spearman's correlation results for 4 IC-based measures (either scalar or vector).

Table 2. Spearman's correlation results of IC-based methods

Semantic Similarity	Spearman's Rank Correlation
Resnik measure	0.4534
Jiang & Conrath measure	0.2852
Lin measure	0.4365
IC vector-based measure	**0.5852**

With observing the results of similarity measures in Fig. 2, we can see IC-based measures perform not as satisfactory as we expected. Even path-based and depth-based functions results yield better correlation with human judgment. This is due to the poorness of computed ICs for the concepts used in the conventional IC-based measures.

5.2   New Approach vs. Other Measures with Add-one Technique

Add-one is an advised strategy that at times helps to improve the result of IC-based similarities. Through consideration of add-one technique, before starting to compute term frequencies of concepts from corpus, it will be assumed that all concepts have occurred at least once in the corpus. By implementing this strategy all the concepts will have IC of their own as the term frequency of them (concept frequency to be exact) is available, and it is either one or more. Fig. 3 represents Spearman's correlation of similarity measures estimations of 64 concept pairs against human scores.

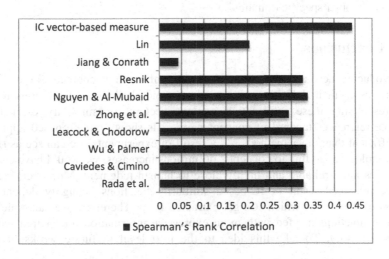

Fig. 3. Spearman's correlation of 64 pairs for similarity measures

From Fig. 3 we can see that even after employing add-one technique, our method using definition-based IC vectors estimates better results. Additionally, add-one does not help to improve the result of IC-based measure to get at least better results from path-based and depth-based approaches.

6   Discussion

The main focus of these experiments is on the information content vector computation as a solution for the conventional IC weaknesses. These IC vectors, for concepts in ontology, are computable by having concepts' definitions, preferably in their extended forms. This is based on the assumptions that for a concept the constructed definition (from the ontology) and summation of all co-occurrence vectors (from a corpus) for the words included in the definition carry the same meaning that the concept does. First, we aimed to extract tacit knowledge hidden in a larger corpus, which was already used in the distributional model of relatedness measures but was intact in the similarity functions. Then, having this information, with considering the definitions and positional information of the concepts in the ontology, we calculated IC vectors for subsequent estimation of similarity for concept pairs. Also, in the experiments done in the study, we showed that these IC vectors help to achieve more accurate estimations of semantic similarity.

Since the measures of semantic similarity play a crucial role in knowledge matching systems, the results provided by this research can be used to develop such matching, mapping and mediation systems by applying semantic similarity algorithms as their fundamental part in order to resolve the problem of data and knowledge heterogeneity. Furthermore, the proposed methods can be applied in different tasks such as information retrieval and word sense disambiguation. While these methods are independent from domain under study, their higher efficiency can be also functional in other specific domains.

7   Conclusions

By introducing new approach for calculating information contents for ontologies' concepts using their definitions, we tried to avoid challenges of previous IC-based measures. Using these information content vectors for similarity estimation of available reference standard of concept pairs, it is shown our proposed approach is more efficient than other methods. As a result of experiments, we can see as richness of the ontology regarding to concepts' definitions increases, we would have estimated similarities more reliable and highly close to human judgment. Moreover, we found that the produced result of IC-based similarity functions is highly dependent on richness of the corpus which ICs get extracted from. Therefore, we can conclude as similarity functions are fed with more expressive information, the output would be more satisfactory. To take this idea to the next level in future works, it can be examined if extending concepts definition from external resources will have even more positive impact on computed outputs. Additionally, our proposed approach can be tested on extrinsic tasks such as ontology alignment to see its real strength in the application.

References

1. Muthaiyah, S., Kerschberg, L.: A Hybrid Ontology Mediation Approach for the Semantic Web. International Journal of E-Business Research 4, 79–91 (2008)
2. Chen, B., Foster, G., Kuhn, R.: Bilingual Sense Similarity for Statistical Machine Translation. In: Proceedings of the ACL, pp. 834–843 (2010)
3. Pucher, M.: WordNet-based Semantic Relatedness Measures in Automatic Speech Recognition for Meetings. In: Proceedings of the ACL, pp. 129–132 (2007)
4. Firth, J.R.: A Synopsis of Linguistic Theory 1930-1955. In: Studies in Linguistic Analysis, pp. 1–32 (1957)
5. Landauer, T.K., Dumais, S.T.: A Solution to Plato's Problem: The Latent Semantic Analysis Theory of the Acquisition. Induction and Representation of Knowledge. Psychological Review 104, 211–240 (1997)
6. Lesk, M.: Automatic Sense Disambiguation Using Machine Readable Dictionaries: How to Tell a Pine Cone from an Ice-cream Cone. In: Proceedings of the 5th Annual International Conference on Systems Documentation, New York, USA, pp. 24–26 (1986)
7. Banerjee, S., Pedersen, T.: An Adapted Lesk Algorithm for Word Sense Disambiguation using WordNet. In: Proceedings of the Third International Conference on Intelligent Text Processing and Computational Linguistics, Mexico City (2002)
8. Patwardhan, S., Pedersen, T.: Using WordNet-based Context Vectors to Estimate the Semantic Relatedness of Concepts. In: Proceedings of the EACL 2006 Workshop, Making Sense of Sense: Bringing Computational Linguistics and Psycholinguistics Together, Trento, Italy, pp. 1–8 (2006)
9. Islam, A., Inkpen, D.: Second Order Co-occurrence PMI for Determining the Semantic Similarity of Words. In: Proceedings of the International Conference on Language Resources and Evaluation, Genoa, Italy (2006)
10. Matveeva, I., Levow, G., Farahat, A., Royer, C.: Terms Representation with Generalized Latent Semantic Analysis. In: Proceedings of the Recent Advances in Natural Language Processing Conference (2005)
11. Evgeniy, G., Shaul, M.: Computing Semantic Relatedness using Wikipedia based Explicit Semantic Analysis. IJCAI, 1606–1611 (2007)
12. Rada, R., Mili, H., Bicknell, E., Blettner, M.: Development and Application of a Metric on Semantic Nets. IEEE Transactions on Systems, Man and Cybernetics 19, 17–30 (1989)
13. Caviedes, J., Cimino, J.: Towards the Development of a ConceptualDistance Metric for the UMLS. Journal of Biomedical Informatic 372, 77–85 (2004)
14. Wu, Z., Palmer, M.: Verb Semantics and Lexical Selections. In: Proceedings of the 32nd Annual Meeting of the Association for Computational Linguistics (1994)
15. Leacock, C., Chodorow, M.: Combining Local Context and WordNet Similarity for Word Sense Identification in WordNet: An Electronic Lexical Database, pp. 265–283 (1998)
16. Zhong, J., Zhu, H., Li, J., Yu, Y.: Conceptual Graph Matching for Semantic Search. In: Proceedings of the 10th International Conference on Conceptual Structures, p. 92 (2002)

17. Nguyen, H.A., Al-Mubaid, H.: New Ontology-based Semantic Similarity Measure for the Biomedical Domain. In: Proceedings of IEEE International Conference on Granular Computing, GrC 2006, pp. 623–628 (2006)
18. Resnik, P.: Using Information Content to Evaluate Semantic Similarity in a Taxonomy. In: Proceedings of the 14th International Joint Conference on Artificial Intelligence, pp. 448–453 (1995)
19. Jiang, J.J., Conrath, D.W.: Semantic Similarity based on Corpus Statistics and Lexical Taxonomy. In: International Conference on Research in Computational Linguistics (1997)
20. Lin, D.: An Information-theoretic Definition of Similarity. In: 15th International Conference on Machine Learning, Madison, USA (1998)
21. Pakhomov, S., McInnes, B., Adam, T., Liu, Y., Pedersen, T., Melton, G.: Semantic Similarity and Relatedness between Clinical Terms: An Experimental Study. In: Proceedings of AMIA, pp. 572–576 (2010)

Enrichment of BOW Representation with Syntactic and Semantic Background Knowledge

Rayner Alfred[1], Patricia Anthony[2], Suraya Alias[1], Asni Tahir[1], Kim On Chin[1], and Lau Hui Keng[1]

[1] Center of Excellence in Semantic Agents, School of Engineering and Information Technology, Universiti Malaysia Sabah, Jalan UMS, 88400, Kota Kinabalu, Sabah, Malaysia
{Ralfred,suealias,asnieta,kimonchin,hklau}@ums.edu.my
[2] Department of Applied Computing, Faculty of Environment, Society and Design, Lincoln University, Christchurch, New Zealand
patricia.anthony@lincoln.ac.nz

Abstract. The basic Bag of Words (BOW) representation, that is generally used in text documents clustering or categorization, loses important syntactic and semantic information contained in the documents. When the text document contains a lot of stop words or when they are of a short length this may be particularly problematic. In this paper, we study the contribution of incorporating syntactic features and semantic knowledge into the representation in clustering texts corpus. We investigate the quality of clusters produced when incorporating syntactic and semantic information into the representation of text documents by analyzing the internal structure of the cluster using the Davies-Bouldin (DBI) index. This paper studies and compares the quality of the clusters produced when four different sets of text representation used to cluster texts corpus. These text representations include the standard BOW representation, the standard BOW representation integrated with syntactic features, the standard BOW representation integrated with semantic background knowledge and finally the standard BOW representation integrated with both syntactic features and semantic background knowledge. Based on the experimental results, it is shown that the quality of clusters produced is improved by integrating the semantic and syntactic information into the standard bag of words representation of texts corpus.

Keywords: clustering, bag of words, syntactic features, semantic back-ground knowledge, automatic text categorization, knowledge management.

1 Introduction

Text document clustering represents a challenging problem to text mining and machine learning communities due to the growing demand for automatic information retrieval systems [17]. With a large volume of text documents accessible to users, it hinders user accessibility to useful information buried in disorganized, incomplete, and unstructured text messages. In order to enhance user accessibility, we propose an algorithm that clusters text documents based on the standard bag of words

S.A. Noah et al. (Eds.): M-CAIT 2013, CCIS 378, pp. 283–292, 2013.

representation of texts corpus integrated with syntactic features and semantic background knowledge. Traditionally, text documents clustering process is based on a *Bag of Words* (BOW) approach, in which each document is represented as a vector with a dimension for each term of the dictionary containing all the words that appear in the corpus, as shown in Fig. 1. The row section represents a document and the column section represents the unique terms exist in documents. For instance, based on Fig. 1, there are n documents that share p unique terms in the BOW representation.

$$\begin{bmatrix} W_{11} & \cdots & W_{1f} & \cdots & W_{1p} \\ \vdots & \vdots & \vdots & \vdots & \vdots \\ W_{i1} & \cdots & W_{if} & \cdots & W_{ip} \\ \vdots & \vdots & \vdots & \ddots & \vdots \\ W_{n1} & \cdots & W_{nf} & \cdots & W_{np} \end{bmatrix}$$

Fig. 1. Standard Bag of Words Representation

The value TF-IDF (Term Frequency – Inverse Document Frequency) associated to a given term represents the weight of the term itself and it is computed based on its frequency of occurrence within the corresponding document (Term Frequency, or TF), and within the entire corpus (Inverse Document Frequency, or IDF). A lot of works conducted related to preprocessing of documents in order to improve the text document representation that includes stemming, removal of stop words and normalization [18]. However, the improvement is still limited due to three major drawbacks: (1) the order to the terms occurrence is not maintained or considered in clustering the text documents; (2) synonymous words are considered different component; and (3) ambiguous words are grouped as a single component (*e.g.*, bank of rivers, financial bank). It is therefore essential to further embed the semantic and syntactic information in order to enhance the quality of clustering.

This paper is organized as follows. Section 2 discusses about some of the related works. Section 3 presents the proposed modified BOW representation used to cluster text documents. Section 4 describes the experimental design and discusses the results obtained and finally this paper is concluded in Section 5.

2   Related Works

In traditional document clustering methods, a document is considered as a bag of words, with no relations between words. The feature vector representing the document is made by using the frequency count of terms in a document. Weights calculated from techniques like Inverse Document Frequency (IDF) and Information Gain (IG) are applied to the frequency count associated with the term. Some works conducted related to preprocessing of documents in order to improve the text document representation that includes semantic and syntactic analysis. Choudhary and Bhattacharyya have proposed a new method to create document vectors in order to improve clustering results by using a *Self Organizing Map* (SOM) technique [1]. This approach uses the Universal Networking Language (UNL) representation of a document. The UNL represents the document in the form of a graph with universal words as nodes and the relation between them as links. Instead of considering the documents as a bag of words they use the information given by the UNL graph to construct the vector. The proposed method managed to improve the clustering accuracy by using the semantic information of the sentences representing the document. Siolas [7] introduced semantic knowledge in Automatic Text Categorization (ATC) by building a kernel that takes into account the semantic distance: first between the different words based on *WordNet*, and then using Fisher metrics in a way similar to Latent Semantic Indexing (LSI). Zelikovitz and Hirsh [8] have also proven that the ATC accuracy can be improved by adding extra semantic knowledge into the LSI representation extracted from unclassified documents.

Caropreso and Matwin [5] and Moschitti and Basili [9] have studied the usefulness of including semantic knowledge in the text representation for the selection of sentences from technical genomic texts [5]. Both works agree that word senses are not adequate to improve ATC accuracy. They have shown that using hierarchical technical dictionaries together with syntactic relations is beneficial for our problem when using state of the art machine learning algorithms. Yamakawa *et al.* have studied a technique for incorporating the vast amount of human knowledge accumulated in Wikipedia into text representation and classification [17]. The aim is to improve classification performance by transforming general terms into a set of related concepts grouped around semantic themes. They proposed a unique method for breaking the enormous amount of extracted Wikipedia knowledge (concepts) into smaller pieces (subsets of concepts). The subsets of concepts are separately used to represent the same set of documents in a number of different ways, from which an ensemble of classifiers is built. The experiments conducted show that the method provides improvement in the classification performance, when compared with the results of a classifier trained on a regular term-document matrix.

Compared to semantics, less works have been done on the syntactic side. Cohen and Singer [3] have conducted experiments that study the importance of introducing the order of the words in the text representation. This can be done by defining position related predicates in the proposed ILP system. This has been extended by Goadich *et al.* [4] in the Information Extraction area, incorporating the order of noun phrases into the representation. Lewis compared different representations using either words or syntactic phrases (but not a combination of both) for Information Retrieval (IR) and Automatic Text Categorization (ATC) [11]. In text documents classification task, Caropreso and Matwin have proposed a method that uses bi-grams representation of text documents together with their single words in the BOW representation. It is shown that syntactic bi-grams (formed by using words that are syntactically linked) provide extra information that improves the classification performance compared to the traditional BOW representation of texts document [5]. In this work, we investigate the effects of integrating both the semantic and syntactic information into the BOW representation of documents.

3   Enrichment of Text Representation for Text Documents Clustering

In this research, the syntactic and semantic knowledge of a text document will be incorporated into the BOW representation.

3.1   Syntactic Enrichment of Text Representation with n-gram

Firstly, the syntactic knowledge of a text document is incorporated into the BOW representation by using *n-gram*. An *n-gram* is a subsequence of *n* items from a given sequence. The items in question can be phonemes, syllables, letters, words or base pairs according to its particular application. An *n-gram* of size 1 is referred to as a *unigram*; size 2 is a *bigram* (or, less commonly, a *digram*); size 3 is a *trigram*; and size 4 or more is simply called an *n-gram*. In this study, the syntactic enrichment of BOW representation is done by using *bi-gram* model which is n-gram of size 2. There are two main steps involved in the integration of syntactic features into the standard BOW representation of a text document. First, all possible *bi-gram* combinations or the likelihood of two words combined together are calculated. Then the frequencies of all the possible *bi-gram* combinations are computed and only the high frequency *bi-gram* terms will be taken into consideration as a new single item introduced into the standard BOW representation.   For instance, when a document *D1* is represented by a vector of *n* terms, where $D1 = < T_1, T_2,..., T_n >$, then the possible bi-gram combinations are $T_1T_2, T_2T_3, T_3T_4, ..., T_{n-1}T_n$. Only the high frequency bi-gram terms will be taken into consideration as a new single item introduced into the standard BOW representation.

3.2   Semantic Enrichment of Text Representation with *WordNet*

Semantic matching is a technique used in computer science to identify information which is semantically related. Given any two graph-like structures, *e.g.*, classifications, database or XML schemas and ontologies, matching is an operator which identifies those nodes in the two structures which semantically correspond to one another. An example applied to file systems is, it can identify that a folder labeled *car* is semantically equivalent to another folder *automobile* because they are synonyms in English. This information can be taken from a linguistic resource like *WordNet* [14].

Semantic matching represents a fundamental technique in many applications in areas such as resource discovery, data integration, data migration, query translation, peer to peer networks, agent communication, schema and ontology merging. In fact, it has been proposed as a valid solution to the semantic heterogeneity problem, namely managing the diversity in knowledge. Interoperability among people of different cultures and languages, having different viewpoints and using different terminology has always been a huge problem. Especially with the advent of the Web and the consequential information explosion, the problem seems to be emphasized. People face the concrete problem to retrieve, disambiguate and integrate information coming from a wide variety of sources. It has been shown that if words are indexed with their WorNet synset or sense then it improves the information retrieval performance [15].

WordNet is a large lexical database of English. Nouns, verbs, adjectives and adverbs are grouped into sets of cognitive synonyms (synsets), each expressing a distinct concept. Synsets are interlinked by means of conceptual-semantic and lexical relations. *WordNet* is also freely and publicly available for download. In this online Lexical Database, English nouns, verbs, and adjectives are organized into synonym sets, that representing one concept that linked by different relations.

In this study, the semantic knowledge of a text document is incorporated into the BOW representation based on *WordNet*. This is done by introducing a new concept that can be used to represent or describe several items that carry similar meaning based on the references obtained from *WordNet*. The basic idea to incorporate semantic background into the text representation is by introducing a new concept that represents two or more words that are semantically related or have similar meaning. Once the semantically related words are grouped under one concept these words will be replaced with the new concept introduced earlier in the BOW representation. Every time a new mutually exclusive semantically relationship is found between words, a new concept will be introduced. In our approach, the method only takes into consideration hyponyms and synonyms related terms. While in the hyponyms, only verbs are considered.

4   Experimental Evaluations

The experiment is designed in order to investigate and compare the effectiveness of clustering text documents based on the following FOUR types of text documents representations;

1. Standard BOW representation (SBOW)
2. Enriched BOW representation with syntactic background knowledge (SYBOW)
3. Enriched BOW representation with semantic background knowledge (SEBOW)
4. Enriched BOW representation with syntactic and semantic background knowledge (SSBOW).

The flow of the experiment for all types of text document representations is shown in Fig. 2. The first step performed in the experiment is Tokenization. Tokenization is the process of breaking a stream of text up into words, phrases, symbols, or other meaningful elements called tokens. Next the stop words removal process is conducted in order to remove irrelevant items in the text document. The stemming process is also conducted in order to turn words into their basic forms.

In this experiment, the Portel stemmer is not used in the stemming process [10]. Instead, the *WordNet* library morphological process is used to remove the suffix and prefix of a word found in the text document. The *WordNet* dictionary morphological process is used because the *WordNet* semantics information will be used to enrich the BOW representation. Thus, by using the *WordNet* stemmer method, the stemmed word produced can be easily found in the *WordNet* application for its synonym and hyponym. After the stemming process, the BOW representation is enriched with syntactic background knowledge by using the n-gram model. The BOW representation is further enriched with semantic background knowledge by using the *WordNet* application. Next, the TF * IDF weight (*Term Frequency–Inverse Document Frequency*) is computed. TF * IDF is a numerical statistic which reflects how important a word is to a document in a collection or corpus [11].

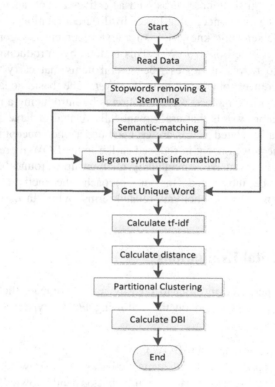

Fig. 2. A data transformation process for data stored in multiple tables with one-to-many relations into a vector space data representation

Once the magnitude of each document is computed, then the similarity between two different documents is computed by using the Cosine similarity distance [12]. All documents are clustered by using the partitional *k*-means clustering method. The clustering results obtained are evaluated by using the Davies-Bouldin Index (DBI) [13]. DBI is used to measure the quality of clusters produced because DBI uses both the within-cluster and between clusters distances to measure the cluster quality. The DBI's description is shown in equation (1) through (4). Let $d_{centroid}(Q_k)$ in (3), denote the centroid distances within-cluster Q_k, where $x_i \in Q_k$, N_k is the number of samples in cluster Q_k, c_k is the center of the cluster and $k \leq K$ clusters. Let $d_{between}(Q_k, Q_l)$, defined in (5), denote the distances between-clusters Q_k and Q_l, where c_k is the centroid of cluster Q_k and c_l is the centroid of cluster Q_l.

$$d_{centroid}(Q_k) \quad = \quad \frac{\sum_i \|x_i - c_k\|}{N_k} \tag{1}$$

$$c_k \quad = \quad \frac{1}{N_k}\left(\sum_{x_i \in Q_k} x_i\right) \tag{2}$$

$$d_{between}(Q_k, Q_l) \quad = \quad \|c_k - c_l\| \tag{3}$$

$$\text{DBI} \quad = \quad \frac{1}{K}\sum_{k=1}^{K}\max_{l \neq k}\left\{\frac{d_{centroid}(Q_k) + d_{centroid}(Q_l)}{d_{between}(Q_k, Q_l)}\right\} \tag{4}$$

Therefore, given a partition of the N points into K-clusters, DBI is defined in (4). This cluster dispersion measure can be incorporated into any clustering algorithm to evaluate a particular segmentation of data.

In the experiment conducted, firstly, the number of clusters, k, is set 20 clusters. Although, usually k is set as square root of the number of document, n, by rule of thumb, we want to investigate the effect of varying the number of documents on the clusters results when we fixed the number of clusters. In this experiment, the number of text documents are varies from 500 to 1500 with an increment of 250 for each experiment. There are two types of text documents representation used in the first part of the experiment which includes SBOW and SSBOW. Secondly, the number of documents is fixed to 1000 documents only but the number of clusters formed is fixed to 20, 25 and 30 respectively. There are four types of text documents representation used in the second part of the experiment which includes all four SBOW, SYBOW, SEBOW and SSBOW representations. Table 1 shows the results of clustering the text documents ranging from 500 to 1500 with a fixed number of clusters by using the standard BOW representation (SBOW) and the enriched BOW representation with syntactic and semantic background knowledge (SSBOW).

Based on the results obtained, the quality of clusters produced is better for documents which are represented by the enriched BOW representation with syntactic and semantic background knowledge (SSBOW). This is shown in Table 1, in which the DBI measurements are lower when documents are clustered by using the enriched BOW representation with syntactic and semantic background knowledge (SSBOW). Thus, the results show the importance of integrating the syntactic and semantic background knowledge into the standard BOW representation of a text documents. Noted that the average value of DBI for the clustering results using the SBOW representation is quite high since more items or words are considered when computing the similarity distance between two different text documents. In the SSBOW representation of text documents, less items or words are used in computing the similarity distance between two different documents.

Table 1. Comparison of DBI Values for The Clustering Results When Using SBOW and SSBOW with Different Number of Documents Clustered

Experiment	Text Documents	SBOW (DBI)	SSBOW (DBI)
1	500	31.1	7.2
2	750	41.7	1.6
3	1000	27.2	2.1
4	1250	44.3	4.4
5	1500	16.1	4.4

Table 2 shows the results of clustering 1000 text documents ranging with three different number of clusters, 20, 25 and 30, by using the standard BOW representation (SBOW), the enriched BOW representation with syntactic background knowledge (SYBOW), the enriched BOW representation with semantic background knowledge (SEBOW) and finally the enriched BOW representation with syntactic and semantic background knowledge (SSBOW).

Table 2. Comparison of DBI Values for the Clustering Results When Using SBOW, SYBOW, SEBOW and SSBOW with Different Number of Clusters Formed in Each Experiment

Number of Clusters	Number of Text Documents	DBI Values			
		SBOW	SYBOW	SEBOW	SSBOW
20	1000	4.91	11.1	4.4	3.1
25	1000	27.2	22.8	6.9	2.1
30	1000	9.5	3.5	3.1	4.1

The results also show that the quality of clusters produced is better for documents which are represented by the enriched BOW representation with syntactic and semantic background knowledge (SSBOW), regardless of the number of clusters produced. Notice that the best quality of clusters is obtained when the number of clusters is 20 with SSBOW representation. However, when the number of clusters is 20, the quality of clusters produced is not consistent, in which the standard BOW representation provides better quality of clusters compared to the enriched BOW representation with syntactic background knowledge (SYBOW). These results show that the syntactic background knowledge alone is not enough in order to improve the quality of clustering produced. It is also shown in Table 2 that the quality of clusters produced is better when the number of clusters requested is larger. When the number of clusters is larger, then the clusters formed are more compact and they are separated well among themselves. On the contrary, the enriched BOW representation with syntactic background knowledge is not always producing better quality of clusters compared to the results obtained when using SBOW representation. The syntactic background knowledge alone is not enough to get better clustering results. Similarly, the semantic background knowledge alone is not good enough to get better clustering results. In short, a better quality of clusters produced when both syntactic and semantic background knowledge can be integrated into the standard bag of word (BOW) representation for text documents.

5   Conclusion

In this paper we have presented the comparisons of clustering results when using different type of bag of words (BOW) representation for text documents. Besides the standard BOW representation, there are three other different types of BOW representation are investigated, which include the enriched BOW representation with syntactic background knowledge (SYBOW), the enriched BOW representation with semantic background knowledge (SEBOW) and finally the enriched BOW representation with syntactic and semantic background knowledge (SSBOW). It is shown that the contributions of having additional semantic and syntactic background knowledge in the standard BOW are very important in order to get better clustering results. We have empirically showed that this syntactic and semantic knowledge is useful for improving the clustering results.

Acknowledgments. This work has been supported by the Long Term Research Grant Scheme (LRGS) project funded by the Ministry of Higher Education (MoHE), Malaysia under Grants No. LRGS/TD/2011/UiTM/ICT/04.

References

1. Choudhary, B., Bhattacharyya, P.: Textclustering using Universal Networking Language representation. In: Eleventh International World Wide Web Conference (2003)
2. Zelikovitz, S., Hirsh, H.: Improving Text Classification with LSI Using Background Knowledge. In: Proceedingsof CIKM 2001,10th ACM International Conference on Information and Knowledge Management (2001)
3. Cohen, W.W., Singer, Y.: Context-sensitive learning methods for text categorization. ACM Trans. Inf. Syst. (1999)
4. Goadrich, M., Oliphant, L., Shavlik, J.: Learning Ensembles of First-Order Clauses for Recall-Precision Curves: A Case Study in Biomedical Information Extraction. In: Proceedings of the Fourteenth International Conference on Inductive Logic Programming, Porto, Portugal (2004)
5. Maria, F.C., Stan, M.: Incorporating Syntax and Semantics in the Text Representation for Sentence Selection. In: Recent Advances in Natural Language Processing, Borovets, Bulgaria (2007)
6. Lewis, D.D.: Representation and Learning in Information Retrieval, Ph.D. dissertation, University of Massachusetts (1992)
7. Siolas, G.: Modèles probabilistes et noyaux pour l'extraction d'informations à partir de documents. Thèsede doctorat de l'Université Paris (2003)
8. Zelikovitz, S., Hirsh, H.: Improving Text Classification with LSI Using Background Knowledge. In: Proceedings of CIKM 2001,10th ACM International Conference on Information and Knowledge Management (2001)

9. Moschitti, A., Basili, R.: Complex Linguistic Features for Text Classification: a Comprehensive Study. In: McDonald, S., Tait, J.I. (eds.) ECIR 2004. LNCS, vol. 2997, pp. 181–196. Springer, Heidelberg (2004)

10. Porter, M.F.: Analgorithm for suffix stripping. In: Jones, K.S., Willett, P. (eds.) Readings in Information Retrieval, pp. 313–316. Morgan Kaufmann Publishers Inc., SanFrancisco (1997)

11. Salton, G., McGill, M.J.: Introduction to modern information retrieval. McGraw-Hill (1983)

12. Bayardo, R.J., Ma, Y., Srikant, R.: Scaling up all pairssimilarity search. In: WWW 2007 - Proceedings of the 16th International World Wide Web Conference, pp.131–140 (2007)

13. Davies, D.L., Bouldin, D.W.: A Cluster Separation Measure. IEEE Transactions on Pattern Analysis and Machine Intelligence 2, 224 (1979)

14. Miller, G.A., Beckwith, R., Fellbaum, C.D., Gross, D., Miller, K.: WordNet: Anonline lexical database. Int. J. Lexicograph 3(4), 235–244 (1990)

15. Gonzalo, J., Verdejo, F., Chugur, I., Cigarrán, J.M.: Indexing with *WordNet* synsets can improve Text Retrieval, CoRR (1998)

16. Yamakawa, H., Jing, P., Feldman, A.: Semantic enrichment of text representation with Wikipedia for text classification. In: Systems Man and Cybernetics (SMC 2010), pp. 4333–4340 (2010)

17. Alfred, R., Mujat, A., Obit, J.H.: A Ruled-Based Part of Speech (RPOS) Tagger for Malay Text Articles. In: Selamat, A., et al. (eds.) ACIIDS 2013, Part II. LNCS, vol. 7803, pp. 50–59. Springer, Heidelberg (2013)

18. Leong, L.C., Basri, S., Alfred, R.: Enhancing Malay Stemming Algorithm with Background Knowledge. In: Anthony, P., Ishizuka, M., Lukose, D. (eds.) PRICAI 2012. LNCS, vol. 7458, pp. 753–758. Springer, Heidelberg (2012)

Learning Relational Data Based on Multiple Instances of Summarized Data Using DARA

Florence Sia, Rayner Alfred, and Kim On Chin

School of Engineering and Information Technology,
Universiti Malaysia Sabah, Jalan UMS, 88400, Kota Kinabalu, Sabah, Malaysia
florenceSia@yahoo.com, {ralfred,kimonchin}@ums.edu.my

Abstract. DARA (Dynamic Aggregation of Relational Attributes) algorithm is designed to summarize non-target records stored in a non-target table. These records have many-to-one relationships with records stored in the target table. The records stored in the non-target table are summarized and the summarized data is then appended to the target table. With these summarized data appended into the target table, a classifier will be applied to learn this data in order to perform the classification task. However, the predictive accuracy of the classification task is highly influenced by the representation of the summarized data. In our previous works, several types of feature construction methods have been introduced especially for the DARA algorithm in order to improve the descriptive accuracy of the summarized data and indirectly improve the predictive accuracy of the target data. This paper proposes a method that learns relational data based on multiple instances of summarized data that are obtained using different types of feature construction methods. This involves investigating the effect of selecting several sets of summarized data which have been summarized using the feature construction methods and appending these summarized data into the target table before the classification task can be performed. The predictive accuracy of the classification task is expected to be improved when multiple instances of summarized data appended into the target table. The experiment results show that there are some improvements in the predictive accuracy of the classification by selecting multiple instances of summarized data and appending them into the target table.

Keywords: Relational Data Summarization, Relational Data Mining, Feature Construction, Feature Selection.

1 Introduction

In recent years, many fields generate and gather large and complex data which are often reside in a relational database for real life applications [5]. Since a single table structure in an attribute-value form is impractical, relational database has been widely used for real world data storage because it is capable to capture and describe complex data in a relational form [1][2]. The increasing demand for tools to analyze data stored in relational databases in order to extract their useful information has drawn researchers' attention to the works in the field of Knowledge Discovery in Database (KDD),

S.A. Noah et al. (Eds.): M-CAIT 2013, CCIS 378, pp. 293–301, 2013.
© Springer-Verlag Berlin Heidelberg 2013

specifically in relational data mining. KDD is commonly presented as a process that comprises of phases to discover knowledge from data where relational data mining is the core process to extract frequent patterns from a large volume of data by using intelligent algorithms [6]. Data mining task can be categorized into descriptive data mining which finds patterns that describe some relations that present in a data, whereas, predictive data mining finds patterns that describe any relationships between data attributes and target concepts of interest [7]. Traditional data mining techniques are only applied to extract frequent patterns from a single data table, thus, many data mining approaches have been developed to extract pattern from a relational data in the relational database.

One of the approaches is called a Dynamic Aggregation of Relational Attributes (DARA), which is designed to extract frequent patterns from relational databases [8][9][10]. In a relational database, it may have a target table that has one-to-many relationship with a non-target table. A target table consists of rows of unique labeled record and a non-target table consists of rows of record where multiple rows can be linked to a single record in a target table [11]. DARA summarizes records stored in the non-target table that have many-to-one relationships with records stored in the target table by clustering these records into groups associated to each unique record in the target table. The summarized data will be then appended to the target table as a new feature-value and fed into classifier for classification tasks. A classification task is a common predictive data mining task which finds patterns that describe relationships that exist between data attributes and class attribute which can be used to classify or predict new unknown data. The predictive accuracy of a classification task is highly dependent on the representation of the input data. In the DARA data summarization algorithm, the quality of the summarized data appended into the target table influences the predictive accuracy of the classification task [10]. In our previous works, several feature construction methods that include a Fixed Length Feature Construction without Substitution (FLFCWOS), a Fixed Length Feature Construction with Substitution (FLFCWS), a Variable Length Feature Construction without Substitution (VLFCWOS), and a Variable Length Feature Construction with Substitution (VLFCWS) have been introduced to construct relevant features to represent the records stored in the non-target table [10][12][13]. These methods are designed in order to improve the quality of the summarized data for records stored in the non-target table and thus it will indirectly improve the predictive accuracy of the classification task of the target table. This is due to the fact that the improvement in the descriptive accuracy will also improve the classification performance [10].

This paper proposes a method that learns relational data based on multiple instances of summarized data that are obtained using different types of feature construction methods [10][12][13]. This involves investigating the effect of selecting multiple instances of summarized data which have been summarized based on different types of feature construction and appending these summarized data into the target table. This paper is organized as follows: Section 2 presents an overview of a general feature selection method used in data mining and the feature construction process used in DARA algorithm. Section 3 reviews the proposed method that learns relational data based on multiple instances of summarized data. Section 4 presents the experimental setup and design. The comparison of the predictive accuracy of the classification task using the proposed method and the tradition techniques [10] is also presented in this section. Section 5 concludes this paper.

2   Related Works

Many research works have been conducted to address problems related to the input data representation for data mining purpose [3][14][15]. This is important because the performance of the data mining algorithm relies highly on the quality of input data representation. A set of relevant features should be presented in order to have a high descriptive accuracy of any clustering techniques or a high predictive accuracy of any classification techniques. The application of feature selection and features construction techniques, in the pre-processing of the input data, has shown a significant improvement on the descriptive and predictive accuracies in the data mining process.

Feature selection is a process of selecting an informative subset of features from an initial set of features. It attempts to remove redundant and irrelevant features which will cause the degradation of the descriptive and predictive accuracy in the data mining performance [16]. Feature selection methods can be categorized into wrapper, filter, and appended approaches [17]. Filter-based feature selection approach selects features subset according to the intrinsic of data which is independent of the data mining method. While the wrapper-based feature selection approach selects features subset by evaluating the data mining performance. Lastly, the appended approach to select feature integrates the process of selecting features subset in the mining process of a given data mining algorithm.

Feature construction is a process of constructing a new set of features by combining the attributes given in the original set of attributes. It attempts to uncover hidden relationships that exist among features [3]. Feature construction methods can also be categorized into wrapper and filter approaches. The filter-based feature construction approach constructs features independent of the data mining process and the quality of the constructed features are evaluated by assessing the data generated based on some criterions. A wrapper-based feature construction approach constructs features within the data mining process and the quality of the constructed features is evaluated based on the data mining performance. Many varieties of feature selection [15-19] and feature construction methods [14] have been developed to generate a highly descriptive input data representation in order to improve the data mining performance.

In DARA (Dynamic Aggregation of Relational Attributes) algorithm [9], feature construction has been applied in the process of transforming a relational data model into a vector space model for summarization process. DARA algorithm summarizes records stored in a non-target table, where these records have many-to-one relationships with records stored in a target table. The records stored in the non-target table are summarized and then appended to the target table. With these summarized data appended into the target table, a classifier will be applied to learn this data in order to perform the classification task. For instance, in DARA data summarization process (see Fig. 1) [10], each unique record stored in the non-target table that is linked to the records stored in the target table is converted into bag of patterns representation. At this stage, a feature construction process is performed in order to enrich the representation of records stored in the non-target table. Herein, a new set of features is constructed by combining several attributes obtained from the original set of attributes, given in the non-target table. For example, given a set of attributes $\{F_1, F_2, F_3, F_4\}$, the possible sets of constructed features could be $\{F_1F_2, F_3F_4\}$ or $\{F_1F_3, F_2F_4\}$. The set of constructed features will be used to generate patterns that represent records in non-target table. Then, bags of patterns are used to represent the relation between records in non-target table and records in target table in the vectors of patterns representation. A vector of patterns comprises of a target record with a set of patterns, patterns frequencies, and the number of target records having the patterns. The vectors of patterns are used to develop a $(n \times p)$ TF-IDF (term frequency inverse document frequency) matrix where n is the number of target records and p is the number of each unique pattern. This matrix is developed to assign weights to all patterns in each target record in order to measure the degree of relevancy of the patterns in differentiating the target records. By using the information obtained from the TF-IDF matrix, clustering process is performed to group the target records into clusters where records in the same clusters share similar characteristics and vice-verse for records in different clusters. Finally, the identification number of each cluster is appended to the corresponding records stored in target table as a new column or new feature-value that represent records in non-target table.

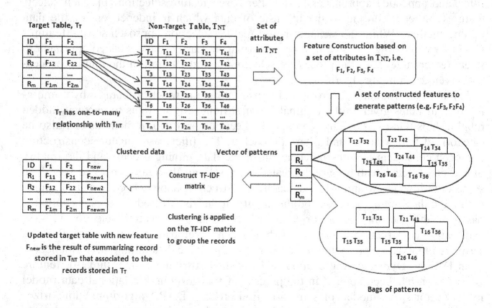

Fig. 1. DARA data summarization with feature construction

Four feature construction methods, namely Fixed Length Feature Construction without Substitution (FLFCWOS), Fixed Length Feature Construction with Substitution (FLFCWS), Variable Length Feature Construction without Substitution (VLFCWOS), and Variable Length Feature Construction with Substitution (VLFCWS) have been applied to construct a set of relevant features in the DARA data summarization process. FLFCWOS constructs a new set of feature by combining the original attributes randomly at fixed length without allowing an individual attribute to be combined more than once [10]. Similar to FLFCWOS, FLFCWS constructs a new set of features by combining the original attributes randomly at fixed length but it allows an individual attribute to be combined more than once [12]. In VLFCWOS, a new set of features is constructed by combining the original attributes randomly at various lengths without allowing an individual attribute to be combined more than once [13]. On the other hand, in VLFCWS, a new set of features is constructed by combining the original attributes randomly at various lengths but it allows an individual attribute to be combined more than once.

Table 1 shows the possible set of features constructed based on the FLFCWOS, FLFCWS, VLFCWOS, and VLFCWS methods. Let $F = \{F_1, F_2, F_3, F_4, F_5, F_6\}$ represents a set of attributes given in a non-target table. By constructing a new set of relevant features using these four approaches, the descriptive accuracy of the summarized data obtained by clustering records stored in the non-target table is also improved. When the summarized data is appended into the target table and fed to a classifier, the predictive accuracy of the classification task can also be improved [10][12][13]. These approaches focus on processing the representation of records stored in the non-target table for summarization process in order to improve the descriptive accuracy of the summarized data. This is performed as an effort to improve the classification performance. However, the representation of the summarized data may also influence the representation of records stored in the target table which is treated as input data for the classification task. In this paper, further investigation works are conducted where several summarized data are selected and appended into the target table in order to improve the predictive accuracy of the target table.

Table 1. Set of Features Constructed Based On Existing Feature Construction Methods

Feature Construction Method	Set of Constructed Feature
FLFCWOS	F_1F_2, F_3F_4, F_5F_6
FLFCWS	$F_1F_2F_3, F_2F_4F_3, F_4F_5F_6$
VLFCWOS	$F_1F_2, F_3F_4F_5, F_6$
VLFCWS	$F_1F_2F_3, F_4F_3, F_4F_5F_6$

3   Learning Relational Data Based on Multiple Instances of Summarized Data

This paper extends our previous works [10][12][13] by selecting multiple instances of summarized data and appending these summarized data into the target table before the classification task can be performed. These summarized data are generated based on four feature construction methods, which includes Fixed Length Feature Construction without Substitution (FLFCWOS), Fixed Length Feature Construction with Substitution (FLFCWS), Variable Length Feature Construction without Substitution (VLFCWOS), and Variable Length Feature Construction with Substitution (VLFCWS). The main purpose of introducing this method is to have a subset of relevant features to represent the summarized data in order to enrich the representation of the records stored in the target table as an input data for classification task. It attempts to find features that contain high discriminative capability to distinguish patterns that belong to a class but not to other classes [16][17][18]. Besides that, there are some features when combined with other features may become a relevant features subset [16].

This method selects any features randomly from the four different features represent the summarized data since there are four different feature construction methods, namely FLFCWOS, FLFCWS, VLFCWOS, and VLFCWS are used to perform the summarization process. For example, given a set of the four features $\{F_A, F_B, F_C, F_D\}$ where F_A, F_B, F_C, and F_D designate features of the data summarized by using FLFCWOS, FLFCWS, VLFCWOS, and VLFCWS respectively. From these features, we could have subset of selected features according to the extended feature construction method such as F_A, F_B or F_A, F_C or F_A, F_D, and so on. The summarized data of each feature in the selected features subset will be used to be appended to the target table as a new feature value. Table 2 shows list of summarized data derived from the features subset. Let $F = \{F_A, F_B, F_C, F_D\}$ with $domain(F_i) = (F_{i,1}, F_{i,2}, F_{i,3}, \ldots, F_{i,n})$ represents the domain of feature F_i, with n different values where $F_i \in F$.

Table 2. Summarized Data Produced Based On The Extended Feature Construction Method

Initial Constructed Features	Selected Constructed Features	Summarized Data
F_A, F_B, F_C, F_D	F_A, F_B	$F_{A,1}, F_{B,n}$
F_A, F_B, F_C, F_D	F_A, F_B, F_C	$F_{A,1}, F_{B,3}, F_{C,n}$

Fig. 2 illustrates the selection of features performed by using the extended feature construction method applied in DARA algorithm to represent the outcome of summarizing relational data for classification task. In the DARA data summarization process, the best set of features constructed is selected from each feature construction methods in order to generate the summarized data. These summarized data are then appended as a new feature into the target table as shown in Fig. 2. Then, a wrapper feature selection approach is used again to select a set of relevant summarized data from the target table before it is fed to a classifier to perform the classification task. The effectiveness of the selected features (summarized data) to represent the target table is evaluated based on the performance of the classification task. These processes iterate to select other new features subset to generate new summarized data representation for the classification task. Since only four features (summarized data) that are produced based on four feature construction methods, the process stops when all possible combination of features have been selected to process the representation of the summarized data. Lastly, the subset of selected features (summarized data) with the highest predictive accuracy of the classification task is used to represent the target table for the final classification task.

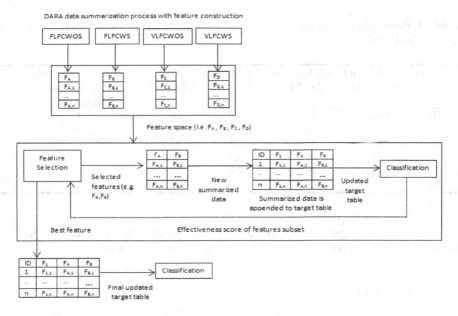

Fig. 2. Extended feature construction process in DARA data summarization algorithm for classification

4   Experimental Design and Evaluations

In these experiments, we evaluate the effects of selecting a set of relevant summarized data produced by the DARA algorithm. Mutagenesis (B1, B2, B3) [10] and Hepatitis (H1, H2, H3) datasets obtained from PKDD 2005 discovery challenge are used for the evaluation of the proposed method. The best summarized data of these datasets based on each FLFCWOS, FLFCWS, VLFCWOS, and VLFCWS are collected before another process of feature selection is performed. A standard *J*48 classifier of WEKA is applied on the newly generated summarized data to perform the classification task. The predictive accuracy of a 10 fold-cross validation is used to evaluate the classification performance. Each experiment is conducted ten times with ten different possible features subset. The effectiveness of the proposed method is analyzed by comparing the predictive accuracies of the classification task based on multiple instances of summarized data with the predictive accuracies of the classification task based on a single summarized data.

Table 3 shows the predictive accuracies of the *J*48 classifier on the target table with multiple instances of summarized data and a single summarized data for the mutagenesis and hepatitis datasets.

Based on the results shown in Table 3, the predictive accuracies of the classification task by using multiple instances of summarized data show comparable results when compared to the predictive accuracies of the classification task by using any single instances of summarized data (FLFCWOS, FLFCWS, VLFCWOS, VLFCWS). The predictive accuracies of the classification task by using multiple instances of summarized data consistently produce good predictive accuracies of the classification

Table 3. Predictive accuracies based on a single instance and multiple instances of the summarized data

Dataset	Using a Single Instance of Summarized Data				Using Multiple Instances of Summarized Data
	FLFCWOS	*FLFCWS*	*VLFCWOS*	*VLFCWS*	
B1	83.5	83.0	83.0	86.7	86.7
B2	83.5	85.1	85.1	87.2	86.7
B3	81.4	81.4	81.1	82.5	81.9
H1	76.7	76.5	76.7	76.5	76.7
H2	78.0	77.2	77.8	78.4	78.2
H3	76.7	76.3	77.9	81.4	75.1

task by using any single instances of summarized data in 9 out of 12 datasets for the Mutagenesis datasets and in 5 out of 12 datasets for the Hepatitis datasets.

5   Conclusion

In this paper, it has shown that the representation of the input data is important in order to get a higher predictive accuracy of the classification task consistently. The proposed method learns the relational data by summarizing records stored in a non-target table using several feature construction methods and then selects a set of multiple instances of summarized data that best describes the data stored in the non-target table. The set of summarized data that is selected is then appended to the target table as a new feature-value and thus, *J*48 classifier is applied on the target table to perform the classification task. The experimental results showed that the predictive accuracy of the classification task performed on the summarized data processed using the proposed method present high predictive accuracies consistently when compared to the existing feature construction methods. Therefore, the proposed method is capable of selecting multiple instances of summarized data that enrich the representation of all the records stored in the non-target table that are treated as an input data for the classification task.

Acknowledgments. This work has been supported by the Research Grant Scheme project funded by the Ministry of Higher Education (MoHE), Malaysia, under Grants No. RAG0007-TK-2012.

References

1. Kavurucu, Y., Senkul, P., Toroslu, I.H.: A Comparative Study on ILP-Based Concept Discovery Systems. Expert Systems with Applications 38(9), 11598–11607 (2011)
2. Xavier, J.C., Canuto, A.M.P., Freitas, A.A., Goncalves, L.M.G., Silla, C.N.: A Hierarchical Approach to Represent Relational Data Applied to Clustering Tasks. In: The 2011 International Joint Conference on Neural Networks (IJCNN), pp. 3055–3062 (2011)

3. Tian, Y., Weiss, G.M., Hsu, D.F., Ma, Q.: A Combinatorial Fusion Method for Feature Construction (2009)
4. Alfred, R.: Summarizing Relational Data Using Semi-Supervised Genetic Algorithm-Based Clustering Techniques. Journal of Computer Science 6(7), 775–784 (2010)
5. Kavurucu, Y., Senkul, P., Toroslu, I.H.: Concept Discovery on Relational Databases: New Techniques for Search Space Pruning and Rule Quality Improvement. Knowledge-Based Systems 23(8), 743–756 (2011)
6. Maimon, O., Rokach, L.: Introduction to Knowledge Discovery and Data Mining. The Data Mining and Knowledge Discovery HandBook, pp. 1–5. Springer, Heidelberg (2010)
7. Choudhary, A.K., Harding, J.A., Tiwari, M.K.: Data Mining in Manufacturing: A Review Based on the Kind of Knowledge. Journal of Intelligent Manufacturing, 501–521 (2009)
8. Alfred, R.: The Study of Dynamic Aggregation of Relational Attributes on Relational Data Mining. In: Alhajj, R., et al. (eds.) ADMA 2007. LNCS (LNAI), vol. 4632, pp. 214–226. Springer, Heidelberg (2007)
9. Alfred, R.: Optimizing Feature Construction Process for Dynamic Aggregation of Relational Attributes. J. Comput. Sci. 5, 864–877 (2009)
10. Alfred, R.: Feature Transformation: A Genetic-Based Feature Construction Method for Data Summarization. Computational Intelligence 26(3), 337–357 (2010)
11. Alfred, R.: Summarizing Relational Data Using Semi-Supervised Genetic Algorithm-Based Clustering Techniques. Journal of Computer Science 6(7), 775–784 (2010)
12. Sia, F., Alfred, R.: Evolutionary-Based Feature Construction With Substitution For Data Summarization Using DARA. In: The 4th 2012 Conference on Data Mining and Optimization (DMO 2012), Langkawi, Malaysia (2012)
13. Sia, F., Alfred, R.: A Variable Feature Construction Method For Data Summarization Using DARA. In: The 3rd International Conference on Advancements in Computing Technology (ICACT 2012), Soeul, Korea (2012)
14. Shafti, L.S., Perez, E.: Evolutionary Multi-Feature Construction for Data Reduction: A Case Study. Appl. Soft Comput. 9, 1296–1303 (2009)
15. Guan, Y., Dy, J.G., Jordan, M.I.: A Unified Probabilistic Model for Global and Local Unsupervised Feature Selection. In: Proc. ICMC (2011)
16. Wong, C., Versace, M.: CARTMAP: A Neural Network Method for Automated Feature Selection in financial Time Series Forecasting. J. Neural Computing and Applications, 969–977 (2012)
17. Vinh, L.T., Lee, S.Y., Park, Y.T., d'Auriol, B.J.: A Novel Feature Selection Method Based On Normalized Mutual Information. J. Applied Intelligence, 100–120 (2012)
18. Pal, M., Foody, G.M.: Feature Selection for Classification of Hyperspectral Data by SVM. IEEE Transactions on Geoscience and Remote Sensing 48(5), 2297–2307 (2010)
19. Song, L.A., Smola, A., Gretton, A., Bedo, J., Borgwardt, K.: Feature Selection Via Dependence Maximization. Journal of Machine Learning Research 13, 1393–1434 (2012)
20. Estevez, P.A., Tesmer, M., Perez, C.A., Zurada, J.M.: Normalized Mutual in-Formation Feature Selection. IEEE Transactions on Neural Networks 20(2), 189–201 (2009)

Author Index